Plasma Catalysis

Plasma Catalysis

Special Issue Editor
Annemie Bogaerts

MDPI • Basel • Beijing • Wuhan • Barcelona • Belgrade

MDPI

Special Issue Editor
Annemie Bogaerts
Universiteit Antwerpen
Belgium

Editorial Office
MDPI
St. Alban-Anlage 66
4052 Basel, Switzerland

This is a reprint of articles from the Special Issue published online in the open access journal *Catalysts* (ISSN 2073-4344) from 2018 to 2019 (available at: https://www.mdpi.com/journal/catalysts/special_issues/plasma_catalysis)

For citation purposes, cite each article independently as indicated on the article page online and as indicated below:

LastName, A.A.; LastName, B.B.; LastName, C.C. Article Title. *Journal Name* **Year**, *Article Number*, Page Range.

ISBN 978-3-03897-750-6 (Pbk)
ISBN 978-3-03897-751-3 (PDF)

Contents

About the Special Issue Editor

Annemie Bogaerts was born in 1971. She obtained her M.Sc. in Chemistry in 1993 and her Ph.D. in Sciences in 1996, both from the University of Antwerp, Belgium. After some postdoc years, she became a professor at the University of Antwerp in 2003, and is full professor since 2014. She is head of the research group PLASMANT, which she started in 2004, based on her own Ph.D. work. Currently, the group has 37 members (2 professors, 15 postdocs, 18 Ph.D. students, and 2 technical coworkers).

She has over 470 peer-reviewed publications since 1995, and over 11,500 citations, with a H-index of 52 (Web of Science) (over 16,000 citations and H-index of 63 in Google Scholar). Furthermore, she has more than 140 invited lectures at international conferences (since 1998) and more than 60 invited seminars at universities/institutes (since 1995), in various countries. She was the supervisor of 29 finished Ph.D. theses (since 2005), and is now supervising 23 Ph.D. students (incl. joint Ph.D. students with other universities), and 12 postdocs.

She has received many scientific awards, including the Prize of the Research Council of the University of Antwerp in Exact Sciences (1998), the Prize of the "Koninklijke Vlaamse Academie van België voor Wetenschappen en Kunsten" in de category Exact Sciences (2001), the Alumni Prize of the "Belgian-American Educational Foundation" (2003), the "Lester W. Strock Award of the New England Section of the Society for Applied Spectroscopy", in recognition of "Outstanding contributions in the areas of plasma and surface modeling" (2008) and the "Winter Plasma Award", in recognition of "Outstanding contributions in the field of laser ablation modeling" (2009). Recently she obtained an ERC Synergy Grant (2019–2025), on plasma catalysis for small molecules conversion, together with G. Centi, V. Hessel and E. Rebrov.

She is a member of the Royal Flemish Academy of Belgium for Sciences and the Arts (since 2012), the Academia Europaea (since 2011), and the Solvay Local Scientific Committee for Chemistry (since 2013). In 2013–2016, she was Francqui Research Professor.

She is a member of the editorial or advisory board of several journals. She was editor of Spectrochimica Acta Part B, responsible for the review papers, from 2002 to 2018. Since 2018, she is Topics Editor of the Journal of Physics D: Applied Physics, for a program "Advances in Plasmas for a Sustainable Future". She also acted more than 10 times as guest editor for a Special Issue. Moreover, she is on the international scientific committee of several international conferences, and was chair of the "International Symposium of Plasma Chemistry" (ISPC), organized in Antwerp in 2015, which attracted almost 600 participants. She is also the Chair of the International Scientific Committee of the "International Workshop on Plasmas for Cancer Treatment" (IWPCT).

She is a world-leading expert in modeling and simulation of reactive plasmas, mainly for environmental/energy (gas conversion) and medical applications (cancer treatment). This includes plasma chemistry and plasma reactor design modeling, as well as plasma–surface interaction simulations, e.g., for catalyst surfaces. She is also working on experimental plasma chemistry, with a special focus on plasma-based CO_2 conversion.

catalysts

MDPI

Editorial

Editorial Catalysts: Special Issue on Plasma Catalysis

Annemie Bogaerts

Research Group PLASMANT, Department of Chemistry, University of Antwerp, Universiteitsplein 1, BE-2610 Wilrijk-Antwerp, Belgium; annemie.bogaerts@uantwerpen.be

Received: 14 February 2019; Accepted: 20 February 2019; Published: 21 February 2019

Plasma catalysis is gaining increasing interest for various gas conversion applications, such as CO_2 conversion into value-added chemicals and fuels, N_2 fixation for the synthesis of NH_3 or NO_x, and CH_4 conversion into higher hydrocarbons or oxygenates [1,2]. In addition, it is widely used for air pollution control (e.g., volatile organic compound (VOC) remediation) and waste gas treatment [3–6]. Plasma allows thermodynamically difficult reactions to proceed at an ambient pressure and temperature because the gas molecules are activated by energetic electrons created in the plasma. Plasma is indeed very reactive, being a cocktail of many different types of reactive species (electrons, various ions, radicals, excited species, besides neutral gas molecules), but for this reason, it is not really selective. Therefore, a catalyst is needed to improve the selectivity towards the production of targeted compounds.

In spite of the growing interest in plasma catalysis, the underlying mechanisms of the (possible) synergy between plasma and catalyst are not yet fully understood [7]. Indeed, these mechanisms are quite complicated, as the plasma will affect the catalyst and vice versa [1,7,8]. Moreover, due to the reactive plasma environment, and the fact that these reactive plasma species can interact at the catalyst surface, the most suitable catalysts for plasma catalysis will probably be different from thermal catalysts. Hence, more research is needed to better understand the plasma–catalyst interactions, in order to further improve the applications.

This special issue gives an overview of the state-of-the-art of plasma catalysis research, for various applications, including VOC abatement, tar component removal, NO_x conversion, CO_2 splitting, dry reforming of CH_4 (DRM), H_2S removal, NH_3 synthesis and NH_3 decomposition into H_2. Moreover, it also contains some papers that provide more insight into the underlying mechanisms of plasma catalysis and packed bed plasma catalysis reactors, by either experiments or modeling.

We have one review paper in this special issue, by Veerapandian et al., presenting an excellent overview of plasma catalysis for VOC abatement in flue gas, applying zeolites as an adsorbent and a catalyst [9]. The authors illustrate that zeolites are ideal packing materials for VOC removal, by cyclic adsorption plasma catalysis, due to their superior surface properties and excellent catalytic activity upon metal loading. The zeolites can be regenerated by plasma, allowing to reduce the energy cost per decomposed VOC molecule.

To better understand the plasma behavior in a packed bed dielectric barrier discharge (DBD), which is the most common configuration of plasma catalysis, Gao et al. developed a two-dimensional (2D) particle-in-cell—Monte Carlo collision (PIC-MCC) model, to study the mode transition from volume to surface discharges in a packed bed DBD operating in various N_2/O_2 mixtures [10]. The calculations reveal that a higher voltage can induce this mode transition from hybrid (volume + surface) discharges to pure surface discharges. Indeed, a higher voltage yields a stronger electric field, so the charged species can escape more easily to the beads and charge them, leading to a strong electric field along the dielectric bead surface, which gives rise to surface ionization waves. The latter enhances the reactive species concentrations on the bead surface, which will be beneficial for plasma catalysis. In addition, changing the N_2/O_2 gas mixing ratio affects the propagation speed of the surface ionization waves, which become faster with increasing N_2 content.

Indeed, a higher O_2 content yields more electron impact attachment, and thus loss of electrons, causing less ionization. Furthermore, different N_2 and O_2 contents result in different amounts of electrons and ions on the dielectric bead surface, which might also affect the performance of plasma catalysis.

Although DBDs are the most convenient and widely studied plasma reactors for plasma catalysis, due to their simplicity, convenient catalyst integration, and easy upscaling, they suffer from limited energy efficiency. To identify the reactions in a DBD that might be responsible for this limited energy efficiency, Navascués et al. propose a method based on isotope labeling [11]. They applied this method to study wet reforming of CH_4, using D_2O instead of H_2O, as well as for NH_3 synthesis, using a $NH_3/D_2/N_2$ mixture. By analyzing the evolution of the labelled molecules as a function of power, they could obtain useful information about exchange events (of H by D atoms and vice versa) between the plasma intermediate species. This isotope labeling technique thus appears to be very appropriate for studying plasma reaction mechanisms.

As mentioned above, the most suitable catalysts for plasma catalysis might not necessarily be the same as for thermal catalysis, due to the presence of many different reactive plasma species. Hence, more research is needed to identify the different mechanisms related to plasma chemistry and thermal effects. Giammaria et al. developed a method to distinguish between both effects and applied it to $CaCO_3$ decomposition in argon plasma [12]. They prepared $CaCO_3$ samples with different external surface area (determined by the particle size), as well as different internal surface area (determined by the pores). As the internal surface area is not exposed to plasma, it only relates to thermal effects, while both plasma and thermal effects take place at the external surface area. The authors concluded that this application is dominated by thermal decomposition, as the decomposition rates were only affected by the internal surface changes, and slow response in the CO_2 concentration (of typically 1 min) was detected upon changes in discharge power. The authors measured a temperature rise within 80 °C for plasma power up to 6 W. In addition, they also studied the mechanism of CO_2 conversion into CO and O_2, which was found to be controlled by the plasma chemistry, as indicated by the fast response (within a few seconds) of the CO concentration upon changing plasma power. Indeed, this reaction is thermodynamically impossible without plasma. This methodology is very interesting to distinguish between thermal and plasma effects, and it would be nice to apply it also to other plasma catalysis reactions, in more reactive plasmas, which the authors indeed plan for their future work.

The other papers in this special issue focus on a particular application, and illustrate the broad applicability of plasma catalysis, for pollution control, gas conversion and destruction.

Zhou et al. studied CO_2 conversion in a packed bed DBD, using a water-cooled cylindrical DBD reactor with ZrO_2 pellets or glass beads of 1–2 mm diameter, to control the temperature [13]. Especially the ZrO_2 pellets provided good results, yielding a maximum CO_2 conversion around 50% (slightly higher for the smaller beads), compared to ca. 33% for the glass beads. The CO selectivity was up to 95%, while the energy efficiency was 7% (compared to 3% without ZrO_2 packing). The authors attributed the improved performance to the stronger electric field, and thus higher electron energy, along with the lower reaction temperature.

Michielsen et al. investigated dry reforming of methane (DRM) in a packed bed DBD, as compared to pure CO_2 splitting [14]. They reported that the packing materials, even when not catalytically activated, can already significantly affect the conversion and product selectivity. This is important to realize because the effect of the packing material is often not taken into account. α-Al_2O_3 packing yielded the highest total conversion (28%), with a high product fraction towards CO and ethane, as well as a high CO/H_2 ratio around 9. γ-Al_2O_3 gave a slightly lower total conversion (22%), but a more pronounced selectivity towards certain products. On the other hand, $BaTiO_3$ resulted in a lower conversion, in contrast to its performance in pure CO_2 splitting. In general, the trends of different packing materials obtained for DRM were different from those obtained for CO_2 splitting. Thus, it is clear that the packing materials can have a vast influence of the reaction performance, and thus, they also need specific attention.

In general, plasma-catalytic DRM is still in its infancy, because up to now, mostly thermal catalysts have been applied, which do not fully exploit the potential of plasma catalysis. Hence, more research is needed to design catalysts tailored to the plasma environment, to make profit of the reactive plasma species and their interactions with the catalyst surface, and to selectively produce value-added chemicals. On the other hand, the application of air pollution control, and specifically VOC removal, by plasma catalysis is already more advanced, as indicated by the vast amount of literature (cf. also the excellent reviews mentioned above [3–6,9]).

Jia et al. investigated toluene oxidation with CeO_2 as an adsorbent and they compared in-plasma catalysis (IPC) and post-plasma catalysis (PPC) [15]. The total, reversible and irreversible adsorbed fractions were quantified. The authors investigated the effect of relative humidity on the toluene adsorption and ozone formation, as well as the effect of specific energy input (SEI) on the mineralization yield and efficiency. The best results were obtained for IPC at the lowest SEI, i.e., lean conditions of ozone. The paper stresses the key role of ozone in the mineralization of toluene and the possible detrimental effect of moisture.

Likewise, Kong et al. studied toluene, nathalene and phenanthrene destruction (as model tar compounds) in humid N_2, in a rotating gliding arc reactor with fan-shaped swirling generator [16]. Tar destruction is one of the greatest technical challenges in commercial gasification technology. The authors studied the effect of tar, CO_2 and moisture concentrations, discharge current, and Ni/γ-Al_2O_3 catalyst on the destruction efficiency. The latter reached 95%, 89% and 84%, for toluene, nathalene and phenanthrene, respectively, at a tar content of 12 g/Nm^3, 15% CO_2, 12% moisture and 6 NL/min flow rate, yielding an energy efficiency of 9.3 g/kWh. The presence of the Ni/γ-Al_2O_3 catalyst significantly improved the destruction efficiency. The major liquid by-products were also identified.

Plasma-catalytic air pollution control also involves NO_x destruction, which was reported by Gao et al. [17]. The authors inserted Mn-based bimetallic nanocatalysts, i.e., Mn-Fe/TiO_2, Mn-Co/TiO_2, and Mn-Ce/TiO_2, in a DBD and demonstrated a clear improvement in the plasma-catalytic conversion compared to plasma alone and nanocatalyst alone. The Mn-Ce/TiO_2 catalyst was found to give the highest catalytic activity and superior selectivity, yielding a maximum NO_x conversion of about 99.5%. The authors applied various surface characterization methods, which revealed that the plasma-catalytic performance was greatly dependent on the phase compositions, explaining the superior performance of the Mn-Ce/TiO_2 catalyst.

H_2S removal is another application of plasma catalysis, which was studied by Xuan et al., for non-stoichiometric La_xMnO_3 perovskite catalysts (x = 0.9, 0.95, 1, 1.05 and 1.1) in a packed bed DBD reactor [18]. The plasma-catalytic performance was found to be much better than the results when only using plasma, reaching a maximum H_2S removal of 96%, producing mainly SO_2 and SO_3, for the $La_{0.9}MnO_3$ catalyst. The sulfur balance was 91%, with the remaining fraction probably deposited sulfur on the catalyst surface. The authors reported that the non-stoichiometric La_xMnO_3 catalyst had a larger specific surface area and smaller crystallite size than the $LaMnO_3$ catalyst and that the non-stoichiometric effect changes the redox properties of the catalyst. Indeed, a lower La/Mn ratio favored the transformation of Mn^{3+} to Mn^{4+}, generating oxygen vacancies on the catalyst surface, yielding a higher concentration of surface-adsorbed oxygen, and a lower reduction temperature.

An emerging application, gaining increasing interest in recent years, is NH_3 synthesis by plasma catalysis. This is attributed to the growing worldwide population and the associated demand for fertilizer production, in combination with the need to find alternatives for the energy-intensive Haber-Bosch process for NH_3 synthesis, which can comply with renewable energy sources. Although plasma catalysis might never become competitive with the current (large-scale) Haber-Bosch process, which has been optimized in industry for so many years, plasma-catalytic NH_3 synthesis might find some niche applications, for the decentralized fertilizer production based on renewable energy, due to the easy on-off switching of plasma, and thus its high potential as turnkey process. While most papers in literature apply DBD reactors for NH_3 synthesis, Shah et al. explored the possibility of an inductively coupled radiofrequency plasma, using Ga, In and their alloys as catalysts [19]. Ga-In alloys with 6:4 or 2:8 ratio at 50 W yielded the highest energy yield (0.31 g-NH$_3$/kWh) and lowest energy cost (196 MJ/mol). The authors tried to explain the results by means of optical emission spectroscopy of the plasma and scanning electron microscopy of the catalyst surface. They reported granular nodes on the catalyst surface, indicating the formation of intermediate GaN.

Finally, Wang et al. studied the opposite process, i.e., NH$_3$ decomposition for H$_2$ production [20]. The authors showed that vacuum-freeze drying and plasma calcination can improve the conventional preparation methods of the catalysts, and thus the performance of plasma-catalytic NH$_3$ decomposition. They reported an enhanced NH$_3$ conversion by 47%, and a rise in energy efficiency from 2.3 to 5.7 mol/kWh, compared to conventional catalyst preparation methods. At optimal conditions, they obtained 98% NH$_3$ conversion with 1.9 mol/kWh energy efficiency. The authors attributed this significant improvement to the creation of more active sites because the Co species can be highly dispersed on the fumed SiO$_2$ support, as well as to the stronger interaction of Co with fumed SiO$_2$ and the stronger acidity of the catalyst, as revealed by their experiments. This improved catalyst preparation method thus seems very promising and might also give inspiration for other plasma catalysis application.

It is obvious that excellent research is being performed worldwide on plasma catalysis for various types of reactions, including VOC decomposition, tar component removal, NO$_x$ conversion, CO$_2$ splitting, DRM, H$_2$S removal, NH$_3$ synthesis, as well as NH$_3$ decomposition into H$_2$. We particularly note numerous activities by various Chinese groups, but also by groups in the US, UK, France, Spain, the Netherlands and Belgium. We can conclude that plasma catalysis is a very active field of research, with promising results for various applications. On the other hand, further research is highly needed, especially to obtain better insight in the underlying plasma-catalyst interactions, in order to develop catalysts that are tailored to the reactive plasma conditions, and to fully exploit the promising plasma catalysis synergy.

Finally, we sincerely thank all authors for their valuable contributions, as well as the editorial team of Catalysts for their kind support and fast responses. Without them, this special issue would not have been possible.

Conflicts of Interest: The author declares no conflicts of interest.

References

1. Neyts, E.C.; Ostrikov, K.; Sunkara, M.K.; Bogaerts, A. Plasma catalysis: Synergistic effects at the nanoscale. *Chem. Rev.* **2015**, *115*, 13408–13446. [CrossRef] [PubMed]
2. Chen, H.L.; Lee, H.M.; Chen, S.H.; Chao, Y.; Chang, M.B. Review of plasma catalysis on hydrocarbon reforming for hydrogen production - Interaction, integration, and prospects. *Appl. Catal. B Environ.* **2008**, *85*, 1–9. [CrossRef]
3. Kim, H.H. Nonthermal plasma processing for air-pollution control: A historical review, current issues, and future prospects. *Plasma Process. Polym.* **2004**, *1*, 91–110. [CrossRef]
4. Chen, H.L.; Lee, H.M.; Chen, S.H.; Chang, M.B.; Yu, S.J.; Li, S.N. Removal of volatile organic compounds by single-stage and two-stage plasma catalysis systems: A review of the performance enhancement mechanisms, current status, and suitable applications. *Env. Sci. Technol.* **2009**, *43*, 2216–2227. [CrossRef]
5. Vandenbroucke, A.M.; Morent, R.; De Geyter, N.; Leys, C. Non-thermal plasmas for non-catalytic and catalytic VOC abatement. *J. Hazardous Mater.* **2011**, *195*, 30–54. [CrossRef] [PubMed]
6. van Durme, J.; Dewulf, J.; Leys, C.; Van Langenhove, H. Combining non-thermal plasma with heterogeneous catalysis in waste gas treatment: A review. *Appl. Catal. B Environ.* **2008**, *78*, 324–333. [CrossRef]
7. Whitehead, J.C. Plasma-catalysis: the known knowns, the known unknowns and the unknown unknowns. *J. Phys. D Appl. Phys.* **2016**, *49*, 243001. [CrossRef]
8. Neyts, E.C.; Bogaerts, A. Understanding plasma catalysis through modeling and simulation—A review. *J. Phys. D Appl. Phys.* **2014**, *47*, 224010. [CrossRef]

9. Veerapandian, S.K.P.; De Geyter, N.; Giraudon, J.-M.; Lamonier, J.-F.; Morent, R. The use of zeolites for VOCs abatement by combining non-thermal plasma, adsorption and/or catalysis. *Catalysts* **2019**, *9*, 98. [CrossRef]

10. Gao, M.; Zhang, Y.; Wang, H.; Guo, B.; Zhang, Q.Z.; Bogaerts, A. Mode transition of filaments in packed-bed dielectric barrier discharges. *Catalysts* **2018**, *8*, 248. [CrossRef]

11. Navascués, P.; Obrero-Pérez, M.; Cotrino, J.; González-Elipe, A.R.; Gómez-Ramírez, A. Isotope labelling for reaction mechanism analysis in DBD plasma processes. *Catalysts* **2019**, *9*, 45. [CrossRef]

12. Giammaria, G.; van Rooij, G.; Lefferts, L. Plasma Catalysis: Distinguishing between Thermal and Chemical Effects. *Catalysts* **2019**, *9*, 185. [CrossRef]

13. Zhou, A.; Chen, D.; Ma, C.; Yu, F.; Dai, B. DBD plasma-ZrO_2 catalytic decomposition of CO_2 at low temperatures. *Catalysts* **2018**, *8*, 256. [CrossRef]

14. Michielsen, I.; Uytdenhouwen, Y.; Bogaerts, A.; Meynen, V. Altering conversion and product selectivity of dry reforming of methane in a dielectric barrier discharge by changing the dielectric packing material. *Catalysts* **2019**, *9*, 51. [CrossRef]

15. Jia, Z.; Wang, X.; Foucher, E.; Thevenet, F.; Rousseau, A. Plasma-catalytic mineralization of toluene adsorbed on CeO_2. *Catalysts* **2018**, *8*, 303. [CrossRef]

16. Kong, X.; Zhang, H.; Li, X.; Xu, R.; Mubeen, I.; Li, L.; Yan, J. Destruction of toluene, napthalene and phenanthrene as model tar compounds in a modified rotating gliding arc discharge reactor. *Catalysts* **2019**, *9*, 19. [CrossRef]

17. Gao, Y.; Jiang, W.; Luan, T.; Li, H.; Zhang, W.; Feng, W.; Jiang, H. High-efficiency catalytic conversion of NO_x by the synergy of nanocatalyst and plasma: Effect of Mn-based bimetallic active species. *Catalysts* **2019**, *9*, 103. [CrossRef]

18. Xuan, K.; Zhu, X.; Cai, Y.; Tu, X. Plasma oxidation of H_2S over non-stoichiometric La_xMnO_3 perovskite catalysts in a dielectric barrier discharge reactor. *Catalysts* **2018**, *8*, 317. [CrossRef]

19. Shah, J.R.; Harrison, J.M.; Carreon, M.L. Ammonia plasma-catalytic synthesis using low melting point alloys. *Catalysts* **2018**, *8*, 437. [CrossRef]

20. Wang, L.; Yi, Y.H.; Guo, H.C.; Du, X.M.; Zhu, B.; Zhu, Y.M. Highly dispersed Co nanoparticles prepared by an improved method for plasma-driven NH_3 decomposition to produce H_2. *Catalysts* **2019**, *9*, 107. [CrossRef]

Review

The Use of Zeolites for VOCs Abatement by Combining Non-Thermal Plasma, Adsorption, and/or Catalysis: A Review

Savita K. P. Veerapandian [1,*], Nathalie De Geyter [1], Jean-Marc Giraudon [2], Jean-François Lamonier [2] and Rino Morent [1]

[1] Research Unit Plasma Technology, Department of Applied Physics, Faculty of Engineering and Architecture, Ghent University, Sint-Pietersnieuwstraat 41 B4, 9000 Ghent, Belgium; Nathalie.DeGeyter@UGent.be (N.D.G.); Rino.Morent@UGent.be (R.M.)

[2] Univ. Lille, CNRS, Centrale Lille, ENSCL, Univ. Artois, UMR 8181-UCCS-Unité de Catalyse et Chimie du Solide, F-59000 Lille, France; jean-marc.giraudon@univ-lille1.fr (J.-M.G.); jean-francois.lamonier@univ-lille.fr (J.-F.L.)

* Correspondence: savita.kaliyaperumalveerapandian@ugent.be

Received: 12 December 2018; Accepted: 13 January 2019; Published: 17 January 2019

Abstract: Non-thermal plasma technique can be easily integrated with catalysis and adsorption for environmental applications such as volatile organic compound (VOC) abatement to overcome the shortcomings of individual techniques. This review attempts to give an overview of the literature about the application of zeolite as adsorbent and catalyst in combination with non-thermal plasma for VOC abatement in flue gas. The superior surface properties of zeolites in combination with its excellent catalytic properties obtained by metal loading make it an ideal packing material for adsorption plasma catalytic removal of VOCs. This work highlights the use of zeolites for cyclic adsorption plasma catalysis in order to reduce the energy cost to decompose per VOC molecule and to regenerate zeolites via plasma.

Keywords: VOC abatement; air pollution; zeolites; adsorption-plasma catalysis

1. Introduction

According to the European Union (EU), volatile organic compounds (VOCs) are defined as organic compounds or substances with a low boiling point ($\leq$523 K) at atmospheric pressure [1]. Volatile organic compounds such as toluene, benzene, and xylene are widely present in the environment due to the fact that they are used in various industries such as in semiconductors, automobiles, and even as domestic cleaning agents. The emission of VOCs are either from static sources (e.g., from production of products such as coal, oil, organic chemicals, plywood, artificial leather, synthetic materials, cosmetics, printing, paint, tobacco smoke, and cleaning products; and from composting units/plants, electroplating, chemical coating, incineration plants, and landfills) or from mobile sources (e.g., petrol and diesel exhaust emissions) [2,3]. In the presence of UV-light, some VOCs can react with nitrogen oxide and form photochemical smog which is harmful to human health and the environment [4,5]. Due to its negative impact on human and environmental health, environmental policies for the emission control of VOCs are becoming more and more stringent.

Volatile organic compounds are the main components of indoor air pollution (IAP) and they can cause carcinogenic, mutagenic, and teratogenic health problems such as skin allergies, dizziness, vomiting, damage to the liver, kidney, and central nervous system [6–8], and are suspected to be the main reason for sick building syndrome (SBS) [9,10]. Several techniques which have been investigated for the removal of VOCs from air [11] including thermal decomposition [12,13], catalytic decomposition/oxidation [14,15], bio-filtration [16–18], adsorption [19], non-thermal plasma [20–22], photo-catalysis [23,24], and plasma catalysis [20,25–27]. Generally, the exhaust gas has a large volume (high flow rate required) and low VOC concentrations (10–1000 ppm) [28]. For example, an indoor gas exhaust from a certain printing factory has a flow rate and VOC concentration of 800 Nm^3/min and 17 ppm, respectively [29]. To treat these flue gases directly (for example, by incineration), huge operating facilities with high energy consumption are required which increase both the instillation and operating costs [28].

Among the various technologies which have been proposed and used for the decomposition of VOCs, plasma catalysis (PC) which is the combination of non-thermal plasma (NTP) and catalysis has been proven efficient; particularly, for the removal of low concentration of VOCs (<1000 ppm) [30,31]. Non-thermal plasma is generated by applying a sufficiently high electric field which produces electrons, excited gas molecules, and free radicals that are suitable to convert environmental pollutants to ideal products. However, the commercialization of this technique has the following disadvantages such as: (a) formation of un-wanted toxic by-products, (b) low energy efficiency, and (c) incomplete oxidation of VOCs. Thus, the combination of NTP and catalysts takes advantage of the ability of plasma to activate catalysts at lower temperature and the high selectivity of the catalysts [32]. In most of the cases, active metals are supported on the surface of substrates such as γ-Al_2O_3, zeolite, and activated carbon which have a large surface area. The area of active sites on the surface of the support material can be improved either by increasing the amount of metal loading or by decreasing the size of metal catalyst (thus, less metal will be required to obtain certain surface area of active metal); of which the latter is more beneficial as precious metals are expensive [32]. On the other hand, from an engineering point of view, the rapid start up and turn off of the plasma devices makes this technique more suitable for small- and medium-scale applications. But the main drawback of using plasma catalysis for the treatment of very low concentrations of VOCs (<100 ppm) in flue gas is that most of the discharge energy will be utilized for the excitement of oxygen and nitrogen [33]. This emphasize the necessity to explore different techniques which are more suitable for the removal of very low concentration of VOCs from a large volume of flue gas. In order to treat the large volume of gas with low VOC concentration and to improve the energy efficiency, the combination of adsorption and plasma catalysis has been proposed and investigated [34–36]. In such an adsorption–plasma catalysis (APC) process, the NTP discharge is either continuous [37] or cyclic [28,38,39].

Figure 1 shows the schematic diagram of the continuous process, where the catalyst/adsorbent material is placed either in the plasma discharge region (in-plasma catalysis, IPC) or downstream of the plasma discharge (post plasma catalysis, PPC) and the plasma is ignited permanently. The main disadvantage of the continuous treatment is that the plasma discharge is applied continuously irrespective of the variation in VOC concentration, and thus the energy consumption to decompose per VOC molecule is high.

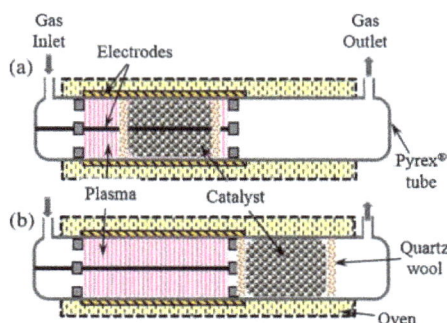

Figure 1. The working principle of the continuous adsorption-plasma catalytic process for volatile organic compounds (VOCs) removal (**a**) in-plasma catalysis and (**b**) post-plasma catalysis. Reprinted from Reference [40], with permission from Elsevier.

Figure 2 shows schematically the working principle of the cyclic APC process for VOC removal. Briefly, in a cyclic APC process, the low-concentration VOCs in flue gas are first stored on catalysts/adsorbents at a storage stage (plasma off) and then the stored VOCs are oxidized to CO_2 by plasma at a discharge stage (plasma on). In a cyclic APC process, adsorption of VOC on an adsorbent for a long time followed by the plasma discharge for the oxidation of adsorbed VOCs to CO_2 during a shorter time improves the energy efficiency [34,41,42]. Thus, the ratio of energy deposited per treated VOC molecule is considerably reduced [43]. Also, the flow rate during the discharge stage of the cyclic APC technology can be chosen to be lower than the storage stage, which leads to higher energy density for the cyclic APC technology for the same discharge power.

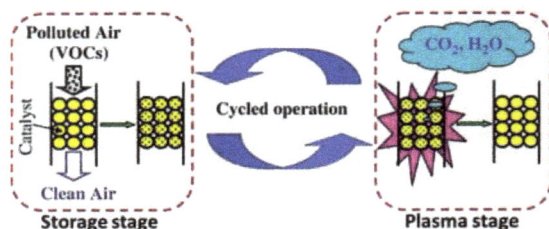

Figure 2. The working principle of the cyclic adsorption–plasma catalytic process for VOC removal. Reprinted from Reference [44], with permission from Elsevier.

The main advantages of the cyclic APC over continuous APC are as follows: (i) during the plasma discharge stage, O_2 plasma can be used instead of air, because O_2 plasma avoids the formation of byproducts such as NO_x, N_2O, etc., and is more efficient in regeneration of the adsorbent/catalyst, (ii) high CO_2 selectivity, (iii) improved carbon balance, (iv) improved energy efficiency and higher power operation is possible, (v) concentration of dilute VOCs (compact system), (vi) adapts to the change in the flow rate and VOC concentration, and (vii) rapid operation.

The adsorbing and catalytic function of the adsorbents can be separated or integrated using a dual-functional material. The most commonly used adsorbents are either (i) physical adsorbents such as alumina, zeolite, and activated carbon or (ii) chemical adsorbents such as alkaline earth metals and metal loaded physical adsorbents. The adsorption of VOCs on adsorbent reduces the chemical barrier by E_b–E_{ads} (E_b is bond energy of the molecule; E_{ads} is the adsorption energy) and the reduction depends on the kind of adsorption (either physisorption or chemisorption) [45]. Physisorption is the Van Der Waals force of attraction on the surfaces and the physically adsorbed molecules can be desorbed by applying heat; whereas chemisorption is because of the chemical reactions that leads to the transfer of electrons and ions between the adsorbent surfaces and molecules. Thus, the presence of adsorbents in the plasma discharge region prolongs the residence time of VOCs, active species, and intermediate by-products resulting in increased collisional probabilities between them and thus enhanced CO_2 selectivity [46].

Another not mentioned yet important advantage of combining adsorption and NTP is the increased lifetime of the used adsorbents. The very low concentration VOCs can be concentrated on the adsorbing material and then desorbed and decomposed to less toxic and/or more useful products. In most of the cases, the adsorbents are discarded or incinerated. With regard to the economic and practical point of view, it is appropriate to decompose the adsorbed VOCs and regenerate the adsorbent [47]. It has been already demonstrated in the literature that some adsorbents can be regenerated by different methods such as heating [48–50], microwave heating [51], pressure and temperature swing adsorption [52], and non-thermal plasma [28,53]. However, the use of techniques such as temperature or pressure swing adsorption and thermal regeneration requires high temperature or vacuum making the regeneration of the adsorbent expensive. For the desorption and decomposition of VOCs adsorbed on the adsorbents, NTP discharge can be used instead of a conventional thermal process in order to reduce both the size of the treatment equipment and the energy consumption [54]. The highly oxidizing environment of plasma promotes the oxidation of the adsorbed VOCs and simultaneously regenerates the adsorbent.

The APC requires an adsorbing/catalytic system which effectively adsorb VOCs and efficiently oxidize the adsorbed VOCs. The selection of adsorbent/catalyst is very crucial to achieve high energy efficiency to decompose diluted VOCs from air. The surface properties such as specific surface area (S_{BET}) and pore size affects the adsorption capacity and plasma discharge. The energy efficiency of APC for the removal of VOC in air is enhanced by using adsorbent with high specific surface area as this increases the adsorbing time [55]. Depending on the pore size, the discharge volume can be effectively increased by packing porous material in the discharge zone of the dielectric barrier discharge (DBD) reactor. The dielectric constant of the adsorbents influences the discharge performance, which affects the number and kind of active species generated. Also, the hydrophobicity of the adsorbent is very important for the selective adsorption of VOCs from the ambient air with humidity [56–58].

Considering the superior and tunable surface properties, hydrophobicity and metal loadability, zeolites are one of the best adsorbents and catalysts for VOC degradation. Zeolites are traditional adsorbents due to their unique micropores, cavities, and channels. The pore diameter in zeolites is in the range of 5–20 Å which is suitable for the adsorption of various VOC molecules. Also, the zeolites can interact with plasma due to their strong natural electric field inside their framework [59,60]. The degradation of VOCs by continuous and cyclic APC using zeolites are accompanied by synergetic effects due to different mechanisms such as ultraviolet (UV) irradiation, generation of electron-hole pairs and their subsequent chemical reactions, changes in work function, adsorption, desorption, direct interaction of gas phase radicals with the catalyst surface and the adsorbed molecules, local heating and activation lattice oxygen and ozone [8,61].

The main target of this work is to review the literature on adsorption and/or plasma catalysis using zeolites for the decomposition of dilute VOCs in flue gas. In this article, the important aspects for the practical use of APC with zeolites for the decomposition of VOCs such as decomposition efficiency, energy efficiency, stability of the adsorbents/catalysts, and post-processing of the by-products obtained after plasma treatment have been discussed. In the first part of the article, the different kind of zeolites which are used in combination with NTP for environmental applications are introduced. In the second part of this article, the pros and cons of using zeolites in combination with non-thermal plasma for VOC abatement and stability of zeolites in APC technique are discussed in detail. Finally, regarding the performance of zeolite adsorption plasma catalytic reactors, the effect of various process parameters such as reactor configuration, humidity, and flow rate of the flue gas, initial concentration of VOCs, and nature of discharge gas is investigated.

2. Zeolites

In recent years, zeolite, a solid acid has been investigated by a number of researchers and used widely in industries due to its high surface area, ordered pore size and structure, thermal stability, shape selectivity, hydrothermal stability, mobility of their cations to act as a catalyst, and the ability to tailor its properties such as wettability and auxiliary mesopore generation in the crystals. Weitkamp and Moshoeshoe et al. [62,63] wrote interesting reviews about the basic principles of zeolite chemistry, structure, properties, and their application in catalysis. Furthermore, a series of zeolite-supported metals as highly efficient catalysts have been well developed for the catalytic combustion of VOCs by combining the selective adsorption of zeolites with catalytically active metal centers [64]. The natural zeolites are known to mankind as aluminosilicate minerals for almost 250 years and most of the natural zeolites are the result of volcanic activities [65]. Zeolites are a three-dimensional framework of SiO_4 and AlO_4 tetrahedra which are bonded by a common oxygen in the corners. Natural zeolite such as Mordenite, after acid treatment which is an essential step in order to eliminate the impurities and to open the pores and thus increasing S_{BET}, can be used for environmental applications [66]. The main disadvantages of natural zeolites are (i) the presence of impurities, (ii) they are not optimized for catalysis, and (iii) a variation in chemical composition which is source/deposit dependent [62].

In order to overcome the shortcomings of natural zeolites, synthetic zeolites with various Si/Al ratios (between 1 and ∞) and well-defined pore structure and dimensions are synthesized. The pioneering work of Barrer, Milton, and their co-workers [67–69] led to the synthesis of the first synthetic zeolites which made this porous material an important commercial heterogeneous catalyst. The general formula of zeolite is $M_{x/n}\left[(AlO_2)_x(SiO_2)_y\right].mH_2$ where M is the metal or hydrogen ion of valency "n" occupying the exchangeable site in the zeolite framework [70]. Figure 3 shows the schematic of the tetrahedral arrangement of SiO_4 and AlO_4 which are the primary building units of zeolites. The overall charge of the aluminosilicate framework, which is formed by polymerization of SiO_4 and AlO_4, is negative due to the +4 charge on silicon and the +3 on aluminum. This negative charge is compensated by the presence of cations such as Na^+, K^+, Ca^{2+}, and Mg^{2+} in pores or in rings, which are the exchangeable cations in the zeolite framework. As shown in Figure 4, the adjacent tetrahedral are linked via the common oxygen at their corners (which are called the secondary building units) which results in a three-dimensional macromolecule with channels and cages which are usually occupied by water or cations [71,72]. Figure 5 depicts the formation of four different commercial zeolites (such as Faujasite type (X and Y), ZSM-5, ZSM-12, and ZSM-22) from the primary and secondary building units and their micropore structure and dimensions.

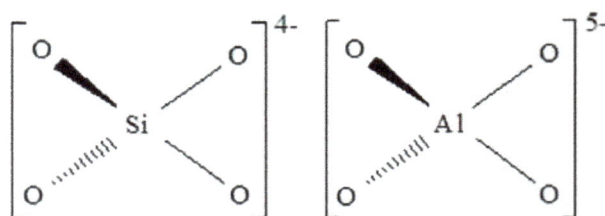

Figure 3. Tetrahedral arrangement of SiO_4 and AlO_4 which forms the primary building units of zeolite. Reproduced from Reference [63] under CC BY 4.0.

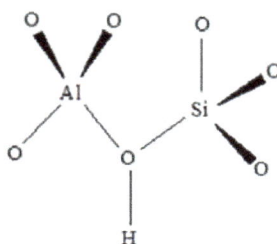

Figure 4. Tetrahedral arrangement of the Si–O and Al–O bonds forming a unit block of zeolite. Reproduced from Reference [63] under CC BY 4.0.

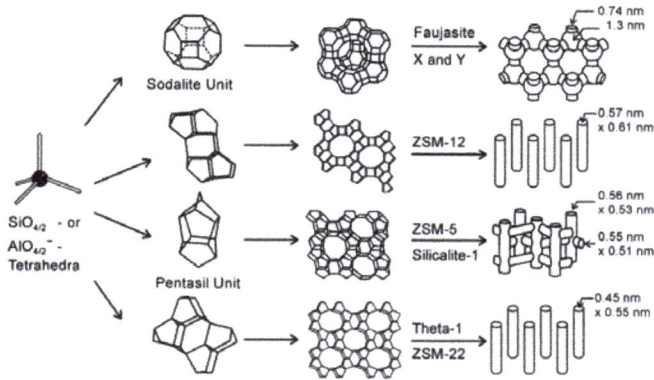

Figure 5. Structures of four commercial zeolites (from top to bottom: Faujasite or zeolites X, Y; zeolite ZSM-12; zeolite ZSM-5 or silicalite-1; zeolite Theta-1 or ZSM-22) and their micropore systems and dimensions. Reprinted from Reference [62], with permission from Elsevier.

Zeolites are classified based on the pore diameter and ring size as these properties have a major role in determining its adsorption properties [73]. Depending on the interconnections between the oxygen bridges and secondary block units, zeolites can be classified as 4-, 5-, 6-, 8-, 10 or 12- membered rings. A ring is made of tetrahedrons and an "n" number ring is made of n-tetrahedrons as shown in Figure 6. Zeolites such as MS-3A, MS-4A, MS-5A, and Mordenite contain eight numbers of rings with a pore diameter in the range of 3–5 Å; whereas ZSM-5 contains 10 rings with the pore diameter of ~3–5 Å. On the other hand, Faujasite type X and Y zeolite have 12 rings with larger pore diameter of 7–8 Å. Zeolites are also classified based on the Si/Al ratio as low silica ($\leq$2, e.g., Na-X Faujasite), intermediate silica (2–5, e.g., Na-Y Faujasite, Mordenite), and high silica zeolites (>5, e.g., ZSM-5, β-zeolites). The ratio of Si/Al influences the hydrophobicity of the zeolite and the hydrophilicity decreases with the increase in this ratio [74].

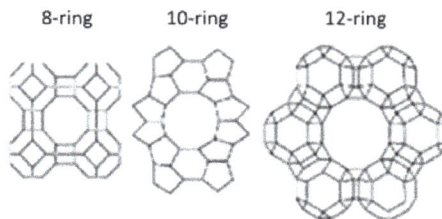

Figure 6. Schematic of "n" rings that are frequently found in the zeolite framework. Reprinted from Reference [75], with permission from Springer nature.

The synthetic zeolites which have been widely used for environmental applications in combination with non-thermal plasma are as follows: (i) Zeolite A, (ii) Zeolite X and Y (Faujasite type), (iii) β-zeolite, and (iv) ZSM-5. Zeolite A is the first industrially used synthetic zeolite which is still in commercial use today. Figure 7 shows the schematic drawing of the structure of zeolite A. The well-known molecular sieves such as MS-4A, MS-5A, and MS-6A are type A zeolite. Zeolite A is formed by the linkage between the sodalite cages which are linked by four ring cages and the nominal Si/Al ratio is 1:1 and the negative charge is balanced by sodium ions.

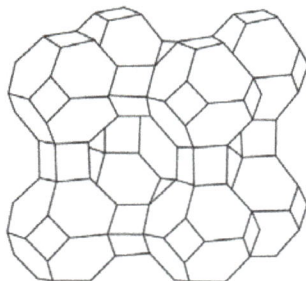

Figure 7. Schematic drawing of the structure of zeolite A. Reproduced from Reference [76] under CC BY 4.0.

Zeolite X and Y belongs to the family of aluminosilicates with Faujasite-type structure. Faujasite is a rare natural zeolite, whereas its synthetic counterpart Linde X and Linde Y are widely used in catalysis. Zeolite X and Y are formed by the linkage over six cornered surfaces and the hexagonal prisms. MS-13X zeolite which is a Faujasite-type zeolite has high adsorption capacity for VOCs due to its high S_{BET}. Figure 8 shows the three-dimensional structure of zeolite 13X and it consists of alumina tetrahedra and silica tetrahedron with oxygen bridge with a certain spatial structure forming a uniform crystal. The basic structural unit of 13X is a beta cage, and adjacent beta cages form the molecular sieve by the six-angle prism connection [77].

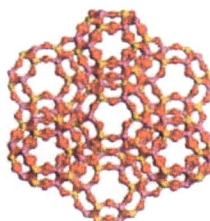

Figure 8. Schematic structure of Faujasite type zeolite. Reproduced from reference [77] under CC BY 4.0.

Zeolite beta is also one of the synthetic zeolites with high silica content and it is formed by the intergrowth hybrid of two different structures (polymorph A and B) with a stacking disorder (shown in Figure 9). Because of its high Si/Al ratio, hydrophobic nature, and higher acidic strength, zeolite beta is usually preferred rather than Faujasite type zeolites in various chemical reactions.

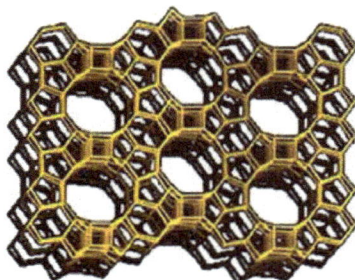

Figure 9. Schematic diagram of beta-zeolite. Reprinted from Reference [78], with permission from Elsevier.

As shown in Figure 10, the ZSM-5 zeolite is formed from the pentasil unit, which are interconnected by an oxygen bridge resulting in 10-membered ring. The Na^+ cations can be removed from the ZSM-5 via ion-exchange and replaced by H^+ ions yields zeolites in its protonic form, i.e., HZSM-5 [79] which is used for molecular sieving [80,81]. The VOC molecules with dynamic diameter smaller than the channel openings of HZSM-5 can access the zeolite's pore network and these adsorbed molecules can be eventually converted on the zeolite acid sites [80–83]. But the bulkier product molecules which cannot diffuse out of the pores being either converted to smaller molecules (resulting in product selectivity) or accumulated as coke (causing deactivation).

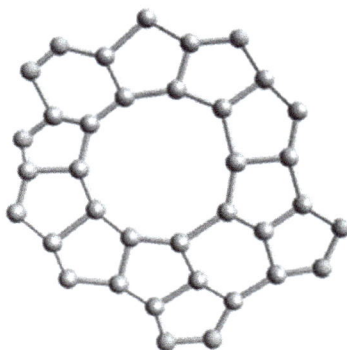

Figure 10. Schematic drawing of synthetic ZSM-5 zeolite. Reprinted from Reference [84], with permission from Elsevier.

3. Adsorption Plasma Catalysis for VOC Abatement

3.1. Adsorption of VOCs on Zeolite

The adsorption of VOCs on zeolite plays an important role in the complete oxidation of VOCs by increasing the residence time of VOCs in the plasma discharge region. The adsorption of VOCs on the adsorbents depends on the surface properties of the materials such as S_{BET}, pore volume, size, and structure. When the pore diameter is larger than the size of the VOC molecule, the VOCs could be adsorbed and enter the internal pore structure. When the pore diameter of the zeolite is smaller than the dynamic diameter of the molecule, they do not exhibit any adsorption of VOCs [85]. The size of the micropores of MS-3A, MS-4A, MS-5A, and MS-13X are approximately 0.3, 0.4, 0.5, and 1 nm respectively; whereas the dynamic size of benzene molecule is 0.59 nm and it is well adsorbed in the intrinsic pores of MS-13X [86]. The 13X zeolite has a hexagonal system and the supercage structure contained four 12-membered ring orifices with tetrahedral orientation [87] which has a large proportion of micropore with the pore size of 1.03 nm, which results in the excellent adsorption of aromatic VOCs such as benzene and toluene [86,88].

Huang et al. [89] investigated different zeolites such as MS-5A (5 Å), HZSM-5 (5.5 Å), Hβ (6.6 Å) and HY (7.4 Å) and reported that except MS-5A, other zeolites exhibited toluene adsorbing ability. The pore diameter of zeolite 5A is smaller than the dynamic toluene molecule diameter resulting in no adsorption. As shown in Figure 11, the breakthrough time for toluene adsorption on other zeolites (such as HZSM-5, Hβ and HY) follows the same order as that of the pore diameter [89]. The breakthrough time for toluene adsorption of HY zeolite is further increased by metal loading which will discussed in detail in the following section. Another work reported that the adsorption of benzene on different zeolites exhibited the same trend as that of their pore diameter MS-13X (10 Å) ≈ HY (7.4 Å) > MOR (6.7~7 Å) > ferrierite (4.3~5.5 Å) [41].

The effect of both S_{BET} and pore diameter of the zeolites on the adsorption of VOCs has been highlighted in the following works. Shiau et al. [54] reported that the adsorption of isopropyl alcohol on different molecular sieves such as MS-3A, MS-4A, MS-5A and MS-13X follow the same order as that of its specific surface area and pore diameter as shown in Table 1. Yi et al. [88] reported that 13X zeolite has better adsorption capacity for toluene when compared to the other adsorbents such as Al_2O_3 and MS-5A. This is because of the high S_{BET} (626.439 m^2·g^{-1}) and suitable pore size and structure [87] of 13X zeolite for toluene adsorption.

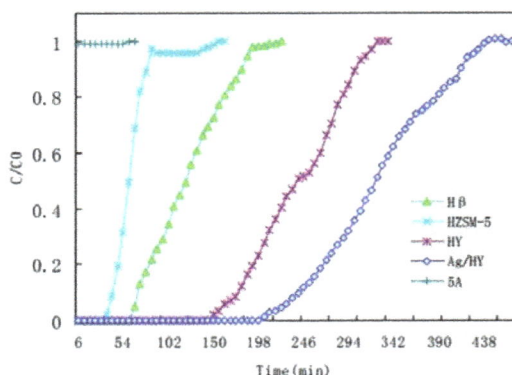

Figure 11. The breakthrough curves of toluene on different zeolites (flow rate: 300 mL/min and toluene inlet concentration: 100 ppm). Reprinted from Reference [89], with permission from RCS publishing.

Table 1. Specific surface area, pore diameter, and adsorption capacity of different zeolite A [54].

Zeolite	Specific Surface Area (m^2/g)	Pore Diameter (Å)	Adsorption Capacity (mg/g)
MS-3A	224.95	3.52	17.75
MS-4A	290.54	4.79	19.21
MS-5A	391.94	5.33	21.27
MS-10A	439.31	9.66	126.61

Thus, the adsorption capacity of zeolites for VOCs are roughly proportional to the S_{BET} and pore size as follows: NaY > MS-13X (10 Å) $\approx$ HY (7.4 Å) >> MOR (6.7–7.0 Å) > ferrierite (4.3–5.5 Å) [32]. If the pore size of the zeolites are larger than the dynamic molecular size of VOCs, the VOCs are adsorbed in the microchannels of zeolites.

The adsorption capacity of zeolites can be increased or decreased by metal loading. The VOCs adsorption capacity of zeolites is enhanced by the presence of metal cation due to the strong interaction between the cation in the zeolite framework and VOC molecules. On the other hand, the adsorption capacity of zeolites can be lowered by metal loading, because the metal ions reduce the specific surface area and block some pores of zeolites, resulting in increased pore diffusion resistance. Thus, the proper selection of the metal ions for loading is important for the enhanced VOC adsorption on zeolites.

The ethylene adsorption capacity of zeolites is enhanced by the presence Ag cation due to the strong interaction between the charge compensating cation in the zeolite framework and the double bond of ethylene [51,90–94]. When compared to bare 13X zeolite, Ag/13X zeolite exhibited higher ethylene adsorption capacity because of the π-complexation between Ag(I) species and ethylene double bond [93,95,96] in addition to the interaction between the adsorbed oxygen and ethylene. The metals with empty s-orbital and available electrons in the d-orbital could form π-complexation bonding. In case of bare 13X zeolite, the adsorption of ethylene on zeolite is via electrostatic interaction with Na$^+$ ions in 13X-zeolite and this physical adsorption is weak and reversible which is insufficient for the ethylene to be adsorbed to the inner pores of zeolite under flowing condition at atmospheric pressure and room temperature. Ethylene adsorption of zeolite is enhanced two orders of magnitude (from 2.66×10^{-4} to 3.37×10^{-2} g per gram of catalyst) by the incorporation of Ag [46]. When compared to HY, Ag loaded HY can adsorb more toluene which is due to the π-complexation of Ag/HY with toluene [89].

Several studies showed that metal loading on zeolite enhances the VOC adsorption capacity despite the reduction in their specific surface area [38,44,97]. As shown in Table 2, the metal loading on HZSM-5 reduces the S_{BET} as they occupy the cationic sites in zeolites which blocks the pores. Despite their reduced S_{BET}, Ag and Ag–Mn loaded HZSM-5 have a longer adsorption time when compared to bare HZSM-5 due to the ability of Ag to form π-complexation with toluene molecule [38]. Generally, the adsorption of toluene on HZSM-5 reduces with the loading of Mn. But, the mixed loading of Ag/Mn shifts the toluene adsorption site to Ag due to the strong intermolecular force (π-complexation) between Ag and toluene. Wang et al. [38] reported that during the adsorption of toluene on metal/HZSM-5, the partial oxidation of toluene has been observed on Ag, Mn, and Ag–Mn loaded HZSM-5 and the combination of Ag and Mn leads to further oxidation of these partial oxidation products to form CO_2 as shown in the diffuse reflectance infrared Fourier transform (DRIFT) spectra bands at 2330 cm^{-1} and 2360 cm^{-1} of Ag–Mn/HZSM-5 catalysts with adsorbed toluene (Figure 12) [38].

Table 2. Specific surface area and toluene adsorption capacity of different metal loaded HZSM-5 zeolite (amount of catalyst = 1.6 g) [38].

Catalyst	S_{BET} (m^2g^{-1})	Toluene Adsorption Amount (mmol)
Ag/HZSM-5	367.6	0.29
Mn/HZSM-5	398.5	0.22
Ce/HZSM-5	380.6	0.21
Ag–Mn/HZSM-5	350.0	0.27
Ce–Mn/HZSM-5	344.0	0.21
HZSM-5	486.6	0.26

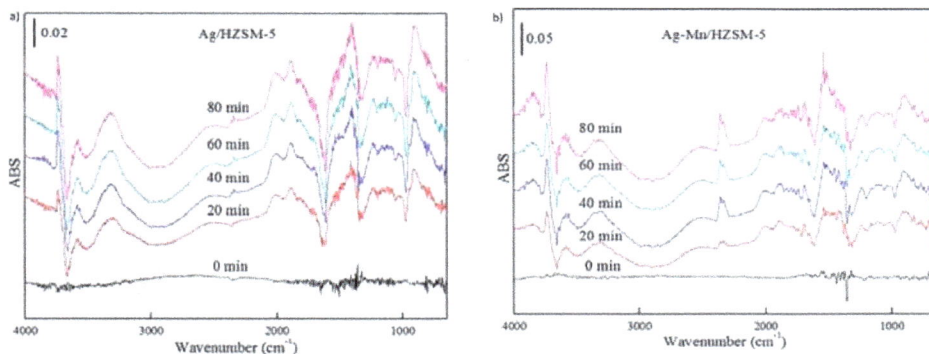

Figure 12. The diffuse reflectance infrared Fourier transform spectra (DRIFTS) of catalyst surface with adsorbed toluene (**a**) Ag/HZSM-5 and (**b**) Ag–Mn/HZSM-5 catalysts. Reprinted from Reference [38], with permission from Elsevier.

Among different transition metal loaded (Cu, Co, Ce, and Mg) 13X zeolites, Co/13X zeolite showed less decrease in the toluene adsorption capacity (as shown in Figure 13). Although the same of amount of metal has been loaded, the excellent adsorption of toluene on Co/13X is due to the ability of Co^{3+} to form π-complexation with toluene which is stronger when compared to other metal species such as Mg, Ce, and Cu [88]. Another work reported that among the different packing materials (CuO/MnO_2, $CeO_2/HZSM-5$ and Ag/TiO_2) tested for the adsorption of chlorobenzene, $CeO_2/HZSM-5$ exhibited rather high adsorption due to its high S_{BET} and pore volume of the support [97].

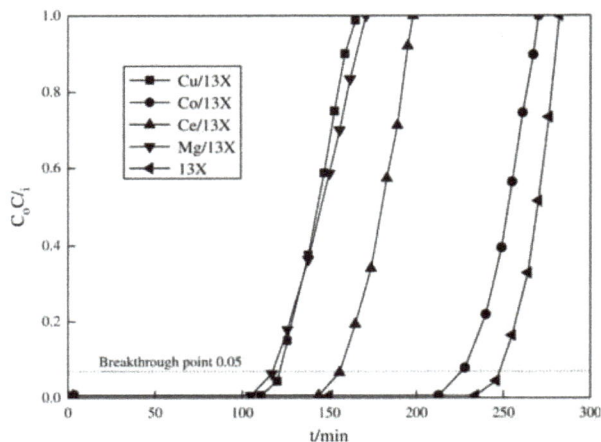

Figure 13. Adsorption breakthrough curves of 13X supported catalysts. Reprinted from Reference [88], with permission from John Wiley and Sons.

The co-incorporation of Ag and transition metal oxide in zeolites substantially increases the VOC adsorption capability due to the new active sites provided by the transition metals. Trinh et al. [95] reported that except Fe_xO_y, the addition of other transition metals (such as Cu, Mn, and Co) enhances the ethylene adsorption capacity of Ag/13X zeolite due to the formation of π-complexation [98]. Among the different bimetallic catalysts, Ag–Co/13X showed the maximum enhancement in ethylene adsorption. Zhao et al. [44] reported that among the different packing materials tested (such as HZSM-5, Ag/HZSM-5, Cu/HZSM-5, and AgCu/HZSM-5), AgCu/HZSM-5 showed higher breakthrough capacity for formaldehyde (as shown in Table 3) due to the synergistic effect of Ag and Cu. The adsorption of formaldehyde happens on both the metal sites (due to π-complexation) and zeolite on Ag/HZSM-5 and AgCu/HZSM-5; while the adsorption happens only on zeolite in Cu/HZSM-5 and HZSM-5. The co-incorporation of Ag and Cu increases the amount of Cu^{3+} ions and the formation of Ag–Cu species on AgCu/HZSM-5 catalyst which increases formaldehyde breakthrough capacity [99]. Qin et al. [100] investigated the metal supported HZSM-5/Al_2O_3 for the toluene decomposition in a cyclic APC system. The S_{BET} of the HZSM-5/Al_2O_3 is higher than the metal loaded HZSM-5/Al_2O_3. However, the breakthrough capacity of Ag–Mn loaded HZSM-5/Al_2O_3 is better when compared to the bare support due to the presence of Ag^+ active sites which are capable of forming π-complexation with the toluene molecule [96,100].

Table 3. Formaldehyde breakthrough capacity of different metal loaded HZSM-5 zeolites [44].

Sample	Breakthrough Capacity (µmol/mL-cat)
HZSM-5	8.5
Ag/HZSM-5	18.5
Cu/HZSM-5	11.8
AgCu/HZSM-5	38.9

High silica zeolite (HZSM-5 with $SiO_2/Al_2O_3 = 360$) shows preferential adsorption of VOCs in humid air due to its hydrophobic nature [39,101,102]. Liu and Fan et al. [39,101] investigated different metal loaded HZSM-5 for the adsorption of dilute benzene in humid air (1.5 vol% H_2O) and the breakthrough capacity are as follows: AgMn/HZSM-5 (131 µmol/cat) > Ag/HZSM-5 (94 µmol/cat) > Mn/HZSM-5 (78 µmol/cat) > HZSM-5 (45 µmol/cat). Ag/HZSM-5 has a longer breakthrough time due to the π-complexation of Ag with benzene [39,101] and the adsorption capacity of AgMn/HZSM-5 is much stronger due to the presence of more adsorption sites [39].

A wide range of humidity tolerance of the adsorbent is also one of the important factors for the cyclic APC process. Zhao et al. [44] reported that the HCHO breakthrough capacity of AgCu/HZSM-5 was almost constant over a wide range of relative humidity (20–93%) due to the hydrophobic nature of high silica HZSM-5 which results in the selective adsorption of formaldehyde on AgCu/HZSM-5 catalyst; whereas the HCHO breakthrough capacity is slightly higher in dry conditions. HiSiv zeolites are high silica Faujasite zeolites (SiO_2/Al_2O_3: HiSiv 1000 < 6.5, HiSiv < 10) known for their hydrophobic nature due to the high silica content and they are used for environmental applications due to their selectivity for organics in the presence of water. Kuroki et al. [47] reported that the mixture of HiSiv 1000 and HiSiv 3000 zeolite enhances the adsorption of a xylene mixture (o-, p-, and m-xylene) due to the capability of HiSiv 1000 zeolite to adsorb larger molecules (0.6–0.9 nm). However, HiSiv 3000 alone is not suitable for the adsorption of the xylene mixture as HiSiv 3000 zeolites are known for the adsorption of molecules which are smaller (<0.6 nm).

3.2. Plasma Catalysis of Adsorbed VOCs on Zeolite

One of the key issues finding an appropriate catalyst for cyclic APC technique is to find a catalyst which is not only good for VOC adsorption and has good humidity tolerance but also a catalyst which has the ability to completely oxidize the adsorbed VOCs. An overview of published papers on abatement of VOCs using zeolites in combination with non-thermal plasma discharge is given in Table 4. The decomposition of O_3 to form highly reactive $O^{\bullet}$ is one of the important reactions which governs the decomposition efficiency of adsorbed VOCs on zeolites because the rate of reaction of $O^{\bullet}$ with VOC is much higher than that of ozone. The decomposition of VOCs by ozone has been known to follow one of the following two mechanisms:

(i)   Langmuir-Hinshelwood (L-H) mechanism [103]:

$$O_3 + * \rightarrow O_2 + *O_{ads} \tag{1}$$

$$*O_{ads} + VOC_{ads} \rightarrow CO_x + H_2O + * \tag{2}$$

where * is the active site on the catalyst and $*O_{ads}$ is the active oxygen species adsorbed on the catalyst. Metal loaded zeolites (for example, MS-13X) have the ability to decompose ozone to form active oxygen species which react with the VOCs adsorbed on zeolite. In this model, both the reactants should be adsorbed on the surface and followed by migration to active site.

(ii)   Eley-Rideal (E-R) mechanism:

$$*VOC_{ads} + O_3 \rightarrow CO_x + H_2O + * \tag{3}$$

where $*VOC_{ads}$ is the VOC adsorbed on the catalyst. Metal unloaded zeolite (bare zeolite) has no ability to decompose ozone to form active oxygen (for example, H-Y zeolite). Unlike the VOC adsorbed on zeolite, which can be oxidized by ozone, ozone on bare zeolite has less ability to oxidize CO to form CO_2 and thus resulting in lower CO_2 selectivity [74]. In this model, only one reactant is adsorbed on the surface and the other exists in the gas phase. Also, the diffusion length of short-lived species such as O and OH and electrons are as small as 100 µm at atmospheric pressure. But ozone has a longer diffusion length of 10 mm due to it long lifetime and can thus reach the inner area of the micropores in zeolite [74].

The adsorption of VOCs on zeolite increases the residence time of VOCs in the plasma discharge region resulting in enhanced mineralization efficiency of cyclic APC technique. Yi et al. [61] investigated three different plasma reactor configurations such as (i) NTP alone, (ii) plasma catalysis (PC), and (iii) cyclic APC for the decomposition of toluene using Co/13X zeolite as a packing material. Toluene removal efficiency of cyclic APC was 92.7%; whereas a removal efficiency of 100% was achieved using NTP alone and PC. However, for APC, the mineralization efficiency and CO_2 selectivity were improved by 23.4% and 35.3%, respectively, when compared to NTP alone; whereas, the improvement was only 18% and 10%, respectively, for PC. Since the product selectivity towards the gaseous products CO_x ($CO + CO_2$) directly corresponds to the extent of mineralization, which is an important evaluation parameter for air pollution treatment. Thus, the higher mineralization efficiency of APC is due to the increase in the residence time of toluene and plasma generated active species in the plasma discharge region which completely oxidizes the adsorbed toluene and the by-products formed show that cyclic APC is suitable for air pollution treatment [104,105].

Yi et al. [88] investigated the cyclic APC using different metal loaded (Cu, Mg, Ce, Co) 13X zeolite for the decomposition of toluene. The metal loading on 13X zeolite enhances the CO and CO_2 yield, CO_x selectivity, and carbon balance. This is due to the dissociation of O_3 produced by the NTP discharge on the metal active sites (*) which played a significant role on the complete oxidation of toluene following the equations [103,106]:

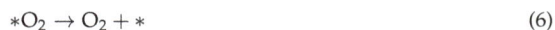

$$O_3 + * \rightarrow O_2 + *O \tag{4}$$

$$O_3 + *O \rightarrow O_2 + *O_2 \tag{5}$$

$$*O_2 \rightarrow O_2 + * \tag{6}$$

Of the different metal loaded 13X zeolites studied, Cu and Co showed excellent CO_2 selectivity, because the active sites of Co and Cu catalysts are Co_3O_4 and CuO_2, respectively, which have large oxygen adsorption capacity and easily convert oxygen to O^- and O^{2-} ion [88]. Oda et al. [107,108] investigated the combination of NTP and metal loaded ZSM-5 zeolite for the decomposition of TCE. They reported that the TCE removal efficiency using zeolites such as Na-ZSM-5 and Cu-ZSM-5 was 80% without the plasma. Due to the large specific surface area of ZSM-5 zeolites, this removal is mainly due to adsorption. As the plasma decomposition of TCE is also associated with its desorption, the TCE removal efficiency is never 100% even at higher discharge power [108]. A good absorbance of TCE at room temperature was exhibited by Cu-ZSM-5 which enhanced the removal efficiency of TCE to >95% even in the presence of weak plasma [107].

Nishimura et al. [46] investigated the use of Ag/13X zeolite in a packed bed dielectric barrier discharge (PBDBD) reactor for the removal of ethylene [46] and reported 100% removal efficiency with and without plasma discharge. Thus, the ethylene removal in the absence of plasma is attributed to the adsorption of ethylene on zeolite (π-complexation) and the enhanced ethylene removal in the presence of plasma is because of the widened range of plasma generation due to Ag loading on zeolite [32,41]. Also, the formation of unwanted by-products such as HCHO, O_3, and NO_x is greatly suppressed by the Ag/13X. The concentration of CO in the outlet is also reduced either due to the adsorption of CO on Ag/13X or via the decomposition reaction with the surface reactive oxygen produced by O_3 (Equations (4)–(6)). Similarly, another work reported that the formation of O_3 is reduced in the Ag/β-zeolite (~20 ppm), when compared to the bare zeolite (350 ppm in transient state and ~30 ppm in steady state) due to the decomposition of O_3 in metal active sites which increased the rate of acetone degradation when compared to the bare zeolite [109]. Kim et al. [8,110] reported that the metal loaded H-Y zeolite also enhances the carbon balance by retarding the formation of carbonaceous products on the surface of the catalyst and the largest enhancement of carbon balance (about 30%) is observed by supporting 2% Ag on H-Y zeolite.

In Ag/H-Y zeolite, ozone was adsorbed and decomposed both by the active sites of the zeolite substrate and by metal loaded on zeolite, and thus exhibiting an enhanced toluene removal efficiency and carbon balance. For example, Ag loading on H-Y zeolite increases the carbon balance from 83% to 98.1% [89]. Also, the formation of organic byproducts was retarded using Ag/H-Y zeolite due to the complete toluene oxidation by the active oxygen species obtained by the decomposition of O_3. Thus, the decomposition of toluene can be achieved by the impact of plasma active species in the gas phase. But the further oxidation of intermediates and by-products formed are effective only by the active $O^\bullet$ species produced by the decomposition of ozone on the surface of the catalyst.

Hu et al. [111] investigated the use of a series surface/packed bed discharge reactor (SSPBD) powered by a bipolar pulsed power supply in combination with metal oxide loaded 13X zeolite (TiO_2, MnO_2, and MnO_2–TiO_2) for the decomposition of benzene. The highest benzene decomposition efficiency and CO_x selectivity of 83.7% and 68.1% were obtained using MnO_2–TiO_2/zeolite packed reactor (10.33 W) because of the generation of OH radicals due to the charge transformation between Ti^{4+} and Mn^{4+} on the surface of MnO_2–TiO_2/zeolite catalyst which resulted in the separation of the photogenerated electron and hole. Also, the concentration of O_3 is greatly reduced due to the O_3 decomposition ability of MnO_2 [112–115]. Also, the formation of some intermediate by-products such as CO, HCOOH, and N_2O were also significantly suppressed.

It is well known from literature that MnO_x efficiently oxidizes O_3 [112–115] and Mn/HZSM-5 enhances the rate of benzene decomposition in cyclic APC [39]; whereas CO_2 selectivity of ~100% was achieved using Ag/HZSM-5 as packing material. This shows that Ag is necessary to promote the complete benzene oxidation [39,101,102]. Thus, the promotional effect of Mn for O_3 decomposition and Ag for complete benzene oxidation is obtained using AgMn/HZSM-5 zeolite [39]. Similar work has been performed for the decomposition of toluene by Wang et al. [38] and they reported that during the NTP regeneration of the toluene adsorbed catalyst, the mineralization efficiency of Ag-Mn/HZSM-5 is higher than the individual metal loaded HZSM-5 (such as Ag/HZSM-5, Mn/HZSM-5, Ce/HZSM-5).

The affinity of metal oxides which are loaded on zeolites towards the adsorbed VOCs plays an important role in determining the time required for a plasma discharge which in turn influences the energy cost of the process. Zhao et al. [44] studied the abatement of the adsorbed formaldehyde on different metal loaded (Cu, Ag, AgCu) and bare zeolites (HZSM-5). As shown in Figure 14, the evolution of CO_2 on AgCu/HZSM-5 and Ag/HZSM-5 is slower than that of Cu-HZSM-5 and bare HZSM-5 due to the difference in the adsorption sites towards HCHO for different zeolites under investigation. Despite the longer discharge time required for the complete decomposition of the adsorbed formaldehyde, the CO_2 selectivity and carbon balance of AgCu/HZSM-5 are ~100% suggesting that there is no other by-product formation [44].

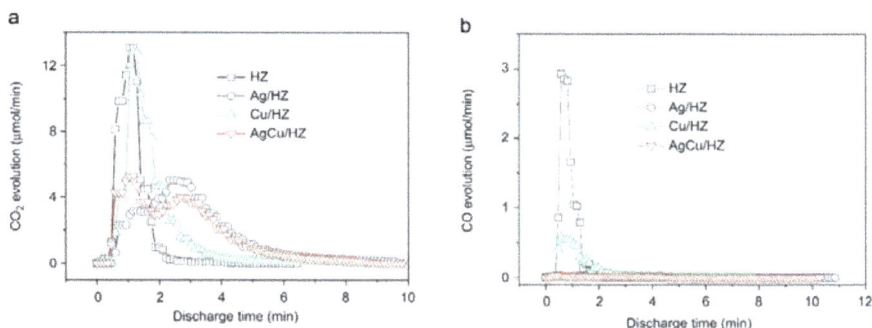

Figure 14. (**a**) CO_2 and (**b**) CO evolutions with discharge time in plasma catalytic oxidation of stored HCHO over HZ, Ag/HZ, Cu/HZ, and AgCu/HZ catalysts (storage stage: simulated air at 300 mL/min, 50% relative humidity (RH), GHSV = 12,000 h^{-1}, C_{HCHO} = 24.4 ppm (HZ), 27.2 ppm (Ag/HZ), 26.7 ppm (Cu/HZ) and 26.6 ppm (AgCu/HZ), t = 40 min; discharge stage: O_2 at 60 mL/min, $P_{discharge}^{APC}$ = 2.3 W). Reprinted from Reference [44], with permission from Elsevier.

Despite the lower affinity towards ethylene, the oxidation of ethylene by AgFe/13X zeolite is faster when compared to other transition metal (Co, Mn, and Cu) loaded Ag/13X zeolite, which is evident from the temporal evolution of CO_2. But, the desorption of ethylene and CO_2 production are the least in AgCo/13X zeolite. Also, the emission of by-products such as O_3 and CH_4 was suppressed by the Ag–Fe/13X zeolite when compared with other bimetallic 13X-zeolite. Thus, the affinity of the metal oxide loaded zeolite towards the adsorbed VOCs plays an important role in the desorption and oxidation of the desorbed VOCs [95] as the rate of oxidation reaction is higher in the gaseous phase [93]. The increase in the amount of adsorbent from 4.8 g (HiSiv 3000) to 8 g (HiSiv3000 = 2.5 g and HiSiv1000 = 5.5 g), decreases the conversion efficiency of p-xylene due to the fact that the conversion is faster in gaseous phase and the increase in the amount of adsorbent reduces the desorption of p-xylene [47], and thus the reduced conversion efficiency.

The metal loading on zeolite not only influences the oxidation state and the crystallinity of the catalyst, but also the pore volume and S_{BET} of the support. Thus, the optimum level of metal loading is critical for the best catalytic performance. Yi et al. [88] reported that the 5% Co/13X zeolite exhibits the best toluene adsorption and good plasma catalytic activity. When the Co loading is too low (1%), the number of active sites available for the π-complexation is too low, resulting in the weak chemical adsorption of toluene. When Co loading exceeds 15%, not only the pores of zeolite are blocked but also the active metal species can be significantly agglomerated which reduces the surface area of active sites, resulting in decreased catalytic performance [88].

The characteristics of the plasma discharge can also be affected by the amount of metal loading on zeolite [32,41,102]. For environmental protection applications, the plasma active region should be widely spread as the localized plasma consumes much energy without enhancing the removal efficiency. Thus, a uniformly distributed plasma is desirable for high energy and conversion efficiency. Kim et al. [32] reported that microscopic ICCD camera snapshots revealed that the metal loading on zeolite increases the number of micro-discharges and reduces the peak current. The number of micro-discharges in a DBD plasma reactor determines the amount of active radicals produced, and thus the VOC decomposition efficiency. As shown in Figure 15, without metal loading, the plasma is confined mainly in the vicinity of zeolite pellets; whereas, with metal loading, the plasma intensity is increased and expanded over a wide area due to the formation of surface streamers. Although the initial decomposition of VOCs occurs on the surface of the support such as zeolites, further decomposition of intermediates occurs on the active sites such as metals on zeolite [110] resulting in different CO_2 selectivity, carbon balance and formation of by-products. Also, the amount of metal loading on zeolite plays a significant role in the complete oxidation of the adsorbed VOCs. Fan et al. [102] reported that the carbon balance during the decomposition of benzene (P = 4.7 W) is ~100% for 0.8 wt% Ag/HZSM-5 and decreases with an increase in Ag loading on HZSM-5 zeolite. This decrease in carbon balance is due to the strong interaction between the adsorbed benzene and Ag^+ which makes it more difficult to desorb and oxidize the adsorbed benzene [102]. Thus, the loading of metals on zeolites showed enhanced VOC removal efficiency and CO_2 selectivity due to the following reasons: (i) enhanced catalytic activity of the metal loaded catalyst (more active sites) and (ii) the formation of a uniform discharge over metal loaded zeolite when compared to the bare zeolite [41].

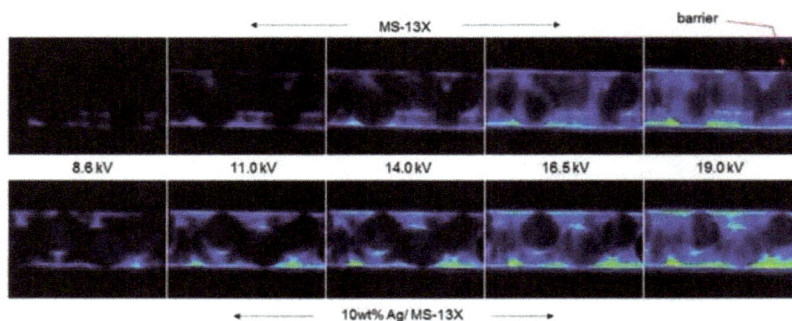

Figure 15. ICCD camera snapshots of the discharge plasma over molecular sieve 13X zeolite (MS-13X) with or without Ag loading; (frequency = 50 Hz, 2× lens, 2 mm gap, exposure time = 100 ms). Reprinted from Reference [32], © EDP sciences, 2011.

It is well known from literature that greater the S_{BET}, greater the catalytic activity [116]. However, Youn et al. [117] reported that Fe/ZSM-5 (S_{BET} = 362.3 m^2/g and pore volume = 0.17 cm^3/g) showed higher toluene oxidation in comparison with Fe/Beta (S_{BET} = 587.8 m^2/g and pore volume = 0.90 cm^3/g) in an APC technique. This is because the number of micro-discharges per pore volume of Fe/ZSM-5 is higher than of Fe/Beta which is evident from the Lissajous figure (as shown in Figure 16). This behavior enhanced both toluene oxidation (100% carbon balance) and CO_x selectivity [117]. With the increase in Fe content, the product selectivity of CO_x increases and ethylene reduces due to the oxidation ability of the Fe/zeolite [117].

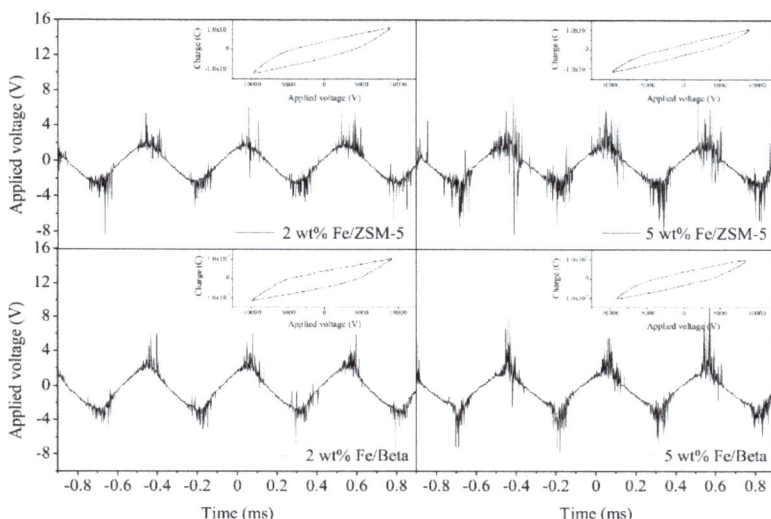

Figure 16. Discharge properties (current pattern and Lissajous plot) of Fe/ZSM-5 and Fe/Beta in toluene oxidation using a dielectric barrier discharge (DBD) plasma–catalyst hybrid system. Reprinted from Reference [117], with permission from Elsevier.

As already mentioned, the number of micro-discharges in a DBD plasma reactor determines the amount of active radicals produced. Apart from the metal loading, the number of micro-discharges also depends on the dielectric constant of a packing material. The discharge characteristics of the zeolite packed plasma reactor can be improved by mixing zeolite with other support materials with higher dielectric constant such as Al_2O_3, TiO_2, and $BaTiO_3$ [100,118]. As the dielectric constant of Al_2O_3 (9~11) is higher than HZSM-5 (1.5~5), the Al_2O_3 packed plasma reactor has a higher electric field strength when compared to a HZSM-5 packed reactor [119,120]. As shown in Figure 17, the number of micro-discharges produced in an Al_2O_3 packed plasma reactor is higher than that of HZSM-5 resulting in an increased number of active species produced [100] which is directly related to the enhanced VOC removal efficiency. The dielectric constant of the zeolites is also dependent on the Si/Al ratio and it is inversely proportional to the Si/Al ratio. Kim et al. [121] studied Ag/HY zeolites with different Si/Al ratio such as 2.6, 15, and 40 and reported that the propagation of the surface streamers are less when the dielectric constant is more than 15, and thus reduced plasma catalytic activity. Qin et al. [100] investigated the mixed packing material (Al_2O_3/HZSM-5) for toluene decomposition in the cyclic APC system. The high S_{BET} and suitable pore size of HZSM-5 keep the adsorbed toluene in the discharge region; whereas, the Al_2O_3 produces more active species for an effective toluene oxidation.

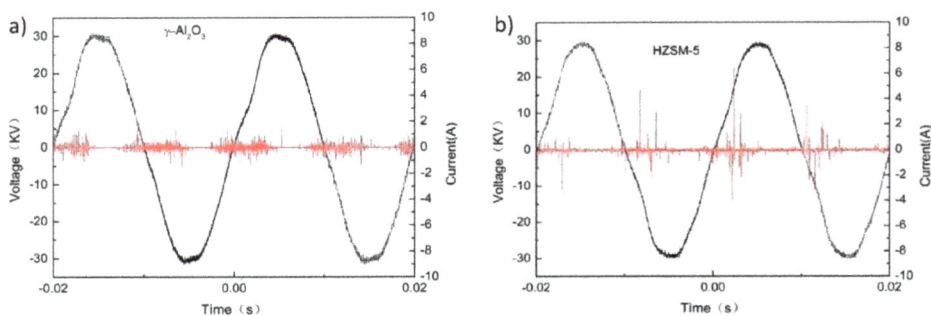

Figure 17. Voltage–current diagrams of (**a**) γ-Al$_2$O$_3$ and a (**b**) HZSM-5 packed NTP reactor. Reprinted from Reference [100], with permission from Elsevier.

Another work investigated the use of mixture of zeolites (such as MS-3A, MS-4A, MS-5A, and MS-13X) and ferroelectric material (BaTiO$_3$) for the decomposition of dilute benzene in flue gas [86]. The zeolite/BaTiO$_3$ hybrid reactor yields 1.4–2.1 times more CO$_2$ than the conventional BaTiO$_3$ reactor except for an MS-13X hybrid reactor which is due to the adsorption of CO$_2$ on MS-13X. In the physical mixture of packing materials, apart from the dielectric constant, the relative size of ferroelectric pellets and adsorbing packing material such as Al$_2$O$_3$ and zeolite also plays an important role in the intensification of plasma. Ogata et al. [55] studied the reaction field and the role of solid surface under plasma for the decomposition of benzene in air. The strong plasma density is necessary to induce catalysis in the plasma media [55] and this phenomenon has been demonstrated by using the combination of BaTiO$_3$ > porous Al$_2$O$_3$ (a mixture of 2 mm BaTiO$_3$ and 1 mm Al$_2$O$_3$) and BaTiO$_3$ < Al$_2$O$_3$ (a mixture of 1 mm BaTiO$_3$ and 2 mm Al$_2$O$_3$) as shown in Figure 18. Since the high-energy plasma is produced around the contact points of BaTiO$_3$ as shown in Figure 18, the combination of BaTiO$_3$ > Al$_2$O$_3$ is more effective for the plasma catalysis when compared to BaTiO$_3$ < Al$_2$O$_3$.

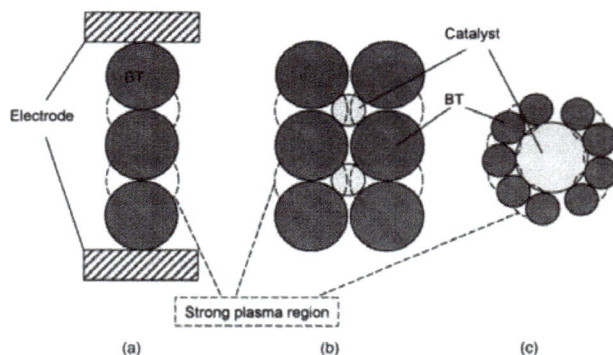

Figure 18. Representation of plasma discharge in (**a**) a conventional reactor packed with BaTiO$_3$ (BT) alone and the reactors packed with a mixture of BaTiO$_3$ (BT) and adsorbent Al$_2$O$_3$ catalyst (**b**) BT > Al$_2$O$_3$ and (**c**) BT < Al$_2$O$_3$. Reprinted from Reference [55], with permission from Elsevier.

The mineralization efficiency can be further improved by loading metal on the mixed packing materials. The highest CO$_2$ selectivity was obtained with AgMn/ZSM-5/BaTiO$_3$ for toluene decomposition [118] because the introduction of Ag–Mn: (i) favors the formation of π-complexation bonds with toluene which aided the reaction pathway to produce CO$_2$ and (ii) favors the oxidation of CO to CO$_2$. The AgMn/ZSM-5/BaTiO$_3$ enhances the toluene removal efficiency to 100%, reduces the concentration of O$_3$ by decomposing O$_3$ to O$_2$ and oxygen active species, which oxidize CO to CO$_2$

and resulting in a 83% mineralization efficiency. AgMn/ZSM-5/BaTiO$_3$ also reduces the production of N$_2$O.

Apart from the dielectric constant and metal loading on zeolites, the plasma discharge characteristics are also influenced by the textural properties of zeolites. When the size of the micropores are much smaller than the dynamic molecular size of VOCs, the adsorption of VOCs is restricted and also the formation of micro-discharges inside the pores is limited resulting in only surface discharges, and thus the decomposition efficiency is reduced [85]. But, the VOCs adsorbed on the external pores of zeolites are easily oxidized when compared to VOCs adsorbed in the inner pores [86]. Ogata et al. [55] reported that the ratio of the surface and the bulk pore volume is important for the decomposition of adsorbed VOCs because it is difficult for the plasma micro-discharges to reach the inner pores when the size of the catalyst is large. On the other hand, when the size of the by-products produced is bigger than the pore diameter, carbon deposition would appear [89].

Huang et al. [89] investigated different zeolites for the plasma driven catalytic abatement of toluene and reported that the mineralization efficiency follows the same order of the pore diameter. Although the pore diameter of MS-5A zeolite is too small to adsorb toluene, it still exhibited high toluene removal efficiency in combination with plasma due to the strong collision of electrons and radicals in the micro pores. For the zeolites that exhibited good toluene adsorption such as H-Y, Hβ and HZSM-5, the toluene removal was due to the combined effect of the collision of electrons and radicals and the adsorption of toluene on the zeolite surface [89]. The highest mineralization was achieved by the zeolite H-Y due to its larger pore size. The natural columbic electric field in the microporous structure of zeolites can strengthen the plasma discharge and enhances the VOC removal efficiency. Also, there was no formation of organic compounds on the surface of MS-5A zeolites after plasma treatment; whereas organic deposits were found on H-Y, Hβ and HZSM-5 zeolites. This is because the pore size of 5A zeolite is too small for the organic intermediates to access the internal pores.

On the other hand, the porous material with high S$_{BET}$ expands the discharge region because the streamers can be generated in the pores [88]. USY zeolite has very high surface area (715 m^{2}g^{-1}) and it is known for its VOC adsorption capacity. Hamada et al. [122] reported that the deposition of Mn$_x$O$_y$ on Y-zeolite enhances the removal efficiency of benzene and avoids the formation of organic by-products which might deactivate the catalyst. This is mainly due to the high S$_{BET}$ of Y-zeolite and O$_3$ decomposing ability of MnO$_2$ in Mn/USY catalyst.

The combination of non-thermal plasma and HZSM-5 has also been investigated for the decomposition of chlorinated VOCs such as dichloromethane (DCM) and chlorobenzene. Wallis et al. [123] investigated the combination of zeolites (such as HZSM-5, calcined HZSM-5, NaZSM-5, NaA, and NaX) and NTP discharge for the destruction of DCM in air. Among the different zeolites tested, HZSM-5 showed the highest DCM destruction efficiency of 36%; whereas the calcined HZSM-5 exhibited lower conversion due to the reduction in Brønsted acid sites as these sites play an important role in the oxidation of chlorinated hydrocarbons [97,124,125]. The HZSM-5 did not exhibit any reduction in NO$_x$ concentration; whereas calcined HZSM-5 reduces the production of NO$_x$ by 15%. The DCM decomposition efficiency of sodium zeolites such as NaZSM-5, NaA, and NaX are similar but not as high as HZSM-5 zeolites [123]. However, NaA is capable of reducing the concentration of unwanted by-products such as CO and HCOCl, and Na-zeolites are also good for deNO$_x$ up to 53% due to its basic sites.

Jiang et al. [125] studied the metal loaded HZSM-5 for the decomposition of chlorobenzene. As shown in Figure 19, for lower specific input energy, the removal efficiency of chlorobenzene was higher in the presence of CeO$_2$/HZSM-5 in the discharge region when compared to NTP alone. The dominant factors that influenced the oxidizability of CeO$_2$/HZSM-5 are its relatively high specific surface area, strong acidity, and redox properties. CeO$_2$/HZSM-5 exhibited better CO$_2$ and CO$_x$ selectivity due to the presence of more oxygen vacancies and reactive oxygen on the surface of deposited CeO$_2$ crystallites, in addition to the Brønsted acid sites of HZSM-5 which oxidize some of the intermediates on the catalyst surfaces [97,125]. The concentration of ozone in the exhaust of

CeO$_2$/HZSM-5 packed reactor is lower than the DBD reactor, because of decomposition of ozone on the catalyst surface to produce active oxygen atom by the following the mechanism mentioned in Equations (1) and (2) [125].

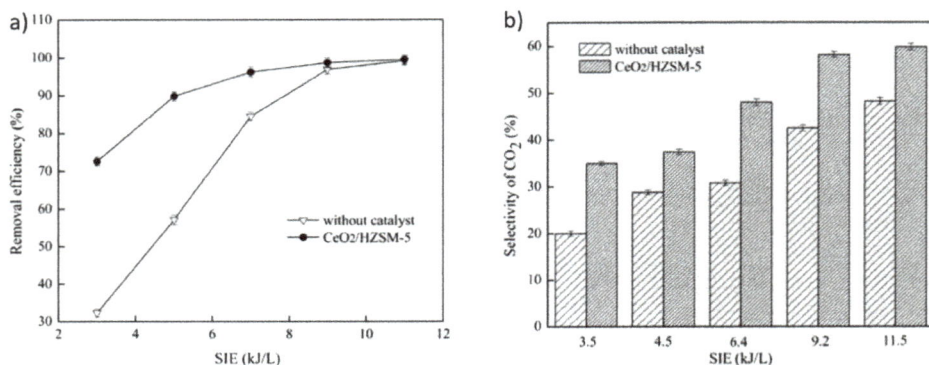

Figure 19. Influence of CeO$_2$/HZSM-5 on (**a**) chlorobenzene removal efficiency and (**b**) selectivity of CO$_2$. Reprinted from Reference [125], with permission from Elsevier.

The plasma catalysis using metal loaded zeolites can also be used as a pretreatment technology for biotrickling filter in order to convert the recalcitrant VOC compounds such as chlorobenzene into water soluble compounds [97]. Zhu et al. [97] reported that the synergy between the NTP and catalysts (CeO$_2$/HZSM-5) increased the chlorobenzene removal efficiency by 40% at low discharge voltages (5–6 kV) [97] which is preferable for a pretreatment technology. The water solubility and the biodegradability of the by-products are important for using the plasma catalysis system as a pretreatment process because the water-soluble carbon-based by-products can be easily used as carbon source by microbes. More water soluble and highly biodegradable by-products were produced by a CeO$_2$/HZSM-5 packed plasma reactor [97].

A number of studies have been conducted on the application of plasma catalysis for the treatment of single VOCs. However, a knowledge gap still exists in understanding the use of plasma catalysis for the treatment of VOC mixtures. Mustafa et al. [126] investigated the removal of a mixture of aliphatic, aromatic, and chlorinated VOCs such as toluene, benzene, ethyl acetate, trichloroethylene, tetrachloroethylene, and carbon disulfide by combining NTP and HZSM-5. A removal efficiency of 100% was achieved for all the compounds except CS$_2$ (80.18%) in the presence of HZSM-5 in the discharge zone at an input power of 16 W. The HZSM-5 has a large adsorbing capacity which extends the retention time of VOCs in the discharge zone and promotes the collisional probability of the adsorbed VOCs and the plasma generated active species, resulting in enhanced VOC removal efficiency. Treatment of VOC mixtures enhances the utilization efficiency of energy and the active species due to the reaction between partial oxidation products of different VOCs. The combination of NTP and HZSM-5 inhibits the formation of certain by-products such as cyclohexane (C$_6$H$_{12}$), pentadecane (C$_{15}$H$_{32}$), benzenonitrile (C$_7$H$_5$N), 2,2-dimethyltetradecane (C$_{16}$H$_{34}$), and 1-propene-1-thiol (C$_3$H$_6$S). These compounds were otherwise found as solid deposits in the plasma reactor.

Typically, the exhaust gas containing VOCs are large in volume with high flow rates and the reactor for treating these effluents should have a low-pressure loss. It has been reported that by using a honeycomb structured zeolite adsorbent in a closed loop DBD reactor, 93% toluene removal efficiency has been obtained (toluene initial concentration = 25 ppm and flow rate = 150 L/min) [33]. Also, the combination of densification of VOCs by adsorption followed by a plasma discharge has the adaptability for the change in the wide range of flow rates and VOC concentrations which is difficult with the flow type reactor (NTP alone or PC). Honeycombs can be produced in different shapes with different zeolite types, allowing customization of their adsorption behavior [127]. The main advantages of honeycomb zeolites are: (i) a lower pressure loss which is suitable for many industrial applications which involve high flue gas flow rate and (ii) an increased surface area for a given volume (as shown in Figure 20) [128].

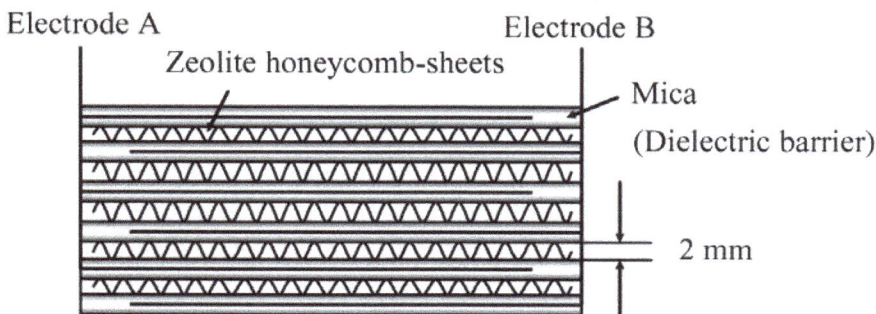

Figure 20. Illustration of the adsorption/plasma combined element with insertion of corrugated honeycomb sheets between discharge electrodes. Reprinted from Reference [37], with permission from Elsevier.

Hydrophobic honeycomb zeolites exhibited high toluene adsorbing capacity in humid air due to their high S_{BET}, pore size and their hydrophobicity retards the adsorption of water which can negatively influence the plasma performance. Inoue et al. [37] investigated the combination of an adsorption/in plasma catalytic reactor filled with hydrophobic dealuminated honeycomb Y-type zeolite as a packing material and a catalytic reactor filled with MnO_x placed in the downstream of the reactor for the decomposition of low concentrations of different VOCs (100 ppm) from humid air at high flow rates (60 m^3/h). In this particular hybrid system, the VOC decomposition takes places due to different reactions such as (i) on the surface of the adsorbent, (ii) in the gas phase, and (iii) on the catalyst due to the active oxygen produced via ozone decomposition. Among the different VOCs studied, the molecules with a C–O bond are easily decomposable with this system. The decomposition efficiency of different VOCs are as follows: alcohol and ether > aromatic and non-aromatic cyclic compounds > ketones.

3.3. Stability of Zeolite in Adsorption-Plasma Catalysis

The main difference between conventional PC and cyclic APC techniques lies in the regeneration of the adsorbents/catalysts. From an economic and application point of view, it is important that the performance of the adsorbents/catalysts is restored completely after plasma assisted regeneration. For the cyclic APC process, high catalytic reaction under the plasma discharge is required for the regeneration of adsorbents. This is achieved by air or O$_2$ plasma which simultaneously decomposes VOCs and regenerates the adsorbent. However, deactivation of the catalysts/adsorbents occurs in continuous usage due to the deposition of carbon containing molecules on zeolites which deteriorates the surface properties of zeolite and poisons the active sites. Thus, it is important that the researchers study the performance of the adsorbents/catalyst in continuous usage for longer time.

The stability of Ag/HZSM-5 for the adsorption of very low concentration of benzene (4.7 ppm) in humid air (RH = 50%) followed by O$_2$ plasma over five cycles of APC technique has been investigated [101]. It has been reported that Ag/HZSM-5 exhibited good stability and the carbon balance and CO$_2$ selectivity were kept around 100% for five cycles of cyclic APC of benzene. The stability of AgCu/HZSM-5 was investigated for the cyclic APC removal of low concentration of formaldehyde (26 ppm) by O$_2$ plasma. As shown in Figure 21, the stability of AgCu/HZSM-5 was maintained for five cycles with the carbon balance and CO$_2$ selectivity of ~100% [44].

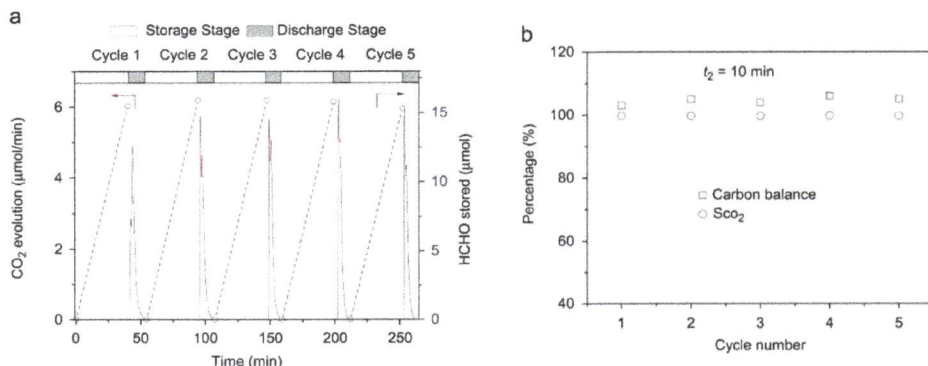

Figure 21. Comparison of (**a**) CO_2 evolution and HCHO storage and (**b**) carbon balance and CO_2 selectivity in five 'adsorption–plasma discharge' cycles. Reprinted from Reference [44], with permission from Elsevier.

The performance of Ag loaded β-zeolite reactor was investigated for four cycles of cyclic APC abatement of acetone (300 ppm) by O_2 plasma and it has been reported that the performance of the reactor was not significantly reduced [109]. However, the concentration peaks of the by-products were getting broader probably due to the oxidation of the low volatile and non-volatile by-products which were deposited on the surface of the catalyst. Kuroki et al. [28] also investigated the stability of honeycomb zeolite for 10 cycles of adsorption of low concentration of toluene (30 ppm) and air plasma for regeneration and reported that a regeneration efficiency of more than 80% was obtained for 10 cycles.

Yi et al. [61] investigated the stability of Co/13X zeolite in cyclic APC and PC technique for the decomposition of toluene (150 ppm) by air NTP discharge. In APC and PC, the toluene removal efficiency was not significantly affected by continuous usage for five adsorption-plasma cycles and 3 h, respectively. But, the adsorption of toluene, mineralization efficiency, and CO_2 selectivity were reduced. For PC, mineralization efficiency and CO_2 selectivity were reduced by 8.1% and 16.4%, respectively. For APC, the adsorption was significantly reduced from 0.51 mmol in the first cycle to 0.382 mmol in the fifth cycle and the mineralization efficiency and CO_2 selectivity were also reduced. The main reasons for the deactivation of Co/13X zeolites in continuous usage are due to the [61]: (i) retention of carbon containing molecules in the zeolite even after plasma discharge which might poison the active sites, (ii) significant reduction in S_{BET} and pore diameter and (iii) retention of H_2O produced during the decomposition of toluene.

Qin et al. [100] investigated a cyclic APC for seven cycles of toluene decomposition (1632 mg·m^{-3}) by air plasma and reported the deactivation of Al_2O_3/HZSM-5 support and the Ag–Mn loaded Al_2O_3/HZSM-5. The main reasons for the deactivation are as follows: (i) formation/residual of organic intermediates on the catalyst surface, which results in reduced S_{BET} and pore volumes, and (ii) change in the active components. For example, the following intermediates are found on the surface of deactivated HZSM-5: undecomposed toluene, 4-benzyl-1,2-dimethylbenzene, 2,2'-bimethylbiphenyl, benzaldehyde, and ethyl(propa-2-yloxl)acetate. The relative amount of Ag$^+$ on the deactivated catalyst is reduced resulting in the deactivation of the catalyst as Ag$^+$ is responsible for the π-complexation with toluene molecule.

During the air plasma catalytic removal of chlorobenzene (1250 mg·m^{-3}), the stability of CeO_2/HZSM-5 was maintained for the first 75 h. After this, the deactivation of CeO_2/HZSM-5 was noticed which may be due to halide and small organic matter deposition which blocks the active sites [125]. The distribution of elements before and after the reaction on the CeO_2/HZSM-5 using energy dispersive X-ray spectroscopy analysis (Figure 22) shows the presence of C and Cl elements after the reaction, confirming that a certain amount of organic and chlorinated matter was deposited or adsorbed on CeO_2/HZSM-5. This is due to the long reaction time and the continuous addition of chlorobenzene that results in insufficient oxidation of intermediates which were deposited on the surface of CeO_2/HZSM-5 covering the active sites, further reducing the decomposition efficiency [125].

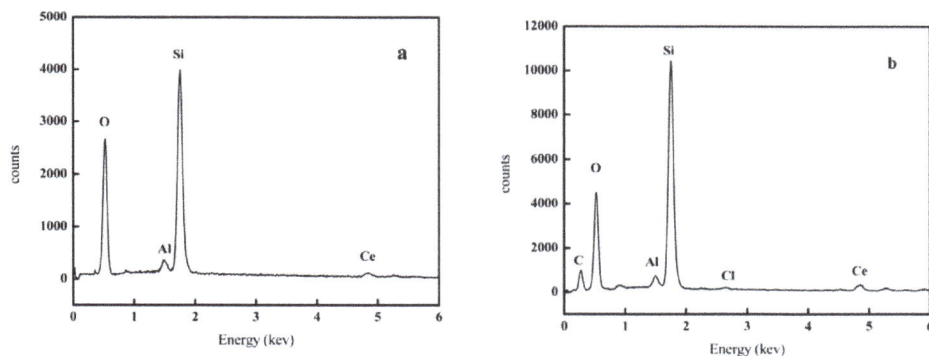

Figure 22. Energy dispersive X-ray spectroscopy spectrum of CeO_2/HZSM-5 (**a**) before and (**b**) after plasma catalytic reaction. Reprinted from Reference [125], with permission from Elsevier.

For the APC process, high catalytic reaction under the plasma is required for the complete regeneration of adsorbents/catalysts and this is achieved by an O_2 plasma that generates more active species and radicals which oxidize and decompose VOCs and simultaneously regenerates packing materials.

4. Effect of Process Parameters

4.1. Effect of Humidity

The presence of water in the flue gas has both a beneficial and detrimental effect: (i) moderate humidity (0.2 vol% to 0.4 vol%) promotes the decomposition of VOCs by the formation of OH• radicals, (ii) however, in the presence of excess humidity (>0.4 vol%) water molecules can adsorb on the surface of the adsorbent/catalyst and block the catalytically active sites and therefore inhibiting the VOC removal [39]. But the effect of presence of water in a flue gas on the adsorption of toluene on a hydrophobic honeycomb zeolite was negligible [29]. Fan et al. [102] reported that the effect of different relative humidity of the flue gas (RH = 0% to 60% at 25 °C) is weak on the decomposition of benzene due to the high hydrophobic nature HZSM-5 zeolite.

The main disadvantage of the presence of H_2O on the on the surface of adsorbent/catalyst and in the plasma discharge region are as follows: (i) water changes the physical and chemical properties of the discharge by quenching the activated chemical species and limits the electron density in the plasma, (ii) water reduces the total charge of a micro-discharge and decreases the plasma volume, and (iii) water covering the catalyst surface results in the hindrance of ozone adsorption and decomposition for the formation of active oxygen species. Hamada et al. [122] investigated the effect of adding water (1%) to the flue gas on the performance of a Mn/USY packed silent discharge reactor and reported that the addition of water has a negative impact on the decomposition of benzene because it inhibits the formation of ozone and lowers the catalytic activity of Mn–USY.

4.2. Effect of Initial Concentration

Inoue et al. [37] investigated the performance of the newly developed hybrid plasma reactor (adsorption on hydrophobic dealuminated honeycomb Y-type zeolite and MnO_2 in the downstream of plasma reactor) in real conditions to treat exhaust gases from painting and adhesive industry. They reported that this hybrid reactor is more suitable for treating the flue gas when the initial VOCs concentration is less than the critical VOC concentration V_{CRT} and it can respond well to temporary rises in VOC concentrations by adsorption and decomposition (e.g., in applications of the printing industry). On the other hand, when concentrations of VOCs are continuously higher than the V_{CRT}, the adsorption function of zeolite is not very effective because the adsorption sites are already saturated by VOCs (e.g., in applications of the adhesive industry). When the initial concentrations of VOCs are as low as the V_{CRT} (~300 ppm), a conversion efficiency of 80% was achieved [37].

4.3. Effect of Discharge Gas

The main difference between the conventional adsorption and a cyclic APC technique lies in the regeneration step [32]. In conventional adsorption, the adsorbent is regenerated either by heating or by supplying hot water vapor and the resulting high concentration of VOCs has to be further oxidized either by thermal or catalytic oxidation. However, for an APC process, high catalytic reaction under the plasma is required for the regeneration of adsorbents/catalysts, and this is achieved by air or O_2 plasma which decomposes VOCs and simultaneously regenerates the adsorbent/catalyst. Kim et al. [129] reported first the complete oxidation of adsorbed benzene on Ag/TiO_2 using oxygen plasma without the formation of by-products such as CO and NO_x and regeneration of the adsorbent.

During discharge, oxygen as background gas enhances the oxidation of adsorbed VOCs and inhibits the formation of unwanted by-products such as NO_x [130]. In air as a background gas, N_2 competes with VOCs for the active oxygen species resulting in reduced VOC conversion efficiency. Kim et al. [8] reported that O_2 driven PC reactor in cyclic operation retards the formation of N_xO_y. Thus, O_2 plasma is suitable for the complete regeneration of the zeolites. The presence of oxygen in a plasma has the following effect on the plasma discharge: photoionization, streamer properties, electron attachment, transition of filamentary to glow discharge in DBD [8], which are more suitable for the complete VOC oxidation. Despite the advantages of using oxygen plasma to avoid the formation of NO_x and to produce highly oxidizing environment, using air as a discharge gas is far more economical in application point of view. Although, the formation of NO_x (such as N_2O and NO_2) are unavoidable while using air plasma, the concentration of NO_x produced can be reduced by minimizing the discharge time and maximizing adsorption time in the APC process [44]. When air was used as the desorption gas, part of toluene has been decomposed before desorption from the hydrophobic honeycomb zeolite because toluene was more easily decomposed in air than N_2 [28,29]. Shiau et al. [54] reported that the desorption of isopropyl alcohol is higher when N_2 is used as a desorbing gas when compared to O_2 because air or O_2 plasma generates more active species and radicals which oxidize and decompose IPA resulting in reduced desorption of IPA and enhanced IPA conversion.

4.4. Effect of Gas Flow Rate

The gas flow rate during the NTP discharge in cyclic APC techniques plays an important role in the desorption and oxidation of the adsorbed VOCs because the flow rate influences the exposure time of the desorbed VOCs to the plasma. On the other hand, increasing the flow rate produces more active species that enhance the VOC decomposition. Kuroki et al. [29] reported that the increase in air flow rate from 1 to 4 L/min increases the toluene desorption ratio from 49% to 72%. This is due to the decrease in exposure time of the desorbed toluene to the plasma discharge and suppressed decomposition of desorbed toluene with increased air flow rate [29]. Nevertheless, the selectivity of CO_2 increases and CO decreases with increase in flow rate.

As shown in Figure 23, the gas flow rate during the plasma discharge process plays an important role determining the plasma operating time and thus the energy efficiency of the process. Youn et al. [117] reported that the time required for toluene desorption and oxidation could be reduced by reducing the air flow rate. This is due to an increased specific energy density with the reduced flow rate for the same input power which provides more energy for desorption and oxidation [117].

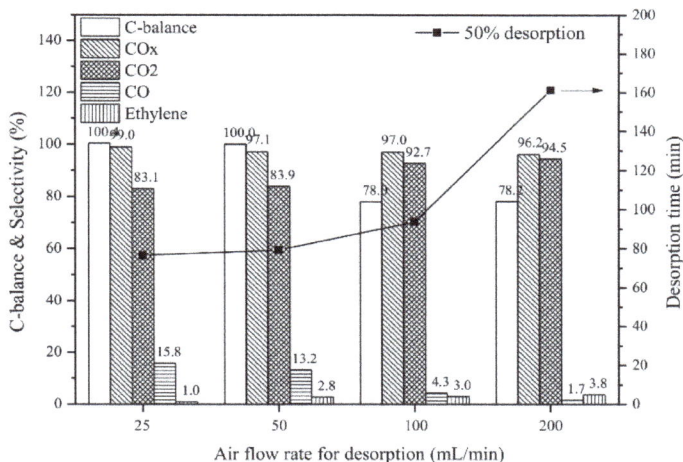

Figure 23. Effect of air flow rate on the performance of 2 wt% Fe/ZSM-5 packed DBD reactor for toluene oxidation. Reprinted from Reference [117], with permission from Elsevier.

4.5. Reactor Configuration

4.5.1. Desorption Method

Yi et al. [131] investigated a closed and ventilated plasma reactor for the removal of toluene in a cyclic APC process. This work reported that the closed reactor is suitable for low concentration of VOCs as the residence time of VOCs in the discharge zone is prolonged; whereas, a ventilated discharge is more suitable for high concentration of VOCs because more reactive species (O$^\bullet$, OH$^\bullet$ and O$_2$$^\bullet$) are generated. For example, the adsorption of toluene on MS-5A is low and the closed reactor during the plasma discharge has enough reactive species per toluene for the conversion, yielding better carbon balance and CO$_x$ selectivity. On the other hand, toluene adsorption on MS-13X zeolite is very high and the average number of reactive species available for every toluene molecule is not enough in a closed reactor resulting in poor oxidation. Thus, a ventilated discharge is more suitable for the higher concentration of VOCs as this produces more oxygen reactive species.

Kuroki et al. [28] investigated the effect of a plasma desorption method (closed and conventional open) on toluene decomposition and regeneration of honeycomb zeolites. This work concludes that the regeneration efficiency and desorption efficiency of the closed system is superior than the conventional open system [28]. The formation of NO$_2$ is suppressed in the closed system. The optimum closing time for the higher toluene desorption and zeolite regeneration efficiency is 1 min; whereas higher closing time is required if complete toluene oxidation is desired [28].

4.5.2. Position of Catalysts

The position of the zeolite in the plasma reactor influences the rate of oxidation reactions, decomposition efficiency and eventually, the mineralization efficiency. Trinh et al. [93] investigated three plasma reactor configurations such as one stage (in-plasma), two stage (post-plasma), and hybrid reactors (zeolites are placed towards the tail of the plasma discharge zone) for the decomposition of ethylene on 13X zeolite (as shown in the Figure 24). During the plasma discharge, the by-products produced in one- and two-stage reactors are mainly CO$_x$; whereas HCHO and desorbed ethylene were also found in the hybrid reactor. The formation of HCHO and ethylene desorption suggests the fast oxidation reaction in the gas phase in the hybrid reactor. As shown in Figure 25, in a two-stage plasma catalytic reactor, there is no formation of CO$_2$ in the first 10 min because the ozone produced in the plasma reactor diffused and decomposed on the catalytic surface to form atomic oxygen which reacts with the adsorbed ethylene to form CO_2. Also, the O_3 production is low in the one stage reactor due to the small gas volume, and thus the oxidation of ethylene is mainly by the diffusion of short-lived species into the micropores of zeolites that have adsorbed ethylene. When compared to the one- and two-stage plasma reactors, fast temporal evolution of CO_2 is observed (Figure 25) with the hybrid plasma reactor due to the synergetic effect of both reactors utilizing the O_3 formed in the blank part of the reactor and the short-lived species formed in the packed part of the reactor.

Oh et el. [36] studied the influence of the amount of toluene adsorbed on a zeolite on the toluene decomposition efficiency depending on the position of the zeolite in the plasma reactor. When the zeolite is placed in the tail of the plasma reactor, the decomposition efficiency of toluene increases with the adsorption capacity; whereas, the decomposition efficiency is not influenced by the adsorption capacity when zeolite is placed in the head of the plasma reactor [36]. These results suggest that the decomposition of toluene adsorbed in the micropores of zeolites by a direct plasma exposure is not easy; while the ozone produced in a plasma forms active atomic oxygen which effectively oxidizes toluene. Other works reported that the positioning of zeolite in the tail of the plasma reactor increases the conversion efficiency between 10% and 20% depending on the type of zeolite (NaY, H-Y, Ferrierite and Mordenite) [36,132].

Teramoto et al. [74] reported that the decomposition efficiency is 3.7 times and 1.2–1.5 times higher when compared to the conventional NTP reactor when a zeolite is placed in the tail (downstream) and in the beginning (upstream) of the plasma reactor, respectively. This is because the amount of ozone increases towards the tail of the reactor and according to the Equation (3), O_3 is important for the decomposition of adsorbed VOCs by metal unloaded zeolites. Thus, the decomposition efficiency is suppressed when a zeolite is placed in the upstream of the plasma reactor [74].

Hamada et al. [122] investigated the use of a surface discharge reactor in combination with Mn loaded USY zeolite for benzene decomposition. For low input power range, the decomposition of ozone can be achieved by placing Mn/USY zeolite in the downstream of the plasma reactor, which enhanced benzene decomposition efficiency. For higher input power range, where the concentration of ozone produced is reduced by heating, it is more effective to place Mn/USY zeolite in the discharge zone as they can make use of reactive species with shorter lifetime in addition to O_3.

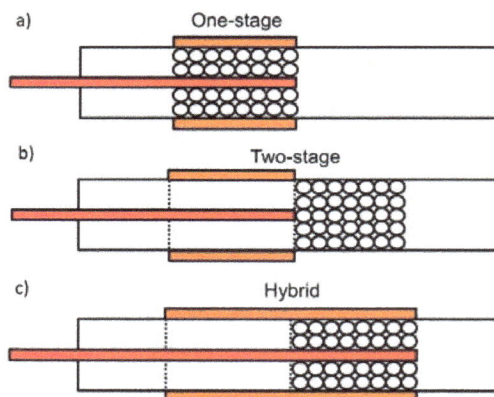

Figure 24. Different configurations of non-thermal plasma reactors. (**a**) One-stage, (**b**) two-stage, and (**c**) hybrid plasma reactors. Reprinted from Reference [93], with permission from Elsevier.

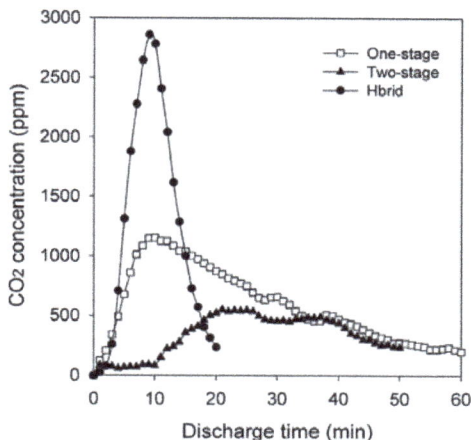

Figure 25. The evolution of CO_2 concentration for different reactor configurations (adsorption time: 100 min, applied voltage: 20 kV). Reprinted from Reference [93], with permission from Elsevier.

4.6. Energy Cost

The energy efficiency of a cyclic APC technique can be three times higher (increased by 1.39 gkw^{-1}h^{-1}) than NTP alone and plasma catalysis under the same experimental conditions for the decomposition of toluene on Co/13X zeolite [61]. This is because the flow rate during the plasma discharge of APC and discharge time were significantly reduced in cyclic APC resulting in enhanced energy density and energy efficiency, respectively. Liu et al. [39] reported that the energy efficiency of APC is almost six times higher (19.72 gkW^{-1}h^{-1}) when compared to the continuous plasma catalysis (3.05 gkW^{-1}h^{-1}) for benzene decomposition using AgMn/HZSM-5 zeolite.

The energy cost (EC) of a cyclic APC technique can be decreased by increasing the ratio of adsorption time (t_1) and discharge time (t_2) according to the following equation [44]:

$$EC^{APC} = \frac{P^{APC}_{discharge} t_2}{t_1} \qquad (7)$$

Fan et al. [101] reported the energy cost of the cyclic APC technique for the decomposition of VOCs was as low as 3.7×10^{-3} kWh m^{-3} for the remediation of 4.7 ppm benzene from humid air (50% RH) using Ag/HZSM-5. In the cyclic treatment, the duration of the plasma discharge for the complete oxidation of adsorbed VOCs depends on two important factors: (i) the amount of VOCs adsorbed during the storage step and (ii) the applied voltage [95]. Zhao et al. [44] studied the effect of the adsorption time of formaldehyde on AgCu/HZSM-5 zeolite on the discharge time for the complete oxidation of adsorbed formaldehyde. As shown in Figure 26, the storage time has no effect on the discharge time and a discharge time of 10 min is required for the complete oxidation of adsorbed formaldehyde to CO_2. The energy cost of the cyclic APC process for a storage time of 690 min has been calculated to be 1.9×10^{-3} kWh m^{-3}; whereas the energy cost can be reduced to the order of 10^{-5} to 10^{-4} kWh m^{-3}, considering the possible long adsorption time of AgCu/HZSM-5 zeolite [44]. On the other hand, another work reported that when the storage period of benzene on Ag/HZSM-5 increased from 1 to 14 h, a small increase in the discharge time from 9 to 24 min was required to achieve ~100% conversion of benzene to CO_2 [102].

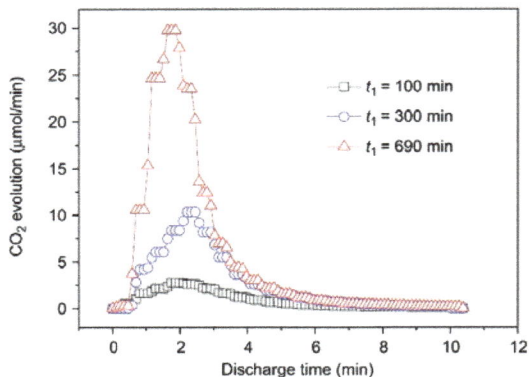

Figure 26. CO_2 evolution with discharge time in cyclic APC of HCHO over AgCu/HZ at various storage periods (100, 300 and 690 min) (storage stage: simulated air at 300 mL/min, 50% RH; discharge stage: O_2 at 6 mL/min, $P_{discharge}^{APC}$ = 2.3 W). Reprinted from reference [44], with permission from Elsevier.

The energy efficiency of a HZSM-5 packed plasma reactor is 29.45% higher when compared to a plasma alone reactor at 16W which can also be attributed to the adsorption capability of HZSM-5 zeolite and outcompetes other reactors [126]. The presence of CeO_2/HZSM-5 in the discharge region enhances the energy efficiency by 1.25 times when compared to NTP alone (at 3.5 kJ/L) for chlorobenzene removal which is an obvious economic benefit [125]. Ogata et al. [55] reported that despite the large adsorption capacity of MS-13X, the energy efficiency of a MS-13X hybrid reactor was lower than the reactors packed with MS-3A, MS-4A, and MS-5A. This is because MS-13X does not completely desorb the large amount of adsorbed benzene to the gas phase and the energy efficiency was calculated based on the CO_x formed and the inlet benzene, resulting in reduced energy efficiency.

The combination of adsorption, catalysis, and non-thermal plasma using zeolites suppresses the formation of by-products such as NO_x, CO, and O_3. Trinh et al. [109] reported that using silver coated β-zeolite for the removal of acetone in a cyclic APC technique improves the energy efficiency and suppresses the formation of un-wanted by-products. But, the formation of unwanted by-products such as CO, N_2O, and NO_2 increases with increase in discharge power which is required for a faster conversion mechanism [39]. The formation of NO_2 may poison the active sites of Ag in AgMn/HZSM-5. Zhao et al. [44] reported that the optimum discharge power for the conversion of adsorbed HCHO to CO_2 is 2.3 W because with the decrease in the discharge power to 1.4 W, the conversion of HCHO reduces to 76%; whereas there is no further improvement in the performance with a further increase of discharge power.

Table 4. Overview of published papers on abatement of VOCs using zeolites in combination with non-thermal plasma discharge.

VOCs	Zeolite	S_{BET} (m²/g)*	Pore Size (Å)*	Type of Discharge	Carrier Gas	Flow Rate (L/min)*	Initial VOC Concentration (ppm)*	Amount of Adsorption/Removal Efficiency	Energy Density and Frequency	By-Products	Ref.
Acetone	3 wt % Ag/β-zeolite (15 g)	-	-	PBDBD (IPC) t_1 = 100 min	N₂	2	300	2.45 mmol	-	-	[109]
				PBDBD (IPC) t_2 = 15 min	O₂	2	-	97%	28 W (400 Hz)	CH₄ (less), HCHO, CH₃CHO	
Benzene	Ferrierite	270	-	SD (IPC) adsorption	N₃	4-5	200	-	-	-	[5]
				SD (IPC) Closed discharge	50% O₂/N₂	5-8	-	57%	130 J/L	-	
	MS-13X	540	-	SD (IPC) adsorption	N₂	4-5	200	-	-	-	
				SD (IPC) Closed discharge	50% O₂/N₂	5-8	-	43%	89 J/L	-	
	H-Y zeolite	-	-	SD (IPC) adsorption	N₂	10	200	-	-	-	
				SD (IPC) Closed discharge	50% O₂/N₂	5-8	-	84%	140 J/L	-	
	2% Ag/H-Y zeolite	520	-	SD (IPC) adsorption	N₂	10	200	-	-	-	
				SD (IPC) Closed discharge	50% O₂/N₂	5-8	-	89%	160 J/L	-	
Benzene	TiO₂-MnO₂/zeolite	-	-	SSPBD (IPC) Continuous	Air	0.5	400	81%	10.33 W (50 Hz; 1 nF)	CO	[111]
Benzene	Mn/USY zeolite	715	-	SD (IPC) Continuous	20% O₂/N₂	1	200	95%	12 W (60 Hz)	CO	[12]
Benzene	Ferrierite	-	-	SD (IPC) Continuous	50% O₂/N₂	5-8	200	58%	130 J/L (500 Hz)	-	[41]
	10% Ag/Mordenite	-	-		50% O₂/N₂			97%	130 J/L (500 Hz)	-	
	10% Ag/HY	-	-		60% O₂/N₂			96%	154 J/L (500 Hz)	-	
	10% Ag/MS-13X	-	-		60% O₂/N₂			88%	154 J/L (500 Hz)	-	

Table 4. Cont.

VOCs	Zeolite	S_{BET} (m²/g)*	Pore Size (Å)*	Type of Discharge	Carrier Gas	Flow Rate (L/min)*	Initial VOC Concentration (ppm)*	Amount of Adsorption/Removal Efficiency	Energy Density and Frequency	By-Products	Ref.
Benzene	Ferrierite	270	4.3–5.3	SD (IPC) Continuous	Synthetic air	4	200	78%	200 J/L (500 Hz)	–	[110]
	2 wt % Ag/H-Y	520	7.4			10		75%			
	2 wt % Ag/H-Y & 0.5 wt % Pt/γ-Al₂O₃	–	–			10		86%			
Benzene	Ag/HZSM-5	334	–	DBD (IPC) t_1=60 min	Humid air (RH = 50%)	0.6	4.7	7.6 µmol	–	–	[101]
				DBD (IPC) t_2=13 min	O₂	0.06	–	100%	4.7 W (2 kHz)	CO	
Benzene	HZSM-5 (0.05 g)	341	–	DBD (IPC) adsorption up to 5% of initial benzene	Humid air (1.5 vol % H₂O)	100 SSCM	20	45 µmol/cat	–	–	[39]
	0.7 wt % Ag/HZSM-5 (0.05 g)	341	–					78 µmol/cat			
	2.4 wt % Mn/HZSM-5 (0.05 g)	334	–					94 µmol/cat			
	0.8 wt % Ag 2.4 wt % Mn/HZSM-5 (0.05 g)	336	–					131 µmol/cat			
Chloro-benzene	12 wt % CeO₂/HZSM-5 (0.3 g)	–	–	DBD (IPC) Continuous	Dry air	–	1250 mg/m³	96%	7 kV (10 kHz)	–	[125]
Chloro-benzene	12 wt % CeO₂/HZSM-5 (0.3 g)	315.8	–	DBD (IPC) Continuous	Air	1	1250 mg/m³	72.6%	5 kV (10 kHz)	CO	[97]
Dichloro-methane	HZSM-5 (140 °C)	338	5.4	PBDBD (PPC) Continuous	Air	1	500	36%	0.9 W (10.25–13.25 kHz)	CO, NOx, HCOCl	[123]
	Calcined HZSM-5 (140 °C)	338	5.4					30%		CO, NOx, HCOCl	
	NaZSM-5 (140 °C)	338	5.4					32%		<NOx, HCOCl, CO	
	NaA (140 °C)	–	4					32%		<HCOCl, CO <NOx	
	NaX (140 °C)	–	10					32%		<NOx, HCOCl, CO	

Table 4. *Cont.*

VOCs	Zeolite	S_{BET} (m²/g)*	Pore Size (Å)*	Type of Discharge	Carrier Gas	Flow Rate (L/min)*	Initial VOC Concentration (ppm)*	Amount of Adsorption/Removal Efficiency	Energy Density and Frequency	By-Products	Ref.
Ethylene	Ag-Fe (1.5–0.5%)/13X (45 g)	–	–	PBDBD (IPC) t_1 = 60 min	21% O₂ + 79% N₂	1	270	662.2 μmol	–	–	[95]
				PBDBD (IPC) t_2 = 30 min			–	100%	15 kV (400 Hz)	–	
Ethylene	Ag/13X (30 g)	699.6	–	PBDBD (IPC) t_1 = 100 min	21% O₂ + 79% N₂	1	200	817.5 μmol	–	–	[94]
				PBDBD (IPC) dist₂ = 20 min			–	42%	20 kV (400 Hz)	CO	
Ethylene	10 wt % Ag/MS-13X (10 g)	–	–	PBDBD (IPC) Continuous	10% O₂, 10% CO₂/N₂	5	200	100%	0–90 J/L (10 kHz)	CO	[46]
Formaldehyde	3.6 wt % Ag 2.1 wt % Cu/HZSM-5	299	–	PBDBD (IPC) t_1 = 40 min	Humid air (RH = 50%)	0.3	26.6	14 μmol	–	–	[44]
				PBDBD (IPC) Open discharge	O₂	0.06	–	100%	2.3 W (2 kHz)	–	
Hexafluro-ethane	BaTiO₃/ zeolite	–	–	PBDBD (PPC) Continuous	N₂	0.025	3000	100%	8 kV (60 Hz)	–	[133]
Iso-propyl alcohol	MS-10A	439.31	9.66	PBDBD (IPC) t_1 = 60 min	N₂	1	400	–	–	–	[54]
				PBDBD (IPC) Open discharge	N₂	2	–	–	9.66 W (300 Hz)	–	
Methane	13X	–	–	PBDBD (IPC) Continuous	–	–	<1000	26%	173–200 Wh/m³ (20 kHz)	NOₓ	[104]
Trichloro-ethylene	Cu-ZSM-5	–	–	DBD (IPC) Continuous	Dry air	0.4	1000	>95%	1 W (50 Hz)	–	[107]
Toluene	ZSM-5-BaTiO₃	165	–	PBDBD (IPC) Adsorption till breakthrough	Humid air (19%)	0.6	1632 mg/m³	6.75 mg/g	–	–	[118]
				PBDBD (IPC) t_2 = 120 min	Air		–	100%	22 kV (50 Hz)	N₂O	
	AgMn-ZSM-5-BaTiO₃ 139			PBDBD (IPC) Adsorption till breakthrough	Humid air (19%)		1632 mg/m³	12.5 mg/g	–	–	
				PBDBD (IPC) t_2 = 120 min	Air		–	100%	22 kV (50 Hz)	<N₂O	

Table 4. Cont.

VOCs	Zeolite	S_{BET} (m²/g)*	Pore Size (Å)*	Type of Discharge	Carrier Gas	Flow Rate (L/min)*	Initial VOC Concentration (ppm)*	Amount of Adsorption/Removal Efficiency	Energy Density and Frequency	By-Products	Ref.
Toluene	Na-Y	750	7.4	DBD (PPC) t_{total} = 160 min t_1 = 30 min t_2 = 10 min	Humid air (0.5% H_2O)	500 cm³/min	200	52	5 W (24 kHz)	CO, HCOOH	[74]
	H-Y (1)	650	7.4					55			
	H-Y (2)	520	7.4					50			
	Mordenite	460	6.7 × 7.0					35			
	Ferrierite	270	4.3 × 5.5					20			
Toluene	NaY	–	–	SD (IPC) t_1 = 160 min	Synthetic humid air (0.5% H_2O)	0.5	200	6.5×10^{-4} mol/g	–	–	[36]
				SD (IPC) Open discharge			–	50%	5 W (50 Hz)	–	
	13X	–	–	DBD (IPC) t_1 = 282 mins	Air	0.4	150	0.57 mmol	–	–	
				DBD (IPC) closed discharge (t_2 = 1 h)	–	–	–	95%	20 W	–	
				DBD (IPC) ventilated discharge (t_1 = 1 h)	Synthetic air (80% N_2 + 20% O_2)	0.1	–	95%	20 W	–	
Toluene	Co/13X	386.214	–	DBD (IPC) ventilated discharge (t_2 = 1 h)	O_2	0.02	–	93.9%	20 W	–	[131]
						0.03	–	93.5%	20 W	–	
					Air (80% N_2 + 20% O_2)	0.1	–	92.7%	20 W	–	
						0.15	–	89.8%	20 W	–	
	MS 5A	–	–	DBD (IPC) t_1 = 30 min	Air	0.4	150	0.021 mmol	–	–	
				DBD (IPC) closed discharge (t_2 = 1 h)	.		–	95%	20 W	–	
				DBD (IPC) ventilated discharge (t_1 = 1 h)	Synthetic air (80% N_2 + 20% O_2)	0.1	–	91%	20 W	–	
Toluene	HZSM-5 (1.6 g)	486.587	–	DBD (IPC) t_1 = 8 h	Humid air (RH 40%)	3	3	0.25 mmol	–	–	[38]
				DBD (IPC) Open discharge	Synthetic air	0.8	–	100%	2.4 W	–	

Table 4. *Cont.*

VOCs	Zeolite	S_{BET} (m²/g)*	Pore Size (Å)*	Type of Discharge	Carrier Gas	Flow Rate (L/min)*	Initial VOC Concentration (ppm)*	Amount of Adsorption/Removal Efficiency	Energy Density and Frequency	By-Products	Ref.
	Ag-Mn/HZSM-5 (1.6 g)	350.051	–	DBD (IPC) t_1 = 8 h	Humid air (RH = 40%)	3	3	0.27 mmol	–	–	[29]
				DBD (IPC) Open discharge	Synthetic air	0.8	–	100%	2.4 W	–	
				DBD (IPC) t_1 = 60 min	Dry air	2	30	3.5 mL	–	–	
	Honeycomb zeolite	–	–		Dry air-forward flow			83%	20 W (420 Hz)	–	
Toluene				DBD (IPC) Open discharge	Dry air-reverse flow	1	–	51%	20 W (420 Hz)	–	[29]
					N₂–forward flow			68%	20 W (420 Hz)	–	
					N₂–reverse flow			37%	20 W (420 Hz)	–	
	MS-13X	626.439	26.6	DBD (IPC) Adsorption till breakthrough	Air	0.4	150	0.57 mmol	–	–	
Toluene				DBD (IPC) Open discharge	Synthetic air	0.1	–	94%	20 W	–	[8]
	5% Co/13X	386.214		DBD (IPC) Adsorption till breakthrough	Air	0.4	150	0.51 mmol	–	–	
Toluene				DBD (IPC) Open discharge	Synthetic air	0.1	–	92.7%	20 W	–	
				DBD (IPC) t_1 = 60 min		2	30	3.6 mL	–	–	
Toluene	Honeycomb zeolite (5.2 g)	–	–	DBD (IPC) t_2 = 5 mins (with 1 min closing time) Reverse flow	Air	1	–	Reg Eff = 82.3%	20 W Pulsed hv (420 Hz)	NO, NO₂	[28]
				DBD (IPC)t_2 = 4 mins Reverse flow				Reg Eff = 78.3%		–	
	Co/13X	386.214	37.78	DBD (IPC) t_1 = till equilibrium	Synthetic air	0.4	150	0.51 mmol	–	–	[6]
Toluene				DBD (IPC) t_2 = 1 h		0.1	–	92.7%	20 W	–	

Table 4. Cont.

VOCs	Zeolite	S_{BET} (m²/g) *	Pore Size (Å) *	Type of Discharge	Carrier Gas	Flow Rate (L/min) *	Initial VOC Concentration (ppm) *	Amount of Adsorption/Removal Efficiency	Energy Density and Frequency	By-Products	Ref.
Toluene	5A	–	5	PBDBD (IPC) Continuous	Dry air	0.3	100	79.8%	3.3–3.7 W (2 kHz)	–	[89]
	HZSM-5	–	5.5					80.8%			
	Hβ	–	6.6					98.4%			
	H-Y	–	7.4					85.2%			
	Ag/H-Y	–	–					97%			
Toluene	CuCeZr/ZSM-5	295	3.7	PBDBD (IPC) Continuous	Dry air	1	1400	77%	43 W	–	[85]
Toluene	Honeycomb zeolite	–	–	DBD (IPC) Continuous	–	150	30	79%	89 W (60 Hz)	–	[33]
				DBD (IPC) APC-closed discharge			25	70% to 93%			
p-xylene	HiSiv 3000 (4.8 g)	–	–	DBD (IPC) t₁ = 130 min	Dry air	2	20	5.2 mL	–	–	[47]
				DBD (IPC) Open discharge	Gas circulation	0.5	–	43%	38 W (60 Hz)	O₃	
Xylene mixture (p-, o-, and m- xylene)	HiSiv 3000 (2.5 g) and HiSiv 1000 (5.5 g)	–	–	DBD (IPC) t₁ = 130 min	Dry air	2	20	5.2 mL	–	–	
				DBD (IPC) Open discharge	Gas circulation	0.5	–	69%	38 W (60 Hz)	O₃	
VOC mixture (C₂Cl₄, C₇H₈, C₂HCl₃, C₆H₆, C₄H₈O₂, CS₂)	HZSM-5 (2 g)	202.82	–	DDBD (IPC) Continuous	N₂	4	100	100%	65.8 W (900 Hz)	SO₂, C₂H₅Cl, C₂H₄O, C₆H₈N₂, CH₄, C₂H₃NO	[126]

5. Conclusions

The main purpose of this article is to provide the readers an insight about the use of non-thermal plasma assisted/driven gas cleaning technology in combination with zeolites for VOC removal. The main conclusions that were obtained from this review are summarized as follows: (i) the adsorbing property of the zeolites enables the use of cyclic-adsorption plasma catalysis for VOC abatement, which is suitable for removal of low concentration of VOCs from large volume of gas, (ii) the energy efficiency of cyclic adsorption-plasma catalysis is improved by maximizing the adsorption time that can be achieved by using zeolites with suitable surface properties (S_{BET} and pore diameter) and/or by metal loading on zeolites, (iii) a proper selection of the nature and amount of metal ions for loading on zeolites is important as it can influence the adsorption of VOCs, characteristics of the plasma discharge, oxidation of adsorbed VOCs, mineralization efficiency and regeneration of zeolites, (iv) the use of oxygen as discharge gas for the complete regeneration of zeolite is economically feasible when the adsorption time is maximized in cyclic-adsorption plasma catalysis, and (v) the shaped zeolites are interesting to treat flue gas with high flow rate due to a reduced pressure drop. From this review, an unambiguous observation can be made that very few works concentrate on the following aspects of this technology: (i) stability of zeolites in continuous usage for longer time, (ii) the lifetime of zeolites and (iii) treating a mixture of VOCs, and (iv) realistic flue gas parameters (presence of humidity, high flow rate and low initial concentration of VOCs). Even though the stability of zeolites in cyclic adsorption-plasma catalysis is studied by some researchers, the number of cycles for which it has been investigated is very low ($\leq$5 cycles), which is not enough to estimate the lifetime of a catalyst. Therefore, more work focusing on the abovementioned aspects must be carried out in order to scale-up and apply this promising technology for real world applications.

Author Contributions: S.K.P.V. wrote the first draft of the article which was then refined by the comments and suggestions from R.M., N.D.G., J.-M.G., and J.-F.L.

Funding: This research was funded by the European Program INTERREG V France-Wallonie-Flanders project "DepollutAir" (grant number 1.1.18).

Acknowledgments: The "DepollutAir" project of the European Program INTERREG V France-Wallonie-Flanders FEDER) is acknowledged for supporting and funding this work.

Conflicts of Interest: The authors declare no conflict of interest.

References

1. EUR-Lex—31999L0013—EN—EUR-Lex. Available online: https://eur-lex.europa.eu/legal-content/EN/TXT/?uri=celex%3A31999L0013 (accessed on 3 September 2018).
2. Liu, Y.; Shao, M.; Fu, L.; Lu, S.; Zeng, L.; Tang, D. Source profiles of volatile organic compounds (VOCs) measured in China: Part I. *Atmos. Environ.* **2008**, *42*, 6247–6260. [CrossRef]
3. Kim, Y.M.; Harrad, S.; Harrison, R.M. Concentrations and sources of VOCs in urban domestic and public microenvironments. *Environ. Sci. Technol.* **2001**, *35*, 997–1004. [CrossRef] [PubMed]
4. Atkinson, R. Atmospheric chemistry of VOCs and NOx. *Atmos. Environ.* **2000**, *34*, 2063–2101. [CrossRef]
5. Finlayson-Pitts, B.J.; Pitts, J.N., Jr. Tropospheric air pollution: Ozone, airborne toxics, polycyclic aromatic hydrocarbons, and particles. *Science* **1997**, *276*, 1045–1052. [CrossRef] [PubMed]
6. Guo, H.; Lee, S.C.; Chan, L.Y.; Li, W.M. Risk assessment of exposure to volatile organic compounds in different indoor environments. *Environ. Res.* **2004**, *94*, 57–66. [CrossRef]
7. Pitten, F.A.; Bremer, J.; Kramer, A. Air pollution by volatile organic compounds (VOC) and health complaints. *Dtsch. Med. Wochenschr.* **2000**, *125*, 545–550. [CrossRef] [PubMed]
8. Kim, H.H.; Ogata, A.; Futamura, S. Oxygen partial pressure-dependent behavior of various catalysts for the total oxidation of VOCs using cycled system of adsorption and oxygen plasma. *Appl. Catal. B Environ.* **2008**, *79*, 356–367. [CrossRef]
9. Jones, A.P. Indoor air quality and health. *Atmos. Environ.* **1999**, *33*, 4535–4564. [CrossRef]
10. Wolkoff, P.; Nielsen, G.D. Organic compounds in indoor air-their relevance for perceived indoor air quality? *Atmos. Environ.* **2001**, *35*, 4407–4417. [CrossRef]

11. Luengas, A.; Barona, A.; Hort, C.; Gallastegui, G.; Platel, V.; Elias, A. A review of indoor air treatment technologies. *Rev. Environ. Sci. Biotechnol.* **2015**, *14*, 499–522. [CrossRef]
12. Katari, V.S.; Vatavuk, W.M.; Wehe, A.H. Incineration techniques for control of volatile organic compound emissions Part I. fundamentals and process design considerations. *JAPCA* **1987**, *37*, 91–99. [CrossRef]
13. Van der Vaart, D.R.; Vatvuk, W.M.; Wehe, A.H. Thermal and catalytic incinerators for the control of VOCs. *J. Air Waste Manag. Assoc.* **1991**, *41*, 92–98. [CrossRef]
14. Kamal, M.S.; Razzak, S.A.; Hossain, M.M. Catalytic oxidation of volatile organic compounds (VOCs)—A review. *Atmos. Environ.* **2016**, *140*, 117–134. [CrossRef]
15. Zhang, Z.; Jiang, Z.; Shangguan, W. Low-temperature catalysis for VOCs removal in technology and application: A state-of-the-art review. *Catal. Today* **2016**, *264*, 270–278. [CrossRef]
16. Iranpour, R.; Cox, H.H.J.; Deshusses, M.A.; Schroeder, E.D. Literature review of air pollution control biofilters and biotrickling filters for odor and volatile organic compound removal. *Environ. Prog.* **2005**, *24*, 254–267. [CrossRef]
17. Guieysse, B.; Hort, C.; Platel, V.; Munoz, R.; Ondarts, M.; Revah, S. Biological treatment of indoor air for VOC removal: Potential and challenges. *Biotechnol. Adv.* **2008**, *26*, 398–410. [CrossRef] [PubMed]
18. Yang, C.; Qian, H.; Li, X.; Cheng, Y.; He, H.; Zeng, G.; Xi, J. Simultaneous removal of multicomponent VOCs in biofilters. *Trends Biotechnol.* **2018**, *36*, 673–685. [CrossRef] [PubMed]
19. Zhang, X.; Gao, B.; Creamer, A.E.; Cao, C.; Li, Y. Adsorption of VOCs onto engineered carbon materials: A review. *J. Hazard. Mater.* **2017**, *338*, 102–123. [CrossRef]
20. Vandenbroucke, A.M.; Morent, R.; De Geyter, N.; Leys, C. Non-thermal plasmas for non-catalytic and catalytic VOC abatement. *J. Hazard. Mater.* **2011**, *195*, 30–54. [CrossRef]
21. Vandenbroucke, A.M.; Morent, R.; De Geyter, N.; Leys, C. Decomposition of toluene with plasma-catalysis: A review. *J. Adv. Oxid. Technol.* **2012**, *15*, 232–241. [CrossRef]
22. Schmid, S.; Jecklin, M.C.; Zenobi, R. Degradation of volatile organic compounds in a non-thermal plasma air purifier. *Chemosphere* **2010**, *79*, 124–130. [CrossRef] [PubMed]
23. Mo, J.; Zhang, Y.; Xu, Q.; Lamson, J.J.; Zhao, R. Photocatalytic purification of volatile organic compounds in indoor air: A literature review. *Atmos. Environ.* **2009**, *43*, 2229–2246. [CrossRef]
24. Huang, Y.; Ho, S.S.H.; Lu, Y.; Niu, R.; Xu, L.; Cao, J.; Lee, S. Removal of indoor volatile organic compounds via photocatalytic oxidation: A short review and prospect. *Molecules* **2016**, *21*, 56. [CrossRef] [PubMed]
25. Kaliya Perumal Veerapandian, S.; Leys, C.; De Geyter, N.; Morent, R. Abatement of VOCs using packed bed non-thermal plasma reactors: A review. *Catalysts* **2017**, *7*, 113. [CrossRef]
26. Thevenet, F.; Sivachandiran, L.; Guaitella, O.; Barakat, C.; Rousseau, A. Plasma-catalyst coupling for volatile organic compound removal and indoor air treatment: A review. *J. Phys. D Appl. Phys.* **2014**, *47*, 224011. [CrossRef]
27. Zhu, T.; Wan, Y.D.; Li, J.; He, X.W.; Xu, D.Y.; Shu, X.Q.; Liang, W.J.; Jin, Y.Q. Volatile organic compounds decomposition using nonthermal plasma coupled with a combination of catalysts. *Int. J. Environ. Sci. Technol.* **2011**, *8*, 621–630. [CrossRef]
28. Kuroki, T.; Fujioka, T.; Kawabata, R.; Okubo, M.; Yamamoto, T. Regeneration of honeycomb zeolite by nonthermal plasma desorption of toluene. *IEEE Trans. Ind. Appl.* **2009**, *45*, 10–15. [CrossRef]
29. Kuroki, T.; Fujioka, T.; Okubo, M.; Yamamoto, T. Toluene concentration using honeycomb nonthermal plasma desorption. *Thin Solid Films* **2007**, *515*, 4272–4277. [CrossRef]
30. Karuppiah, J.; Sivachandiran, L.; Karvembu, R.; Subrahmanyam, C. Catalytic nonthermal plasma reactor for the abatement of low concentrations of isopropanol. *Chem. Eng. J.* **2010**, *165*, 194–199. [CrossRef]
31. Tang, X.; Feng, F.; Ye, L.; Zhang, X.; Huang, Y.; Liu, Z.; Yan, K. Removal of dilute VOCs in air by post-plasma catalysis over Ag-based composite oxide catalysts. *Catal. Today* **2013**, *211*, 39–43. [CrossRef]
32. Kim, H.-H.; Ogata, A. Nonthermal plasma activates catalyst: From current understanding and future prospects. *Eur. Phys. J. Appl. Phys.* **2011**, *55*, 13806. [CrossRef]
33. Yamagata, Y.; Niho, K.; Inoue, K.; Okano, H.; Muraoka, K. Decomposition of volatile organic compounds at low concentrations using combination of densification by zeolite adsorption and dielectric barrier Discharge. *Jpn. J. Appl. Phys.* **2006**, *45*, 8251–8254. [CrossRef]
34. Mok, Y.S.; Kim, D.H. Treatment of toluene by using adsorption and nonthermal plasma oxidation process. *Curr. Appl. Phys.* **2011**, *11*, S58–S62. [CrossRef]

35. Lin, B.Y.; Chang, M.B.; Chen, H.L.; Lee, H.M.; Yu, S.J.; Li, S.N. Removal of C_3F_8 via the combination of non-thermal plasma, adsorption and catalysis. *Plasma Chem. Plasma Process.* **2011**, *31*, 585–594. [CrossRef]

36. Oh, S.M.; Kim, H.H.; Einaga, H.; Ogata, A.; Futamura, S.; Park, D.W. Zeolite-combined plasma reactor for decomposition of toluene. *Thin Solid Films* **2006**, *506–507*, 418–422. [CrossRef]

37. Inoue, K.; Okano, H.; Yamagata, Y.; Muraoka, K.; Teraoka, Y. Performance tests of newly developed adsorption/plasma combined system for decomposition of volatile organic compounds under continuous flow condition. *J. Environ. Sci.* **2011**, *23*, 139–144. [CrossRef]

38. Wang, W.; Wang, H.; Zhu, T.; Fan, X. Removal of gas phase low-concentration toluene over Mn, Ag and Ce modified HZSM-5 catalysts by periodical operation of adsorption and non-thermal plasma regeneration. *J. Hazard. Mater.* **2015**, *292*, 70–78. [CrossRef]

39. Liu, Y.; Li, X.; Liu, J.; Wu, J.; Zhu, A. Cycled storage-discharge (CSD) plasma catalytic removal of benzene over AgMn/HZSM-5 using air as discharge gas. *Catal. Sci. Technol.* **2016**, *6*, 3788–3796. [CrossRef]

40. Pham Huu, T.; Gil, S.; Da Costa, P.; Giroir-Fendler, A.; Khacef, A. Plasma-catalytic hybrid reactor: Application to methane removal. *Catal. Today* **2015**, *257*, 86–92. [CrossRef]

41. Kim, H.H.; Kim, J.H.; Ogata, A. Adsorption and oxygen plasma-driven catalysis for total oxidation of VOCs. *Int. J. Plasma Environ. Sci. Technol.* **2008**, *2*, 106–112.

42. Ohshima, T.; Kondo, T.; Kitajima, N.; Sato, M. Adsorption and plasma decomposition of gaseous acetaldehyde on fibrous activated carbon. *IEEE Trans. Ind. Appl.* **2010**, *46*, 23–28. [CrossRef]

43. Sivachandiran, L.; Thevenet, F.; Rousseau, A. Non-thermal plasma assisted regeneration of acetone adsorbed TiO_2 surface. *Plasma Chem. Plasma Process.* **2013**, *33*, 855–871. [CrossRef]

44. Zhao, D.-Z.; Li, X.-S.; Shi, C.; Fan, H.-Y.; Zhu, A.-M. Low-concentration formaldehyde removal from air using a cycled storage-discharge (CSD) plasma catalytic process. *Chem. Eng. Sci.* **2011**, *66*, 3922–3929. [CrossRef]

45. Rajanikanth, B.S.; Srinivasan, A.D.; Nandiny, B.A. A cascaded discharge plasma-adsorbent technique for engine exhaust treatment. *Plasma Sci. Technol.* **2003**, *5*, 1825–1833. [CrossRef]

46. Nishimura, J.; Kawamura, T.; Takahashi, K.; Teramoto, Y.; Takaki, K.; Koide, S.; Suga, M.; Orikasa, T.; Uchino, T. Removal of ethylene and by-products using packed bed dielectric barrier discharge with Ag nanoparticle-loaded zeolite. *Electron. Commun. Jpn.* **2017**, *100*, 320–327. [CrossRef]

47. Kuroki, T.; Hirai, K.; Kawabata, R.; Okubo, M.; Yamamoto, T. Decomposition of adsorbed xylene on adsorbents using nonthermal plasma with gas circulation. *IEEE Trans. Ind. Appl.* **2010**, *46*, 672–679. [CrossRef]

48. Bagreev, A.; Rahman, H.; Bandosz, T.J. Thermal regeneration of a spent activated carbon previously used as hydrogen sulfide adsorbent. *Carbon* **2001**, *39*, 1319–1326. [CrossRef]

49. Yun, J.-H.; Yun, J.-H.; Choi, D.-K.; Moon, H. Benzene adsorption and hot purge regeneration in activated carbon beds. *Chem. Eng. J.* **2000**, *55*, 5857–5872. [CrossRef]

50. Lee, D.-G.; Kim, J.-H.; Lee, C.-H. Adsorption and thermal regeneration of acetone and toluene vapors in dealuminated Y-zeolite bed. *Sep. Purif. Technol.* **2011**, *77*, 312–324. [CrossRef]

51. Kim, K.-J.; Ahn, H.-G. The effect of pore structure of zeolite on the adsorption of VOCs and their desorption properties by microwave heating. *Microporous Mesoporous Mater.* **2012**, *152*, 78–83. [CrossRef]

52. Ghoshal, A.K.; Manjare, S.D. Selection of appropriate adsorption technique for recovery of VOCs: An analysis. *J. Loss Prev. Process Ind.* **2002**, *15*, 413–421. [CrossRef]

53. Fan, Y.; Cai, Y.; Li, X.; Yin, H.; Chen, L.; Liu, S. Regeneration of the HZSM-5 zeolite deactivated in the upgrading of bio-oil via non-thermal plasma injection (NTPI) technology. *J. Anal. Appl. Pyrolysis* **2015**, *111*, 209–215. [CrossRef]

54. Shiau, C.H.; Pan, K.L.; Yu, S.J.; Yan, S.Y.; Chang, M.B. Desorption of isopropyl alcohol from adsorbent with non-thermal plasma. *Environ. Technol.* **2017**, *38*, 2314–2323. [CrossRef] [PubMed]

55. Ogata, A.; Einaga, H.; Kabashima, H.; Futamura, S.; Kushiyama, S.; Kim, H.-H. Effective combination of nonthermal plasma and catalysts for decomposition of benzene in air. *Appl. Catal. B Environ.* **2003**, *46*, 87–95. [CrossRef]

56. Bahri, M.; Haghighat, F.; Rohani, S.; Kazemian, H. Metal organic frameworks for gas-phase VOCs removal in a NTP-catalytic reactor. *Chem. Eng. J.* **2017**, *320*, 308–318. [CrossRef]

57. Delkash, M.; Bakhshayesh, B.E.; Kazemian, H. Using zeolitic adsorbents to cleanup special wastewater streams: A review. *Microporous Mesoporous Mater.* **2015**, *214*, 224–241. [CrossRef]

58. Zaitan, H.; Manero, M.H.; Valdés, H. Application of high silica zeolite ZSM-5 in a hybrid treatment process based on sequential adsorption and ozonation for VOCs elimination. *J. Environ. Sci.* **2016**, *41*, 59–68. [CrossRef] [PubMed]

59. Liu, C.; Zou, J.; Yu, K.; Cheng, D.; Han, Y.; Zhan, J.; Ratanatawanate, C.; Jang, B.W.-L. Plasma application for more environmentally friendly catalyst preparation. *Pure Appl. Chem.* **2006**, *78*, 1227–1238. [CrossRef]

60. Rhodes, C.J. Electric fields in zeolites: Fundamental features and environmental implications. *Chem. Pap.* **2016**, *70*, 4–21. [CrossRef]

61. Yi, H.; Yang, X.; Tang, X.; Zhao, S.; Xie, X.; Feng, T.; Ma, Y.; Cui, X. Performance and pathways of toluene degradation over Co/13X by different processes based on nonthermal plasma. *Energy Fuels* **2017**, *31*, 11217–11224. [CrossRef]

62. Weitkamp, J. Zeolites and Catalysis. *Solid State Ion.* **2000**, *131*, 175–188. [CrossRef]

63. Moshoeshoe, M.; Nadiye-tabbiruka, M.S.; Obuseng, V. A Review of the Chemistry, Structure, Properties and Applications of Zeolites. *Am. J. Mater. Sci.* **2017**, *7*, 196–221.

64. Zhang, L.; Peng, Y.; Zhang, J.; Chen, L.; Meng, X.; Xiao, F.-S. Adsorptive and catalytic properties in the removal of volatile organic compounds over zeolite-based materials. *Chin. J. Catal.* **2016**, *37*, 800–809. [CrossRef]

65. Khodayar, M.; Franzson, H. Fracture pattern of Thjórsárdalur central volcano with respect to rift-jump and a migrating transform zone in South Iceland. *J. Struct. Geol.* **2007**, *29*, 898–912. [CrossRef]

66. Wang, H.; Cao, Y.; Chen, Z.; Yu, Q.; Wu, S. High-efficiency removal of NO_x over natural mordenite using an enhanced plasma-catalytic process at ambient temperature. *Fuel* **2018**, *224*, 323–330. [CrossRef]

67. Barrer, R.M. Zeolites and their synthesis. *Zeolites* **1981**, *1*, 130–140. [CrossRef]

68. Barrer, R.M. Synthesis of a zeolitic mineral with chabazite-like sorptive properties. *J. Chem. Soc.* **1948**, 127–132. [CrossRef]

69. Breck, D.W.; Eversole, W.G.; Milton, R.M. New synthetic crystalline zeolites. *J. Am. Chem. Soc.* **1956**, *78*, 2338–2339. [CrossRef]

70. Jha, B.; Singh, D.N. A review on synthesis, characterization and industrial applications of flyash zeolites. *J. Mater. Educ.* **2011**, *33*, 65–132.

71. Valdés, M.G.; Pérez-Cordoves, A.I.; Díaz-García, M.E. Zeolites and zeolite-based materials in analytical chemistry. *Trends Anal. Chem.* **2006**, *25*, 24–30. [CrossRef]

72. Hashimoto, S. Zeolite photochemistry: Impact of zeolites on photochemistry and feedback from photochemistry to zeolite science. *J. Photochem. Photobiol. C Photochem. Rev.* **2003**, *4*, 19–49. [CrossRef]

73. Corma, A.; Rey, F.; Valencia, S.; Jordá, J.L.; Rius, J. A zeolite with interconnected 8-, 10- and 12-ring pores and its unique catalytic selectivity. *Nat. Mater.* **2003**, *2*, 493–497. [CrossRef] [PubMed]

74. Teramoto, Y.; Kim, H.-H.; Negishi, N.; Ogata, A. The role of ozone in the reaction mechanism of a bare zeolite-plasma hybrid system. *Catalysts* **2015**, *5*, 838–850. [CrossRef]

75. Xu, M.; Mukarakate, C.; Robichaud, D.J.; Nimlos, M.R.; Richards, R.M.; Trewyn, B.G. Elucidating zeolite deactivation mechanisms during biomass catalytic fast pyrolysis from model reactions and zeolite syntheses. *Top. Catal.* **2016**, *59*, 73–85. [CrossRef]

76. Lutz, W.; Zeolite, Y. Synthesis, modification, and properties—A case revisited. *Adv. Mater. Sci. Eng.* **2014**, *36*, 1389–1404.

77. Guo, Y.; Zhang, H.; Liu, Y. Desorption characteristics and kinetic parameters determination of molecular sieve by thermogravimetric analysis/differential thermogravimetric analysis technique. *Adsorpt. Sci. Technol.* **2018**, *36*, 1389–1404. [CrossRef]

78. Khan, N.A.; Yoo, D.K.; Bhadra, B.N.; Jun, J.W.; Kim, T.-W.; Kim, C.-U.; Jhung, S.H. Preparation of SSZ-13 zeolites from beta zeolite and their application in the conversion of ethylene to propylene. *Chem. Eng. J.* **2018**. [CrossRef]

79. Al-Dughaither, A.S.; De Lasa, H. HZSM-5 zeolites with different SiO_2/Al_2O_3 ratios. Characterization and NH_3 desorption kinetics. *Ind. Eng. Chem. Res.* **2014**, *53*, 15303–15316. [CrossRef]

80. Datka, J.; Tuznik, E. Infrared Spectroscopic Studies of Acid Properties of NaHZSM-5 Zeolites. *J. Catal.* **1986**, *102*, 43–51. [CrossRef]

81. Reitmeier, S.J.; Gobin, O.C.; Jentys, A.; Lercher, J.A. Enhancement of sorption processes in the zeolite H-ZSM5 by postsynthetic surface modification. *Angew. Chem. Int. Ed.* **2009**, *48*, 533–538. [CrossRef]

82. Weber, R.W.; Fletcher, J.C.Q.; Möller, K.P.; O'Connor, C.T. The characterization and elimination of the external acidity of ZSM-5. *Microporous Mater.* **1996**, *7*, 15–25. [CrossRef]

83. Armaroli, T.; Simon, L.J.; Digne, M.; Montanari, T.; Bevilacqua, M.; Valtchev, V.; Patarin, J.; Busca, G. Effects of crystal size and Si/Al ratio on the surface properties of H-ZSM-5 zeolites. *Appl. Catal. A Gen.* **2006**, *306*, 78–84. [CrossRef]

84. GhavamiNejad, A.; Kalantarifard, A.; Yang, G.S.; Kim, C.S. In-situ immobilization of silver nanoparticles on ZSM-5 type zeolite by catechol redox chemistry, a green catalyst for A^3-coupling reaction. *Microporous Mesoporous Mater.* **2016**, *225*, 296–302. [CrossRef]

85. Dou, B.; Liu, D.; Zhang, Q.; Zhao, R.; Hao, Q.; Bin, F.; Cao, J. Enhanced removal of toluene by dielectric barrier discharge coupling with Cu-Ce-Zr supported ZSM-5/TiO$_2$/Al$_2$O$_3$. *Catal. Commun.* **2017**, *92*, 15–18. [CrossRef]

86. Ogata, A.; Ito, D.; Mizuno, K.; Kushiyama, S.; Yamamoto, T. Removal of dilute benzene using a zeolite-hybrid plasma reactor. *IEEE Trans. Ind. Appl.* **2001**, *37*, 959–964. [CrossRef]

87. Makowski, W.; Ogorzałek, Ł. Determination of the adsorption heat of *n*-hexane and *n*-heptane on zeolites beta, L, 5A, 13X, Y and ZSM-5 by means of quasi-equilibrated temperature-programmed desorption and adsorption (QE-TPDA). *Thermochim. Acta* **2007**, *465*, 30–39. [CrossRef]

88. Yi, H.; Yang, X.; Tang, X.; Zhao, S.; Wang, J.; Cui, X.; Feng, T.; Ma, Y. Removal of toluene from industrial gas over 13X zeolite supported catalysts by adsorption-plasma catalytic process. *J. Chem. Technol. Biotechnol.* **2017**, *92*, 2276–2286. [CrossRef]

89. Huang, R.; Lu, M.; Wang, P.; Chen, Y.; Wu, J.; Fu, M.; Chen, L.; Ye, D. Enhancement of the non-thermal plasma-catalytic system with different zeolites for toluene removal. *RSC Adv.* **2015**, *5*, 72113–72120. [CrossRef]

90. Díaz, E.; Ordóñez, S.; Vega, A.; Coca, J. Adsorption characterisation of different volatile organic compounds over alumina, zeolites and activated carbon using inverse gas chromatography. *J. Chromatogr. A* **2004**, *1049*, 139–146. [CrossRef]

91. Yu, Y.; Zheng, L.; Wang, J. Adsorption behavior of toluene on modified 1X molecular sieves. *J. Air Waste Manag. Assoc.* **2012**, *62*, 1227–1232. [CrossRef]

92. Triebe, R.W.; Tezel, F.H.; Khulbe, K.C. Adsorption of methane, ethane and ethylene on molecular sieve zeolites. *Gas Sep. Purif.* **1996**, *10*, 81–84. [CrossRef]

93. Trinh, Q.H.; Mok, Y.S. Effect of the adsorbent/catalyst preparation method and plasma reactor configuration on the removal of dilute ethylene from air stream. *Catal. Today* **2015**, *256*, 170–177. [CrossRef]

94. Van Mao, R.L.; Mclaughlin, G.P. Ethylene recovery from low grade gas stream by adsorption on zeolites and controlled desorption. *Can. J. Chem. Eng.* **1988**, *66*, 686–690.

95. Trinh, Q.H.; Lee, S.B.; Mok, Y.S. Removal of ethylene from air stream by adsorption and plasma-catalytic oxidation using silver-based bimetallic catalysts supported on zeolite. *J. Hazard. Mater.* **2015**, *285*, 525–534. [CrossRef]

96. Takahashi, A.; Yang, F.H.; Yang, R.T. Aromatics/aliphatics separation by adsorption: New sorbents for selective aromatics adsorption by π-complexation. *Ind. Eng. Chem. Res.* **2000**, *39*, 3856–3867. [CrossRef]

97. Zhu, R.; Mao, Y.; Jiang, L.; Chen, J. Performance of chlorobenzene removal in a nonthermal plasma catalysis reactor and evaluation of its byproducts. *Chem. Eng. J.* **2015**, *279*, 463–471. [CrossRef]

98. Kim, H.; Park, J.; Jung, Y. The binding nature of light hydrocarbons on Fe/MOF-74 for gas separation. *Phys. Chem. Chem. Phys* **2013**, *15*, 19644–19650. [CrossRef]

99. Shen, B.; Fan, K.; Wang, W.; Deng, J. Ab initio study on the adsorption and oxidation of HCHO with Ag$_2$ cluster. *J. Mol. Struct.* **1999**, *469*, 157–161. [CrossRef]

100. Qin, C.; Huang, X.; Zhao, J.; Huang, J.; Kang, Z.; Dang, X. Removal of toluene by sequential adsorption-plasma oxidation: Mixed support and catalyst deactivation. *J. Hazard. Mater.* **2017**, *334*, 29–38. [CrossRef]

101. Fan, H.-Y.; Shi, C.; Li, X.-S.; Zhao, D.-Z.; Xu, Y.; Zhu, A.-M. High-efficiency plasma catalytic removal of dilute benzene from air. *J. Phys. D Appl. Phys.* **2009**, *42*, 225105. [CrossRef]

102. Fan, H.-Y.; Li, X.-S.; Shi, C.; Zhao, D.-Z.; Liu, J.-L.; Liu, Y.-X.; Zhu, A.-M. Plasma catalytic oxidation of stored benzene in a cycled storage-discharge (CSD) process: Catalysts, reactors and operation conditions. *Plasma Chem. Plasma Process.* **2011**, *31*, 799–810. [CrossRef]

103. Einaga, H.; Futamura, S. Catalytic oxidation of benzene with ozone over Mn ion-exchanged zeolites. *Catal. Commun.* **2007**, *8*, 557–560. [CrossRef]
104. Xia, J.F.; Gao, X.X.; Kong, J.Y.; Hui, H.X.; Cui, M.; Yan, K.P. By-products NOx control and performance improvement of a packed-bed nonthermal plasma reactor. *Plasma Chem. Plasma Process.* **2000**, *20*, 225–233. [CrossRef]
105. Xu, X.; Wang, P.; Xu, W.; Wu, J.; Chen, L.; Fu, M.; Ye, D. Plasma-catalysis of metal loaded SBA-15 for toluene removal: Comparison of continuously introduced and adsorption-discharge plasma system. *Chem. Eng. J.* **2016**, *283*, 276–284. [CrossRef]
106. Bahri, M.; Haghighat, F. Plasma-Based Indoor Air Cleaning Technologies: The State of the Art-Review. *Clean-Soil Air Water* **2014**, *42*, 1667–1680. [CrossRef]
107. Oda, T.; Takahashi, T.; Kohzuma, S. Decomposition of dilute trichloroethylene by using non-thermal plasma processing—Frequency and catalyst effect. In Proceedings of the Thirty-Third IAS Annual Meeting Conference Record of 1998 IEEE Industry Applications Conference, St. Louis, MO, USA, 12–15 October 1998; pp. 1871–1876.
108. Oda, T. Non-thermal plasma processing for environmental proection: Decomposition of dilute VOCs in air. *J. Electrost.* **2003**, *57*, 293–311. [CrossRef]
109. Trinh, Q.H.; Gandhi, M.S.; Mok, Y.S. Adsorption and plasma-catalytic oxidation of acetone over zeolite-supported silver catalyst. *Jpn. J. Appl. Phys.* **2015**, *54*, 01AG04. [CrossRef]
110. Kim, H.H.; Ogata, A.; Futamura, S. Effect of different catalysts on the decomposition of VOCs using flow-type plasma-driven catalysis. *IEEE Trans. Plasma Sci.* **2006**, *34*, 984–995. [CrossRef]
111. Hu, J.; Jiang, N.; Li, J.; Shang, K.; Lu, N.; Wu, Y. Degradation of benzene by bipolar pulsed series surface/packed-bed discharge reactor over MnO2-TiO2/zeolite catalyst. *Chem. Eng. J.* **2016**, *293*, 216–224. [CrossRef]
112. Li, W.; Gibbs, G.V.; Oyama, S.T. Mechanism of ozone decomposition on a manganese oxide catalyst. 1. In situ Raman spectroscopy and ab initio molecular orbital calculations. *J. Am. Chem. Soc.* **1998**, *120*, 9041–9046. [CrossRef]
113. Li, W.; Oyama, S.T. Mechanism of ozone decomposition on a manganese oxide catalyst. 2. Steady-state and transient kinetic studies. *J. Am. Chem. Soc.* **1998**, *120*, 9047–9052. [CrossRef]
114. Reed, C.; Xi, Y.; Oyama, S.T. Distinguishing between reaction intermediates and spectators: A kinetic study of acetone oxidation using ozone on a silica-supported manganese oxide catalyst. *J. Catal.* **2005**, *235*, 378–392. [CrossRef]
115. Reed, C.; Lee, Y.-K.; Oyama, S.T. Structure and oxidation state of silica-supported manganese oxide catalysts and reactivity for acetone oxidation with ozone. *J. Phys. Chem. B* **2006**, *110*, 4207–4216. [CrossRef] [PubMed]
116. Garcia, T.; Agouram, S.; Sánchez-Royo, J.F.; Murillo, R.; Mastral, A.M.; Aranda, A.; Vázquez, I.; Dejoz, A.; Solsona, B. Deep oxidation of volatile organic compounds using ordered cobalt oxides prepared by a nanocasting route. *Appl. Catal. A Gen.* **2010**, *386*, 16–27. [CrossRef]
117. Youn, J.S.; Bae, J.; Park, S.; Park, Y.K. Plasma-assisted oxidation of toluene over Fe/zeolite catalyst in DBD reactor using adsorption/desorption system. *Catal. Commun.* **2018**, *113*, 36–40. [CrossRef]
118. Qin, C.; Guo, H.; Liu, P.; Bai, W.; Huang, J.; Huang, X.; Dang, X.; Yan, D. Toluene abatement through adsorption and plasma oxidation using ZSM-5 mixed with γ-Al2O3, TiO2or BaTiO3. *J. Ind. Eng. Chem.* **2018**, *63*, 449–455. [CrossRef]
119. Kim, H.-H.; Teramoto, Y.; Negishi, N.; Ogata, A. A multidisciplinary approach to understand the interactions of nonthermal plasma and catalyst: A review. *Catal. Today* **2015**, *256*, 13–22. [CrossRef]
120. Patil, B.S.; Cherkasov, N.; Lang, J.; Ibhadon, A.O.; Hessel, V.; Wang, Q. Low temperature plasma-catalytic NOx synthesis in a packed DBD reactor: Effect of support materials and supported active metal oxides. *Appl. Catal. B Environ.* **2016**, *194*, 123–133. [CrossRef]
121. Kim, H.-H.; Teramoto, Y.; Sano, T.; Negishi, N.; Ogata, A. Effects of Si/Al ratio on the interaction of nonthermal plasma and Ag/HY catalysts. *Appl. Catal. B Environ.* **2015**, *166–167*, 9–17. [CrossRef]
122. Hamada, S.; Hojo, H.; Einaga, H. Effect of catalyst composition and reactor configuration on benzene oxidation with a nonthermal plasma-catalyst combined reactor. *Catal. Today* **2018**. [CrossRef]
123. Wallis, A.E.; Whitehead, J.C.; Zhang, K. The removal of dichloromethane from atmospheric pressure nitrogen gas streams using plasma-assisted catalysis. *Appl. Catal. B Environ.* **2007**, *72*, 282–288. [CrossRef]

124. Fitzsimmons, C.; Ismail, F.; Whitehead, J.C.; Wilman, J.J. The chemistry of dichloromethane destruction in atmospheric-pressure gas streams by a dielectric packed-bed plasma reactor. *J. Phys. Chem. A* **2000**, *104*, 6032–6038. [CrossRef]

125. Jiang, L.; Nie, G.; Zhu, R.; Wang, J.; Chen, J.; Mao, Y.; Cheng, Z.; Anderson, W.A. Efficient degradation of chlorobenzene in a non-thermal plasma catalytic reactor supported on CeO_2/HZSM-5 catalysts. *J. Environ. Sci.* **2017**, *55*, 266–273. [CrossRef] [PubMed]

126. Mustafa, M.F.; Fu, X.; Liu, Y.; Abbas, Y.; Wang, H.; Lu, W. Volatile organic compounds (VOCs) removal in non-thermal plasma double dielectric barrier discharge reactor. *J. Hazard. Mater.* **2018**, *347*, 317–324. [CrossRef] [PubMed]

127. Formisano, B.; Bonten, C. Extruded zeolitic honeycombs for sorptive heat storage. *AIP Conf. Proc.* **2016**, *1779*, 030003.

128. Chang, F.-T.; Lin, Y.-C.; Bai, H.; Pei, B.-S. Adsorption and desorption characteristics of semiconductor volatile organic compounds on the thermal swing honeycomb zeolite concentrator. *J. Air Waste Manag. Assoc.* **2003**, *53*, 1384–1390. [CrossRef] [PubMed]

129. Kim, H.-H.; Oh, S.-M.; Ogata, A.; Futamura, S. Decomposition of gas-phase benzene using plasma—Driven catalyst reactor: Complete oxidation of adsorbed benzene using oxygen plasma. *J. Adv. Oxid. Technol.* **2005**, *8*, 226–233. [CrossRef]

130. Dang, X.; Huang, J.; Cao, L.; Zhou, Y. Plasma-catalytic oxidation of adsorbed toluene with gas circulation. *Catal. Commun.* **2013**, *40*, 116–119. [CrossRef]

131. Yi, H.; Yang, X.; Tang, X.; Zhao, S. Removal of toluene from industrial gas by adsorption–plasma catalytic process: Comparison of closed discharge and ventilated discharge. *Plasma Chem. Plasma Process.* **2018**, *38*, 331–345. [CrossRef]

132. Oh, S.M.; Kim, H.H.; Ogata, A.; Einaga, H.; Futamura, S.; Park, D.W. Effect of zeolite in surface discharge plasma on the decomposition of toluene. *Catal. Lett.* **2005**, *99*, 101–104. [CrossRef]

133. Urashima, K.; Kostov, K.G.; Chang, J.S.; Okayasu, Y.; Iwaizumi, T.; Yoshimura, K.; Kato, T. Removal of C2F6 from a semiconductor process flue gas by a ferroelectric packed-bed barrier discharge reactor with an adsorber. *IEEE Trans. Ind. Appl.* **2001**, *37*, 1456–1463. [CrossRef]

catalysts

MDPI

Article

Mode Transition of Filaments in Packed-Bed Dielectric Barrier Discharges

Mingxiang Gao [1], Ya Zhang [1,2,*] , Hongyu Wang [3,4], Bin Guo [1], Quanzhi Zhang [2] and Annemie Bogaerts [2]

[1]   Department of Physics, Wuhan University of Technology, Wuhan 430070, China;
      mxgao@whut.edu.cn (M.G.); binguo@whut.edu.cn (B.G.)
[2]   Research group PLASMANT Department of Chemistry University of Antwerp, B-2610 Wilrijk-Antwerp,
      Belgium; Quan-Zhi.Zhang@uantwerpen.be (Q.Z.); annemie.bogaerts@uantwerpen.be (A.B.)
[3]   School of Physics Science and Technology, Anshan Normal University, Anshan 114007, China;
      why@btitgroup.com
[4]   Shanghai Bright-Tech Information Technology Co. Ltd., Shanghai 201206, China
*   Correspondence: yazhang@whut.edu.cn

Received: 15 May 2018; Accepted: 13 June 2018; Published: 15 June 2018

check for
updates

Abstract: We investigated the mode transition from volume to surface discharge in a packed bed dielectric barrier discharge reactor by a two-dimensional particle-in-cell/Monte Carlo collision method. The calculations are performed at atmospheric pressure for various driving voltages and for gas mixtures with different N_2 and O_2 compositions. Our results reveal that both a change of the driving voltage and gas mixture can induce mode transition. Upon increasing voltage, a mode transition from hybrid (volume+surface) discharge to pure surface discharge occurs, because the charged species can escape much more easily to the beads and charge the bead surface due to the strong electric field at high driving voltage. This significant surface charging will further enhance the tangential component of the electric field along the dielectric bead surface, yielding surface ionization waves (SIWs). The SIWs will give rise to a high concentration of reactive species on the surface, and thus possibly enhance the surface activity of the beads, which might be of interest for plasma catalysis. Indeed, electron impact excitation and ionization mainly take place near the bead surface. In addition, the propagation speed of SIWs becomes faster with increasing N_2 content in the gas mixture, and slower with increasing O_2 content, due to the loss of electrons by attachment to O_2 molecules. Indeed, the negative O_2^- ion density produced by electron impact attachment is much higher than the electron and positive O_2^+ ion density. The different ionization rates between N_2 and O_2 gases will create different amounts of electrons and ions on the dielectric bead surface, which might also have effects in plasma catalysis.

Keywords: plasma catalysis; mode transition; packed-bed dielectric barrier discharge; particle-in-cell/Monte Carlo collision method; surface filament; gas composition

---

1. Introduction

Plasma catalysis is gaining increasing interest for various environmental applications, such as gaseous pollutant removal, the splitting of CO_2, hydrogen generation and O_3 production [1–9]. Plasma catalysis can be regarded as the combination of a plasma with a catalyst, and often results in improved performance, in terms of selectivity and energy efficiency of the process. Plasma is an ionized gas, consisting of various reactive species, like electrons, positive ions, negative ions and radicals. These reactive species are created by applying a potential difference to a gas. The gas itself remains at room temperature, which is beneficial in terms of energy saving compared to classical thermal catalysis. Plasma-based pollutant removal and gas conversion have recently gained increased attention, being possible alternatives for chemical reactors. The production of new molecules in the plasma can be selective when a catalyst is added to the plasma. Plasma catalysis can be realized by introducing dielectric packing beads (coated with catalyst material) in the discharge gap, forming a packed bed dielectric barrier discharge (PB-DBD) reactor. The DBD generally occurs in a filamentary mode by applying a high driving voltage [10–14], which induces a very fast ionization avalanche, propagating from powered electrode to grounded electrode, i.e., a so-called streamer [12,13,15–19]. Each streamer starts when the driving voltage passes a certain threshold, and will further polarize the dielectric surface [20].

The ionization avalanches can occur either within the bulk plasma (so-called volume discharge) or along the surface of a dielectric (so-called surface discharge), sustained by electron impact surface ionization waves (SIWs) [21]. SIW discharges operating in N_2 at low pressure (5–100 Torr) and created by high voltage (10–15 kV) nanosecond pulses, traveling along a dielectric surface or a liquid surface, have been experimentally investigated by Intensified Charge Coupled Device (ICCD) camera images [22], with measured propagation speeds of $\sim 5 \times 10^5$ m/s. Furthermore, a SIW discharge in hydrogen was experimentally reported, by measuring the time-resolved electric field at special positions, within tens of nanosecond pulse scales, using a picosecond four-wave mixing technique [23]. The results showed that the discharge developed itself as a SIW propagating along the dielectric surface at an average speed of $\sim 10^6$ m/s with a maximum electric field of $\sim 2.3 \times 10^6$ V/m [23].

On the other hand, volume discharges can be sustained by filamentary microdischarges (MDs) in a PB-DBD reactor, by limiting the dimensions at atmospheric pressure, following the famous Paschen law [24–26]. A filamentary MD may be tens of micrometers in space and can exist for a few nanoseconds [27], yielding a high concentration of reactive species in a narrow gap, which is beneficial for pollutant remediation. Furthermore, filamentary MDs can co-exist with SIWs along the surface of the dielectric beads in a PB-DBD reactor under proper conditions [28,29]. This may yield high concentrations of chemically active species and radicals on the surface due to electric field enhancement along the dielectric surface, compared to an unpacked DBD.

The generation of reactive species at catalyst surfaces and in the gaps between the packing beads is very important for plasma catalysis. Therefore, it is crucial to understand the properties of surface and volume discharges and their formation mechanisms. However, only a few theoretical studies have been performed for PB-DBD reactors [11,29–37]. Russ et al. [30] adopted a two-dimensional (2D) hydrodynamic theory to study transient MDs in a PB-DBD reactor for a gas mixture 80% N_2, 20% O_2, and 500 ppm NO, with 23 species and 120 plasma reactions. However, this work was limited to a one-directional discharge, without considering discharge channels in the voids of the packed beads. Kang et al. [31] implemented a 2D fluid model to investigate the typical properties of MDs in a ferroelectric PB-DBD, focusing only on the surface discharge with strong electric field for two packing beads. Chang et al. [32] studied a ferroelectric packed bed N_2 plasma for a spherical void between two pellets based on the 1D Poisson equation and transport equations. This work focused on the electron density, electron temperature, and electric field as a function of the applied voltage, the discharge gap size, and the dielectric constants of the packing beads.

In our previous work, we employed a kinetic 2D particle-in-cell/Monte Carlo collision (PIC/MCC) model to investigate the formation and propagation of a filamentary MD in a PB-DBD, and the results were compared to an unpacked DBD [11]. Van Laer and Bogaerts developed a comprehensive 2D fluid model for a PB-DBD in helium [33–35]. They reported that a packing enhanced the electric field strength and electron energy near the contact points of the dielectric beads. Microplasma discharges in humid air in a PB-DBD reactor were experimentally characterized through ICCD imaging and numerically simulated by the 2D multi-fluid simulator non PDPSIM, with a negative driving voltage at atmospheric pressure [29]. The behavior of three kinds of discharge modes, including positive streamers (restrikes), filamentary MDs and SIWs, were reported in their work. Wang et al. [36] studied the microplasma characteristics by means of fluid modelling and experimental observations using ICCD imaging, for packing materials with different dielectric constants, in dry air at atmospheric pressure. They also studied the same three types of discharges predicted in Ref. [29]. In addition, they reported a transition in discharge mode from surface discharges to local filamentary discharges upon increasing the dielectric constant of the packing from 5 to 1000. Finally, Kang et al. [37] experimentally and numerically investigated surface streamer propagation on an alumina bead, and predicted three distinct phases, i.e., the generation and propagation of a primary streamer with a moderate electric field and velocity, rapid primary streamer acceleration with an enhanced electric field, and slow secondary streamer propagation.

Although several studies have been performed to better understand the microplasma properties in a PB-DBD reactor, as outlined above, the interaction mechanisms between the plasma and the catalyst are still poorly understood. Both physical and chemical effects can play a role in this interaction, and this will affect the discharge types and properties. Therefore, in the present work, we investigate the filament formation and mode transition between volume and surface discharges, as a function of driving voltage and O_2/N_2 mixing ratio at atmospheric pressure, by using a 2D PIC/MCC model in a micro gap PB-DBD reactor. Specifically, we aim to obtain a better understanding of the production mechanisms and concentrations of reactive species sustained by the different modes.

2. Results and Discussion

The spatial and temporal evolutions of the species densities, electric field, and excitation and ionization rates are presented in this section, including the whole process of filament formation and transition to SIWs.

2.1. Effect of Driving Voltage

It has been reported that the SIW properties are dependent on both the direction and value of the driving voltage and the surface properties [21]. In Figures 1–7, we present the effect of driving voltage on the density of the reactive species and the electric field for a dry air discharge (so only a mixture of 80% N_2 and 20% O_2, no other air components).

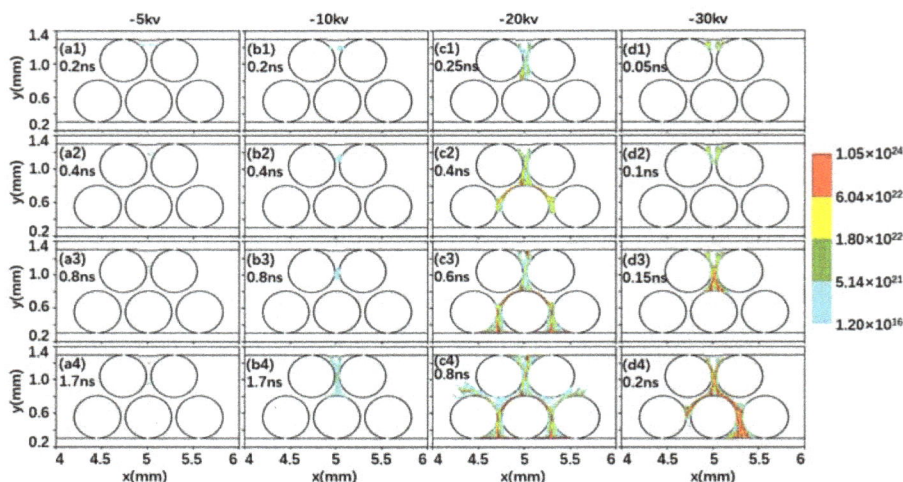

Figure 1. Electron density (m^{-3}) at different times, for different driving voltages, i.e., −5 kV (**a1–a4**), −10 kV (**b1–b4**), −20 kV (**c1–c4**), and −30 kV (**d1–d4**), in an air discharge. The same color scale is used in all panels to allow comparison. The different time intervals for different driving voltages are based on the different filament formation times for the different conditions.

Figure 1 shows the electron density for four different driving voltages, i.e., -5 kV (a1–a4), -10 kV (b1–b4), -20 kV (c1–c4) and -30 kV (d1–d4), and four different time intervals. Note that the different time intervals at various driving voltages are based on the filament formation times under the different conditions. The driving voltages are all above the breakdown voltage of air (~3 kV/mm) in glow mode. Since a filamentary discharge needs higher breakdown voltage than a glow discharge, no filament occurs at -5 kV, so only the seed electron density ($\sim10^{18}$ m^{-3}) is shown in Figure 1a1–a4, which corresponds with the values in Ref. [29]. A limited volumetrically sustained filament appears at -10 kV, with an average speed of 3×10^5 m/s and a maximum electric field of 8.6×10^7 V/m. For -20 kV, the filament is formed in the gap at around 0.25 ns with an average speed of 2×10^6 m/s. Subsequently, the SIWs are developed and start dominating the discharge. The SIWs can be further classified as downward- and upward-propagating modes. The downward-directed SIWs are formed in the time interval of 0.25–0.6 ns with an average velocity of 2.2×10^6 m/s, while the upward-directed SIWs are developed in the time range of 0.6–0.8 ns with an average velocity of 2×10^6 m/s. It is worth to note that a multi-channel filamentary MD also develops with the upward-directed SIWs, i.e., the plasma extends away from the dielectric surface. However, the maximum density in the MD is nearly two orders of magnitude lower than in the SIWs, as shown in Figure 1c4, which indicates that the discharge is mainly governed by the SIWs. For the driving voltage of -30 kV, the filament is formed and sustained in the time range of 0–0.2 ns, as nearly pure SIWs propagating downward along the surface with a high speed of 5×10^6 m/s, while no upward-directed SIW develops. There is obvious asymmetry in the electron density for -30 kV, due to the inherent statistical character of the streamers.

Therefore, as the driving voltage increases, a mode transition will happen, i.e., the filament is mainly sustained by an MD at low driving voltage (-10 kV), it becomes a combination of a limited portion of MD and significant SIW at a moderate driving voltage (-20 kV), and finally it is almost completely dominated by SIWs. The physical mechanism is that the charged species will rapidly escape to the dielectric surface under the effect of high driving voltage (strong electric field) and charge the dielectric. The surface charging will induce a strong tangential component of electric field along the dielectric surface, and lead to the formation of SIWs. The SIWs will in turn yield high concentrations of reactive species, as well as high electron density on the surface, which will further charge the dielectric, and gradually form a filament along the dielectric surface. This filament may increase the surface activity of the beads, which is probably beneficial for plasma catalysis.

Our results are consistent with the theoretical and experimental predictions in Refs. [22,23,29,36]. For -10 kV, the average speed of the volume streamer is 3×10^5 m/s, which is in the same order as in Refs. ($\sim5 \times 10^5$ and $\sim10^5$–10^6 m/s) [29,36]. For -20 kV, both volume and surface filaments co-exist. Therefore, the filament could propagate in the gaps and on the surface of the dielectric beads. This trend is in good agreement with recent experimental observations by fast-camera imaging for numerous packing beads with dielectric constant of 25. Once the SIWs are initiated in the PB-DBD, we calculate a propagation speed of the SIWs around 10^6 m/s, which is in the same order as in literature ($\sim5 \times 10^5$ and $\sim10^6$ m/s) [22,23]. The reactive species are highly concentrated on the surface due to the presence of SIWs along the catalyst surface, which agrees with experimental observations in literature [22,36].

Figures 2–4 show the density of N_2^+, O_2^+ and O_2^- ions, respectively, for the same conditions as in Figure 2.

As shown in Figure 2, the N_2^+ ion density profile is a bit different from the electron density distribution at -20 kV, i.e., with a higher density in the voids between the beads. The maximum N_2^+ ion density is a bit higher than the electron density, due to loss of electrons by attachment to O_2 molecules. However, the N_2^+ ion density profile becomes very similar to the electron density distribution at -30 kV, because the discharge is now dominated by pure surface discharges. This indicates that the electron attachment is not significant for a pure surface discharge.

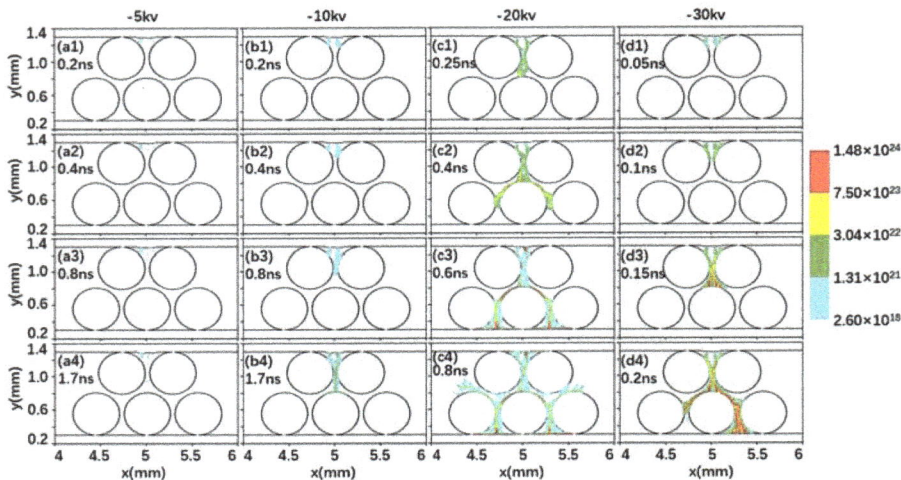

Figure 2. N_2^+ ion density (m^{-3}) for the same conditions as in Figure 1.

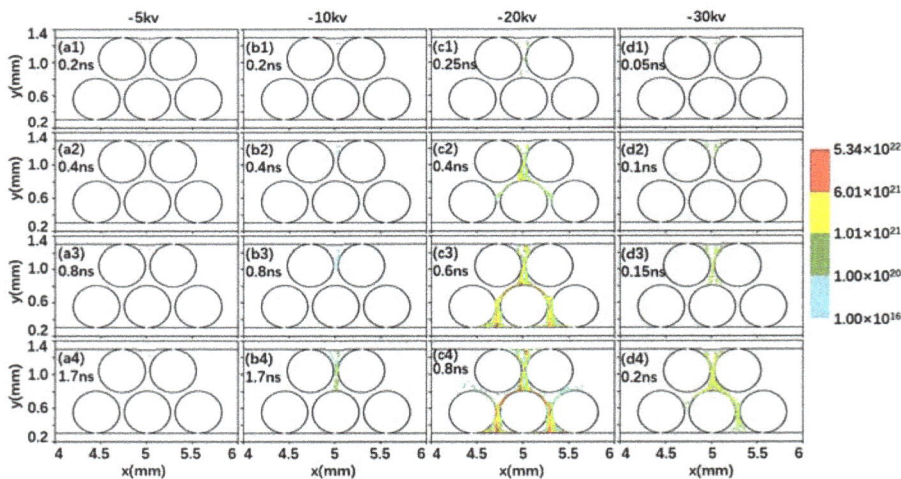

Figure 3. O_2^+ ion density (m^{-3}) for the same conditions as in Figure 1.

Figure 3 clearly shows that the O_2^+ ions are mainly formed along the dielectric surface by the SIWs, and the maximum density is about 20 times lower than the maximum electron density for all driving voltages, because it is more difficult for the discharge to become sustained in O_2 gas [38]. This can be justified by the ionization rate shown later in Figure 5.

Figure 4 presents the O_2^- ion density distribution. The O_2^- ions are generated by electron impact attachment. Thus, the space and time distributions of the O_2^- ion density are nearly the same as for the electron density, but the maximum density is about two times smaller than the maximum electron density due to the smaller fraction of O_2 in air.

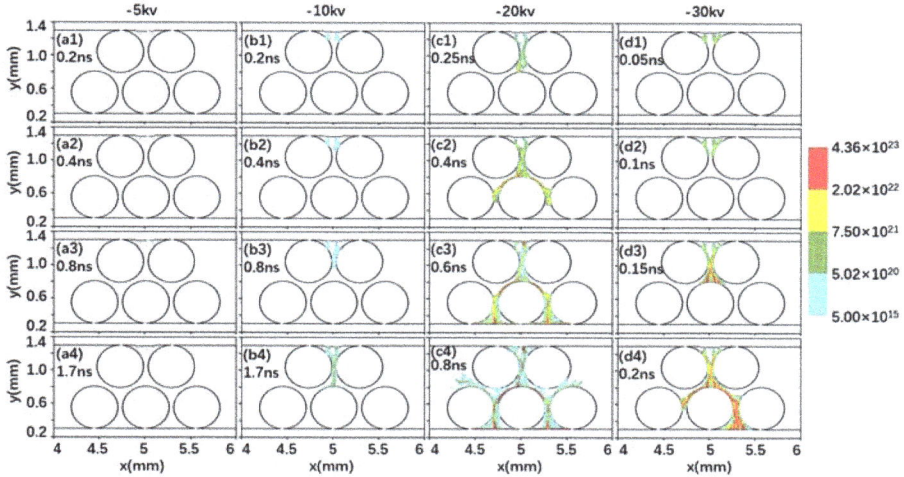

Figure 4. O_2^- ion density (m^{-3}) for the same conditions as in Figure 1.

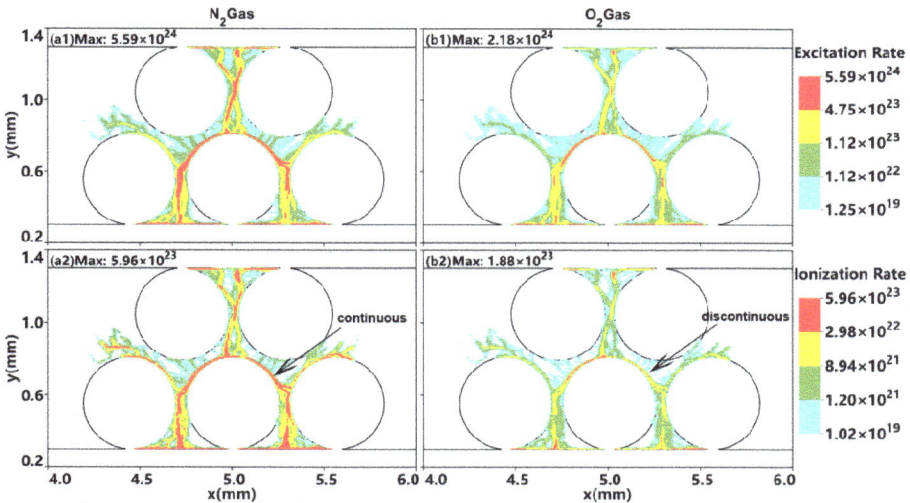

Figure 5. (**a1,b1**) Electron impact excitation rate ($m^{-3}s^{-1}$) and (**a2,b2**) electron impact ionization rate ($m^{-3}s^{-1}$), at 0.8 ns for a driving voltage of -20 kV, in an air discharge, for the nitrogen (**a1,a2**) and oxygen (**b1,b2**) components. The same color scale is used in all panels to allow comparison. The maximum values are noted in the figure.

Therefore, in total, the electrons and N_2^+ ions are present both along the dielectric surface and in the gaps between the beads, whereas the O_2^+ and O_2^- ions only concentrate along the dielectric surface at -20 kV. The difference in spatial distributions of charged reactive species may affect the catalytic selectivity, which is important for plasma catalysis, as the interaction between the charged species and the surface might influence the morphology and work function of the catalyst, and accordingly affect the catalyst performance. However, the exact influence cannot be deduced from our model and can be quite complicated, which is beyond the scope of this work.

It has been reported [11,33,36,39] that the local electric field can be enhanced near the contact points and at the boundaries between beads and dielectric layers. This is because the dielectric beads and plates are strongly polarized by the applied electric field, which reduces the potential drop in the dielectric, but increases the potential drop in the gas gap. However, most of the previous works only focus on local electric field in the gas gap [29,40], but they do not study the variation and influence of the electric field in the dielectric materials.

Figure 6 shows the electric field amplitudes for the same conditions as in Figure 1, and Figure 7 shows the vertical and horizontal electric field components for −20 kV (at 0.8 ns) and −30 kV (at 0.2 ns). As clearly seen from Figures 6 and 7, the presence of dielectric beads and plates induces a sharp boundary with the gas phase, which strongly enhances the displacement electric field inside the dielectric material near this boundary. As a result, the electric field on the surface and in the voids of the packing beads is also enhanced, due to the significant charge accumulation in the narrow gaps (see Figures 1–4 above). Indeed, the local electric field inside the vertical gaps between two beads (or between bead and plate) is enhanced so much that the maximum value is much larger than the average external electric field, even for the lowest driving voltage of −5 kV without breakdown. The maximum values, presented in Figure 6, stay constant when there is no breakdown, but they increase with time when there is a discharge.

The electric field is more enhanced in the vertical gaps between the dielectric beads than inside the materials for the −10 kV case, due to the very limited filamentary MD, as displayed in Figure 1b. On the other hand, at −20 kV and −30 kV, the electric field on and near the surface of the dielectric materials is significantly enhanced, due to the high concentration of the charged species on the catalyst surfaces (see Figures 1–4 above). In addition, as shown in Figure 6c4, the electric field is also obviously enhanced in the head of the upward-directed SIWs (noted in Figure 1c4 above), where the electric field is characterized by an overlap of both vertical and horizontal electric field components. Behind the SIW head, the vertical component rapidly reduces, as shown in Figure 7a1.

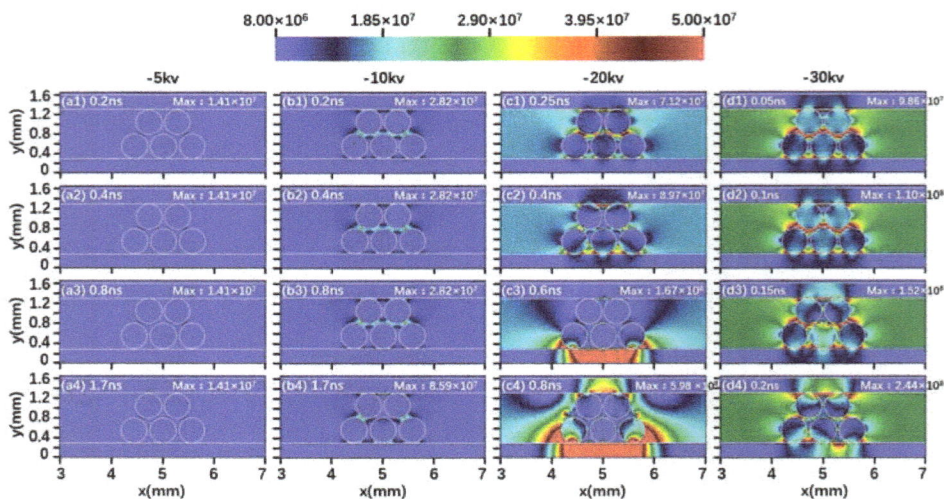

Figure 6. Electric field amplitude $|E|$ (V/m) for the same conditions as in Figure 1. The maximum values are noted in each panel. They occur at the contact points between the beads, and are larger than the color scale, but the color scale is chosen in such a way to better illustrate the general behavior.

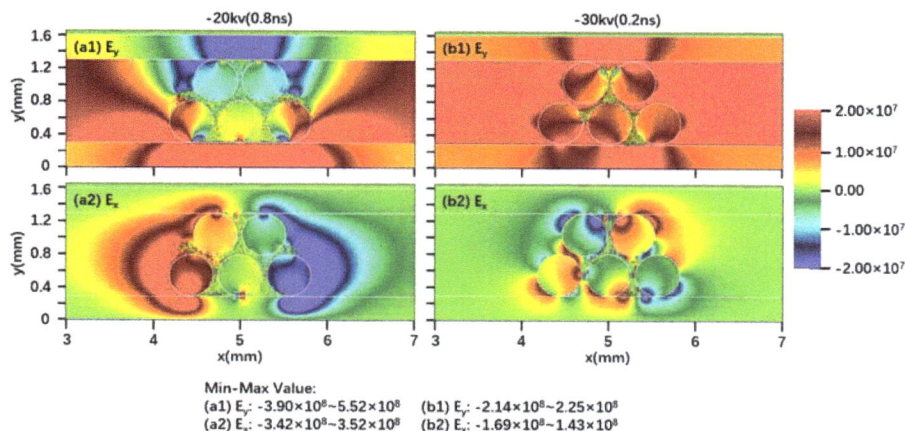

Figure 7. Electric field components in the y and x direction, (**a1,b1**) E_y (V/m) and (**a2,b2**) E_x (V/m), (**a1,a2**) for -20 kV (0.8 ns), and (**b1,b2**) for -30 kV (0.2 ns), in an air discharge. The maximum values are noted under the figure. They occur at the contact points between the beads, and are larger than the color scale, but the color scale is chosen in such a way to better illustrate the general behavior.

Moreover, at -30 kV, the electric field has its maximum values on the surface of the beads and plates, and the values gradually decrease away from the surface boundary, because of the pure surface discharge mode. The high electric field on the dielectric bead surface is important for plasma catalysis [8], because it could induce locally stronger electron impact reactions, thus higher reaction rates and a higher plasma density. The SIW discharges therefore play an important role in a PB-DBD reactor for plasma catalysis applications. The physical mechanism is related to the high concentration of reactive species on the surface of the beads, which yields tangential components of the electric field, and the latter in turn enhance the SIWs.

We calculated the tangential components of the electric field E_x to be around 2×10^7 V/m, as shown in Figure 7b2, which is larger than the value of 6×10^6 V/m from literature [29]. The difference is mainly due to the much smaller discharge gap of 1 mm, and bead size of 508 μm (bead diameter) used in this work, compared to the large discharge gap of 1 cm and bead size of 1.8 mm (bead diameter) in Ref. [29], as the same driving voltage will induce a much stronger electric field in a smaller gap. In addition, we consider ZrO_2 dielectric beads with a dielectric constant of 22, which yields stronger polarization than for quartz material with a dielectric constant of 4, used in Ref. [29]. This stronger polarization also induces a higher electric field in the gap.

In order to elucidate the mechanisms giving rise to the SIWs on the dielectric surfaces, we need to distinguish where the reactive species are generated, either on the dielectric surfaces or in the gap. To clarify this, the electron impact excitation rate (a1–b1) and ionization rate (a2–b2), for N_2 (a1–a2) and O_2 (b1–b2) components in the mixture, are plotted in Figure 8, for 0.8 ns and a driving voltage of -20 kV.

As seen in Figure 5, both excitation and ionization mainly take place on the surface of the beads. The maximum excitation rate is about one order of magnitude higher than the maximum ionization rate, due to the lower threshold energy needed for excitation reactions (see Table 1 below). The excitation rate for the N_2 component is almost uniformly distributed on all surfaces, and fills the gaps between beads 3, 4 and 5, indicating that the excited N_2^* molecules will be fairly uniformly distributed on these surfaces and in the gaps. The ionization rate for the N_2 component is locally enhanced on the top surface of dielectric bead 4 and the bottom dielectric plate, due to the high local electric field there, which corresponds to the higher N_2^+ ion density at these positions, shown in Figure 2c4. Although the excitation and ionization rates for the N_2 component spread out from the surface and cause restrikes [29], which may further give rise to multi-channel filamentary MD (see Figure 1c4 above), the maximum values in the restrikes are about one order of magnitude smaller than on the surface, as shown in Figure 5a1–a2. This again demonstrates that the SIWs are the main discharge mode at the driving voltage of −20 kV, while the MDs sustained by column-like filaments are more or less negligible.

Table 1. Reaction set of a N_2/O_2 gas mixture used in the model, as well as the threshold energy for electron impact ionization and excitation collisions. The cross sections for all reactions are adopted from Refs. [41–44], and are downloaded from the LXCat database [45].

Reactions	Threshold Energy (eV)
Electron-impact ionization	
$e + O_2 \rightarrow 2e + O_2^+$	12.06
$e + N_2 \rightarrow 2e + N_2^+$	15.58
Attachment	
$e + O_2 \rightarrow O_2^-$	
Electron-impact excitation	
$e + O_2 \rightarrow e + O_2^*$	0.98
$e + O_2 \rightarrow e + O_2^*$	1.63
$e + O_2 \rightarrow e + O_2^*$	6.0
$e + O_2 \rightarrow e + O_2^*$	8.4
$e + O_2 \rightarrow e + O_2^*$	10.00
$e + N_2 \rightarrow 2e + N_2^*$	6.169
$e + N_2 \rightarrow 2e + N_2^*$	7.353
$e + N_2 \rightarrow 2e + N_2^*$	7.362
$e + N_2 \rightarrow 2e + N_2^*$	8.165
$e + N_2 \rightarrow 2e + N_2^*$	8.399
$e + N_2 \rightarrow 2e + N_2^*$	8.549
$e + N_2 \rightarrow 2e + N_2^*$	8.89
$e + N_2 \rightarrow 2e + N_2^*$	9.7537
$e + N_2 \rightarrow 2e + N_2^*$	11.032
$e + N_2 \rightarrow 2e + N_2^*$	12.771
$e + N_2 \rightarrow 2e + N_2^*$	13.37
$e + N_2 \rightarrow 2e + N_2^*$	13.382
$e + N_2 \rightarrow 2e + N_2^*$	14.0
Elastic collision	
$e + O_2 \rightarrow e + O_2$	
$e + N_2 \rightarrow e + N_2$	

On the other hand, both the excitation and ionization rates for the O_2 component show discontinuous distributions on the bead surfaces, except for the top surface of central bead 4, as shown in Figure 5b1–b2. This indicates that the production of the excited O_2^* molecules and the O_2^+ ions is not a continuous process, but rather a discrete process, as predicted in Ref. [29]. This discontinuous ionization rate leads to a rather discretely distributed O_2^+ ion density (see Figure 3 above).

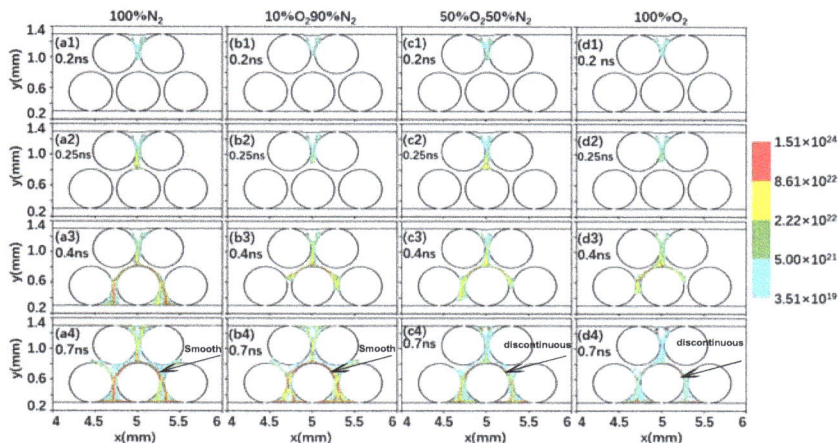

Figure 8. Electron density (m^{-3}) at 0.2, 0.25, 0.4 and 0.7 ns, for a driving voltage of -20 kV, in a gas mixture of nitrogen and oxygen with different composition, (**a1–a4**) 100% N_2, (**b1–b4**) 10% O_2 and 90% N_2, (**c1–c4**) 50% O_2 and 50% N_2, and (**d1–d4**) 100% O_2. The same color scale is used in all panels to allow comparison.

The difference between the ionization rate for N_2 and O_2 gas components may be attributed to the different mean free path for ionization collisions. The mean free path for ionization collisions with N_2 is about 250 μm (hence similar to the radius of the beads, i.e., 254 μm), while it is $\sim$2 mm (similar with the perimeter of the beads) for O_2. A shorter ionization mean free path results in more intensive ionization collisions, whereas an ionization mean free path longer than the bead radius and the gap distance between beads results in weaker ionization and may yield a discrete distribution, as shown in Figure 5b2. Although the densities of the excited species (N_2^* and O_2^*) are not displayed in this work, the electron impact excitation rate reveals that these excited species are highly concentrated on the surface. The electron impact ionization rate profile can be considered as evidence for SIW formation on the surface.

2.2. Effect of the Gas Composition

We vary the O_2/N_2 mixing ratio in this subsection, to obtain better insight in the different character between electropositive and electronegative gases. The effect of gas composition on the production of MDs inside porous ceramics was investigated experimentally by Hensel et al. [38], using photographic visualization and optical emission spectroscopy. It was predicted that a higher O_2 content resulted in the redistribution of the MD channels inside the porousceramics [38], while the total number of MD channels was reduced, since the breakdown voltage was increased. Their results can be partially correlated to the excitation and ionization rates, which are stronger in N_2 than in O_2, as shown in Figure 5 above. Therefore, it is important to understand the effect of gas composition on the mechanisms of filament formation and mode transition between volume and surface filaments in a PB-DBD reactor, which has not been reported before.

In this section, we will demonstrate that the gas composition is a critical parameter for mode transition. Figures 8–11 present the density of charged species (electrons, N_2^+, O_2^+ and O_2^- ions) at 0.2, 0.25, 0.4 and 0.7 ns, for a driving voltage of -20 kV, and different compositions of N_2 and O_2 in the mixture, i.e., 100% N_2 (a1–a4), 10% O_2 and 90% N_2 (b1–b4), 50% O_2 and 50% N_2 (c1–c4), and 100% O_2 (d1–d4).

Figure 8 shows the electron density. The filament is initiated from t = 0 with a few seed electrons. The filaments travel for a similar distance within 0.2 ns, and arrive at the central bead 4 at 0.25 ns, for the different gas compositions, except in the case of 10% O_2 and 100% O_2. However, after the filaments reach the central bead, their propagation speed becomes faster in N_2 gas and gradually becomes slower with increasing O_2 contents, as shown in Figure 8 at 0.4 and 0.7 ns. Indeed, the average speed of the filaments is 2×10^6 m/s during the first 0.25 ns, while during 0.25–0.4 ns, the average speed is 3×10^6 m/s for pure N_2 gas, indicating the fastest SIW, 2.3×10^6 m/s for 10% O_2, 2×10^6 m/s for 50% O_2, and 1.7×10^6 m/s for pure O_2.

The physical mechanism is that the discharge is more difficult to be sustained, and thus it needs a higher breakdown voltage and leads to a slower discharge evolution, for electronegative gas, due to the loss of electrons by attachment to O_2 molecules. This feature was experimentally confirmed by Hensel et al. [38], through measurements of the breakdown voltage in a N_2/O_2 gas mixture. Therefore, for the same driving voltage, the discharge can more easily be created in N_2 than in O_2, resulting in gradually slower propagation speeds of the filaments upon higher O_2 fractions in the mixture.

For pure N_2 gas, our calculations predict a dominated surface discharge with a negligible volume discharge. Indeed, the maximum electron density (1.51×10^{24} m^{-3}) on the surface is approximately five orders of magnitude higher than in the gap (3.51×10^{19} m^{-3}). The electron density is almost uniformly distributed on the bead surfaces after the filament reaches the central bead 4.

The maximum electron density on the bead surfaces is represented by the red color in Figure 8. It is smooth for the first two cases, indicating that the electron density is indeed uniformly distributed, while it becomes discontinuous with increasing O_2 contents. However, the maximum electron density still occurs on the surface and not in the gap, giving rise to a high concentration of electrons on a specific part of the surface in pure O_2. Again it shows that the SIWs can produce a high density of charged species on the surface of the beads, which might be beneficial for gas treatment by plasma catalysis [46]. Finally, while adding more O_2 to the mixture makes the SIWs to become discontinuous, the gas mixture almost does not affect the volume discharge, as seen in Figure 8.

The ion density profiles, for the same conditions as in Figure 8, are plotted in Figure 9 (N_2^+), Figure 10 (O_2^+), and Figure 11 (O_2^-).

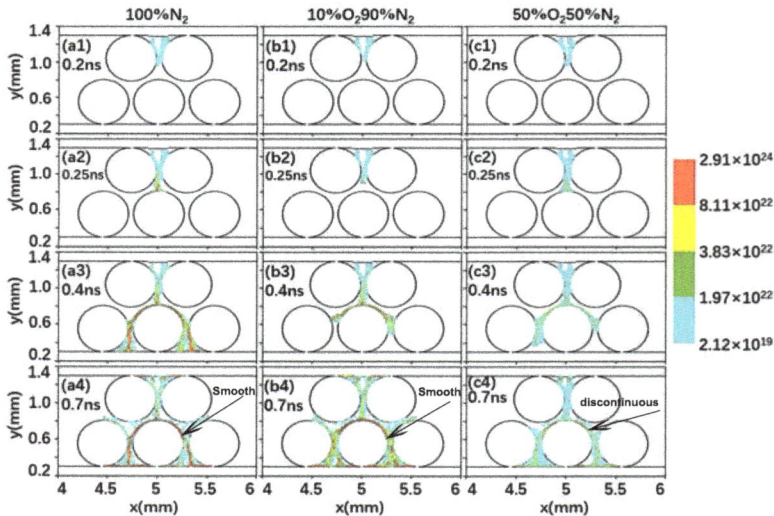

Figure 9. N_2^+ ion density (m^{-3}) for the same conditions as in Figure 8.

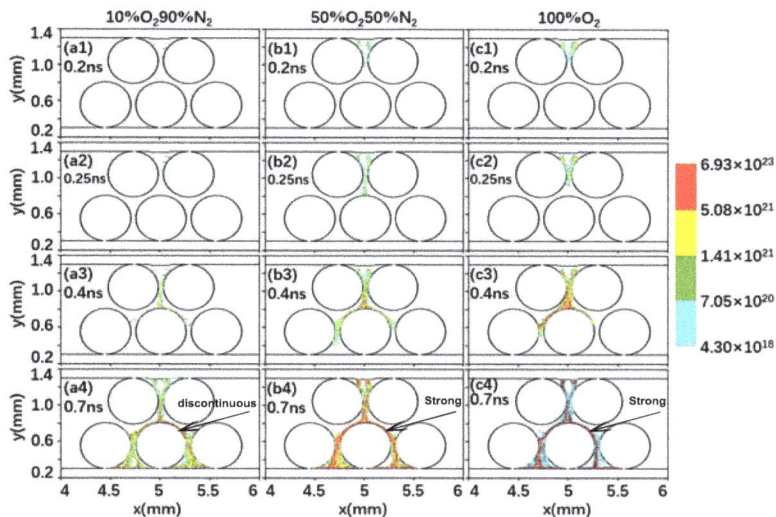

Figure 10. O_2^+ ion density (m^{-3}) for the same conditions as in Figure 8.

The N_2^+ ion density profile is similar to the electron density profile, but the maximum N_2^+ ion density is two times larger than the electron density, again due to the loss of electrons by attachment to the O_2 molecules. The number of lost electrons can be determined by the O_2^- ion density (see Figure 11 below). The maximum N_2^+ ion density $(2.91 \times 10^{24}$ $m^{-3})$ is always found on the surfaces, and the speed of the N_2^+ ion filament is almost the same as for the electron filament discussed above.

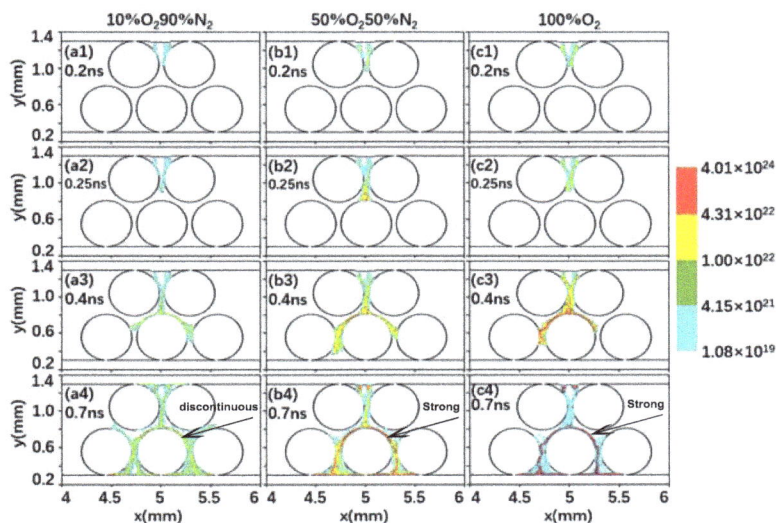

Figure 11. O_2^- ion density (m^{-3}) for the same conditions as in Figure 8.

The O_2^+ ion density exhibits different profiles with increasing O_2 content, as shown in Figure 10. The O_2^+ ion density is at least two times smaller than the electron density, for the same reason as explained above, i.e., it becomes harder for the discharge to be created for high O_2 content at a fixed driving voltage, resulting in a lower O_2^+ ion density. As a result, the O_2^+ ion density is at least four times smaller than the N_2^+ ion density at equal gas fractions (50%/50% mixture). In pure O_2, the propagation speed of the SIW is about two times slower than in pure N_2, and no MD channel is formed above the surface of beads 3 and 5, due to the absence of an upward-directed SIW.

The O_2^- ions are again mainly formed on the surface by the SIWs, with a maximum value of 4.01×10^{24} m^{-3} in 100% O_2, which is 2.6 times higher than the maximum electron density and 5.8 times higher than the maximum O_2^+ ion density, indicating that electron attachment is quite significant. Indeed, attachment is easier than ionization, because it requires no threshold energy, while the ionization threshold is 12.06 eV for O_2 and 15.58 eV for N_2. Therefore, the O_2^- ion density sharply increases with rising O_2 content, as expected.

To summarize, the O_2^+ and O_2^- ion densities are enhanced, while the electron and N_2^+ ion densities drop on the bead surface, for higher O_2 content, as expected. This might affect the surface reactions in plasma catalysis [47,48].

3. Description of the Model

3.1. Model Assumptions

The dimensions of the whole reactor are 1.65 mm in the y direction and 10 mm in the x direction. The discharge is sustained between two parallel plate electrodes covered by two dielectric layers of 0.3 mm thickness, separated by a gap of 1 mm in the y direction. Dielectric beads are inserted in the gap, forming a packed bed reactor. The dielectric constant of the layers and the beads is $\epsilon_r = 22$, characteristic for ZrO_2. A schematic illustration of the reactor is presented in Figure 12, showing in the x direction only the central part including the dielectric beads, but the entire simulation region is taken from 0 to 10 mm in the x dimension. The dielectric layers and beads are colored in blue. We consider five dielectric spheres with diameter d = 508 μm. In order to leave some gas space for the filament formation between the packing beads, these five beads are packed in a non-strict spherical packing manner [11], with distance between adjacent bead centers of $(0.1 + \sqrt{3}/2)d$ in the y dimension and 1.1d in the x dimension. Note that some gaps are present between the dielectric beads in Figure 12. This was intentional to show filament formation. Indeed, streamers cannot propagate in a 2D system without gaps between the beads, while they can propagate either along the bead surfaces or the gaps between the beads in a 3D case. We thus assume a certain gap between the beads, to allow the streamers propagate in our 2D model. Furthermore, even in experimental (3D) packed-bed reactors, there are some gaps between the dielectric beads, i.e., the cross sections between the beads are similar with the 2D model, when a lot of beads are packed together, as the beads cannot touch each other perfectly. This can be observed from Ref. [49]. On the other hand, it needs extremely long calculation times to model a PB-DBD with 5 beads in an entire 3D geometry, owing to severe mesh requirements and very large number of macro particles. Therefore, we consider a 2D model.

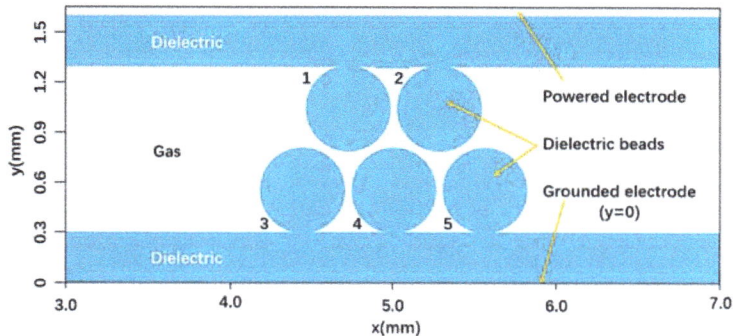

Figure 12. Geometry used in the packed-bed dielectric barrier discharge (PB-DBD) reactor. The entire simulation domain is 10 mm in the x dimension and 1.65 mm in the y dimension, but only a smaller part in the x direction is shown here, for clarity. The discharge gap is 1 mm, and the top and bottom electrodes are covered by 0.3 mm thick dielectric plates. The top electrode is 0.05 mm thick, and it acts as powered electrode. The bottom electrode (x = 0) is grounded. Dielectric packing beads with diameter of 508 μm are placed in the gap. The numbering is used later in the paper.

The upper electrode (y = 1.65 mm) is powered by a pulsed voltage with a rise time of 0.1 ns and then kept constant for the duration of the simulation, considered as the cathode. The lower electrode (anode; y = 0 mm) is grounded. The dielectric surfaces are considered as absorption boundaries for the reactive plasma species, i.e., all reactive species will be removed from the simulation if they hit the surface of the dielectrics and they cannot participate in the discharge anymore. The absorbing boundary condition is a reasonable and widely used boundary condition in PIC model [50]. When charged species are absorbed on the dielectrics, they can emit a secondary electron and they will deposit their charge on the dielectric surface. The deposited charge is determined by the charge of the ion striking the dielectric surface and the secondary electron emission charge and coefficient, accounting for the formalism $Q_D = Q_D + Q_{ion} - Q_{se}$. Here, Q_D is the deposited charge with initial value of zero, Q_{ion} is the charge of the ions striking the dielectric surfaces of the beads or layers, and Q_{se} is the charge of the secondary electron, respectively. The effect of secondary electron emission is self-consistently coupled in the PIC/MCC model, assuming a constant ion impact secondary electron emission coefficient of 0.15 [29].

Photoionization is neglected in this study. Indeed, we found in our previous work [39] that the results are nearly the same with and without photoionization in a gap of tens of μm, which is the typical gap size between the packing beads and between the dielectric layers and the beads in a PB-DBD. Furthermore, the photoionization rate was nearly two or three orders of magnitude lower than the electron impact ionization rate, even in a large gap (hundreds of μm) [29,36]. Indeed, as will be shown below, the filaments in the present model are mainly sustained by SIWs, and the volume discharges will be negligible in a narrow gap of ~10 μm.

The simulations are applied to a mixture of N_2 and O_2 at atmospheric pressure with a temperature of 300 K. Only charged species, i.e., electrons (e) and ions (N_2^+ , O_2^+ and O_2^-), are simulated in this PIC/MCC model, as the total ionization degree is typically less than 10%. Filaments are initiated from initial seeding electrons emitted from the surface of the top dielectric plate in the region $x \in [4.9, 5.1]$ mm. A few seed electrons can be generated by cosmic radiation or external emission. The emission current density is set as 10^5 Am^{-2}. Subsequently, the filament sustains itself through anode-directed avalanches due to the electron impact ionization and secondary electron emission. Dissociation process, and the behavior of atomic ions, are not included in the model, because the cross sections of these reactions are relatively small, thus omitting these reactions will almost not affect the kinetics of the streamer.

3.2. Simulation Method

We developed a 2D PIC/MCC model based on the VSIM simulation software [51], which has been widely used and validated [11,39,51]. This PIC/MCC simulation is based on an explicit and electrostatic method, which was first introduced and described in detail in Ref. [52]. The PIC/MCC model takes advantage of accounting for the detailed kinetic behavior of charged particles, compared to a fluid model. There are four steps in a PIC/MCC model: (1) pushing the particle velocity based on the previous electric field; (2) weighting from the positions of all charged particles to obtain the charge density; (3) summing and extrapolating the charge density to achieve a new electric field based on the Poisson equation; and (4) using a standard MCC method to describe electron impact elastic collisions, excitations, ionizations and electron attachment.

The particle pushing procedure is described by the Newton equations

$$\mathbf{v}_p^{n+1} = \mathbf{v}_p^n + \mathbf{a}_p^n \delta t \tag{1}$$

$$\mathbf{r}_p^{n+1} = \mathbf{r}_p^n + \mathbf{v}_p^{n+1} \delta t \tag{2}$$

where

$$\mathbf{a}_p^n = \frac{Z_p e}{m_p} \mathbf{E^n}. \tag{3}$$

Here n represents the nth time step (δt). The new electric field in time t^{n+1} is solved by the Poisson

$$\nabla \cdot \epsilon \mathbf{E}^{n+1} = \rho^{n+1} + Q_D \tag{4}$$

where $\epsilon = \epsilon_r \epsilon_0$ is the permittivity, and ρ^{n+1} is the total charge density, given by

$$\rho^{n+1} = \sum_p Z_p e S(\mathbf{r} - \mathbf{r}_p^{n+1}). \tag{5}$$

Here, p represents the electrons (e) and ions (N_2^+, O_2^+ and O_2^-), and $\mathbf{r}_p$, $\mathbf{v}_p$, Z_p, m_p and S are the particle position, velocity, charge, mass and shape function in space. The shape function is chosen to be a first order b-spline (b_l) function, $S(\mathbf{r} - \mathbf{r}_p) = b_l(\frac{\mathbf{r}-\mathbf{r}_p}{\triangle_r})$, with $\triangle_r = \delta x \delta y$. The time step δt fulfills the Courant condition

$$c\delta t < \frac{1}{\sqrt{(\frac{1}{\delta x})^2 + (\frac{1}{\delta y})^2}}, \tag{6}$$

with the light speed c, and the space steps in x and y directions δx and δy. As mentioned above, the simulation region is 1.65 mm in the y direction and 10 mm in the x direction, and we use a mesh of 1000 × 500 uniform grid points. Dirichlet boundary conditions are adopted in the y direction, and Neumann conditions are used in the x direction.

A standard MCC method [53] is used to account for the electron impact elastic, excitation, ionization and attachment collisions with N_2 and O_2 gas molecules, as listed in Table 1. The cross sections and threshold energies used for these reactions are adopted from the Refs. [41–44] and downloaded from LXCat database [45].

We consider only three types of ions, since they are the most important ones, with the lowest ionization threshold and largest density compared to other ions. We consider a simulation time up to 0.8 ns, which is enough to develop both the volume and surface discharges in a PB-DBD reactor. Thus, the effect of electron-ion and negative ion-positive ion recombination is negligible. Indeed the recombination reactions have a larger relaxation time up to the microsecond time scale, as predicted by Kruszelnicki et al. [29], as it requires a sequence of two-body reactions or three body reactions.

4. Conclusions

We have applied a 2D PIC/MCC model to study the formation and mode transition of filamentary discharges for a PB-DBD reactor with various driving voltages and N_2/O_2 gas mixtures at atmospheric pressure. The discharge is sustained between two parallel plate electrodes covered by two dielectric plates, separated by a gap distance of 1mm. Dielectric beads are inserted in the gap to form a PB-DBD reactor.

As the driving voltage increases, a pure surface discharge gradually dominates, because the charged species can more easily escape to the beads and charge the bead surface due to the strong electric field at high driving voltage. This significant surface charging will enhance the tangential component of the electric field along the bead surface, yielding SIWs. The SIWs, in turn, yield a high concentration of reactive species on the bead surface, which will affect the chemical reactions (and energy efficiency) in plasma catalysis. Indeed, electron impact excitation and ionization mainly take place on the bead surfaces.

The propagation speed of SIWs becomes faster in N_2 gas and slower with increasing O_2 content, because it is more difficult for the discharge to be created, and it yields a slower discharge evolution in electronegative gas, due to the loss of electrons by attachment to O_2 molecules. This trend can also be understood from the significant difference in ionization rates between N_2 and O_2 gases. These different ionization rates will create different amounts of electrons and ions on the dielectric bead surface, which might have consequences for plasma catalysis.

Author Contributions: Visualization and Simulations, M.G.; Conceptualization, Y.Z.; Formal analysis, Y.Z., B.G., Q.Z. and A.B.; Software, H.W.; Writing—original draft, Y.Z.

Funding: This work was supported by the Natural Science Foundation of China (NSFC) (11775164, 11775090, 11575135) and the Fundamental Research Funds for the Central Universities (WUT: 2018IB011 and 2017IVA79). Ya Zhang gratefully acknowledges the Belgian Federal Science Policy Office for financial support.

Acknowledgments: The authors are very grateful to Wei Jiang for the useful discussions on the particle-in-cell/Monte-Carlo collision model.

Conflicts of Interest: The authors declare no conflict of interest.

References

1. Neyts, E.C.; Ostrikov, K.; Sunkara, M.K.; Bogaerts, A. Plasma Catalysis: Synergistic Effects at the Nanoscale. *Chem. Rev.* **2015**, *115*, 13408–13446.
2. Yi, H.; Du, H.; Hu, Y.; Yan, H.; Jiang, H.L.; Lu, J. Precisely Controlled Porous Alumina Overcoating on Pd Catalyst by Atomic Layer Deposition: Enhanced Selectivity and Durability in Hydrogenation of 1,3-Butadiene. *ACS Catal.* **2015**, *5*, 2735–2739.
3. O'Neill, B.J.; Jackson, D.H.; Lee, J.; Canlas, C.; Stair, P.C.; Marshall, C.L.; Elam, J.W.; Kuech, T.F.; Dumesic, J.A.; Huber, G.W. Catalyst design with atomic layer deposition. *ACS Catal.* **2015**, *5*, 1804–1825.
4. Chen, H.L.; Lee, H.M.; Chen, S.H.; Chao, Y.; Chang, M.B. Review of plasma catalysis on hydrocarbon reforming for hydrogen production—Interaction, integration, and prospects. *Appl. Catal. B Environ.* **2008**, *85*, 1–9.
5. Van Durme, J.; Dewulf, J.; Leys, C.; Van Langenhove, H. Combining non-thermal plasma with heterogeneous catalysis in waste gas treatment: A review. *Appl. Catal. B Environ.* **2008**, *78*, 324–333.
6. Chen, H.L.; Lee, H.M.; Chen, S.H.; Chang, M.B.; Yu, S.J.; Li, S.N. Removal of volatile organic compounds by single-stage and two-stage plasma catalysis systems: A review of the performance enhancement mechanisms, current status, and suitable applications. *Environ. Sci. Technol.* **2009**, *43*, 2216–2227.
7. Whitehead, J.C. Plasma catalysis: A solution for environmental problems. *Pure Appl. Chem.* **2010**, *82*, 1329–1336.
8. Neyts, E.; Bogaerts, A. Understanding plasma catalysis through modelling and simulation—A review. *J. Phys. D Appl. Phys.* **2014**, *47*, 224010.
9. Takashima, K.; Kaneko, T. Ozone and dinitrogen monoxide production in atmospheric pressure air dielectric barrier discharge plasma effluent generated by nanosecond pulse superimposed alternating current voltage. *Plasma Sources Sci. Technol.* **2017**, *26*, 065018.

10. Dong, L.; Xiao, H.; Fan, W.; Yin, Z.; Zhao, H. Temporal symmetry of individual filaments in different spatial symmetry filaments pattern in a dielectric barrier discharge. *Phys. Plasmas* **2010**, *17*, 102314.

11. Zhang, Y.; Wang, H.y.; Jiang, W.; Bogaerts, A. Two-dimensional particle-in cell/Monte Carlo simulations of a packed-bed dielectric barrier discharge in air at atmospheric pressure. *New J. Phys.* **2015**, *17*, 083056.

12. Babaeva, N.Y.; Kushner, M.J. Self-organization of single filaments and diffusive plasmas during a single pulse in dielectric-barrier discharges. *Plasma Sources Sci. Technol.* **2014**, *23*, 065047.

13. Babaeva, N.Y.; Naidis, G.V.; Kushner, M.J. Interaction of positive streamers in air with bubbles floating on liquid surfaces: Conductive and dielectric bubbles. *Plasma Sources Sci. Technol.* **2018**, *27*, doi:10.1088/1361-6595/aaa5da.

14. Fan, W.; Sheng, Z.; Liu, F.; Zhong, X.; Dong, L. Mechanisms of fine structure formation in dielectric barrier discharges. *Phys. Plasmas* **2018**, *25*, 023502.

15. Chanrion, O.; Neubert, T. A PIC-MCC code for simulation of streamer propagation in air. *J. Comput. Phys.* **2008**, *227*, 7222–7245.

16. Papageorghiou, L.; Panousis, E.; Loiseau, J.; Spyrou, N.; Held, B. Two-dimensional modelling of a nitrogen dielectric barrier discharge (DBD) at atmospheric pressure: Filament dynamics with the dielectric barrier on the cathode. *J. Phys. D Appl. Phys.* **2009**, *42*, 105201.

17. Babaeva, N.Y.; Kushner, M.J. Ion energy and angular distributions onto polymer surfaces delivered by dielectric barrier discharge filaments in air: I. Flat surfaces. *Plasma Sources Sci. Technol.* **2011**, *20*, 035017.

18. Akishev, Y.; Aponin, G.; Balakirev, A.; Grushin, M.; Karalnik, V.; Petryakov, A.; Trushkin, N. DBD surface streamer expansion described using nonlinear diffusion of the electric potential over the barrier. *J. Phys. D Appl. Phys.* **2013**, *46*, 464014.

19. Inada, Y.; Aono, K.; Ono, R.; Kumada, A.; Hidaka, K.; Maeyama, M. Two-dimensional electron density measurement of pulsed positive primary streamer discharge in atmospheric-pressure air. *J. Phys. D Appl. Phys.* **2017**, *50*, 174005.

20. Kogelschatz, U. Dielectric-barrier discharges: Their history, discharge physics, and industrial applications. *Plasma Chem. Plasma Process.* **2003**, *23*, 1–46.

21. Vasilyak, L.M.; Kostyuchenko, S.; Kudryavtsev, N.N.; Filyugin, I. Fast ionisation waves under electrical breakdown conditions. *Physics-Uspekhi* **1994**, *37*, 247–268.

22. Petrishchev, V.; Leonov, S.; Adamovich, I.V. Studies of nanosecond pulse surface ionization wave discharges over solid and liquid dielectric surfaces. *Plasma Sources Sci. Technol.* **2014**, *23*, 065022.

23. Goldberg, B.M.; Böhm, P.S.; Czarnetzki, U.; Adamovich, I.V.; Lempert, W.R. Electric field vector measurements in a surface ionization wave discharge. *Plasma Sources Sci. Technol.* **2015**, *24*, 055017.

24. Kushner, M.J. Modeling of microdischarge devices: Pyramidal structures. *J. Appl. Phys.* **2004**, *95*, 846–859.

25. Bruggeman, P.; Brandenburg, R. Atmospheric pressure discharge filaments and microplasmas: Physics, chemistry and diagnostics. *J. Phys. D Appl. Phys.* **2013**, *46*, 464001.

26. Zhang, Y.; Jiang, W.; Zhang, Q.; Bogaerts, A. Computational study of plasma sustainability in radio frequency micro-discharges. *J. Appl. Phys.* **2014**, *115*, 193301.

27. Šimek, M. Optical diagnostics of streamer discharges in atmospheric gases. *J. Phys. D Appl. Phys.* **2014**, *47*, 463001.

28. Tu, X.; Gallon, H.J.; Whitehead, J.C. Transition behavior of packed-bed dielectric barrier discharge in argon. *IEEE Trans. Plasma Sci.* **2011**, *39*, 2172–2173.

29. Kruszelnicki, J.; Engeling, K.W.; Foster, J.E.; Xiong, Z.; Kushner, M.J. Propagation of negative electrical discharges through 2-dimensional packed bed reactors. *J. Phys. D Appl. Phys.* **2017**, *50*, 025203.

30. Russ, H.; Neiger, M.; Lang, J.E. Simulation of micro discharges for the optimization of energy requirements for removal of NO x from exhaust gases. *IEEE Trans. Plasma Sci.* **1999**, *27*, 38–39.

31. Kang, W.S.; Park, J.M.; Kim, Y.; Hong, S.H. Numerical study on influences of barrier arrangements on dielectric barrier discharge characteristics. *IEEE Trans. Plasma Sci.* **2003**, *31*, 504–510.

32. Takaki, K.; Chang, J.S.; Kostov, K.G. Atmospheric pressure of nitrogen plasmas in a ferroelectric packed bed barrier discharge reactor. Part I. Modeling. *IEEE Trans. Dielectr. Electr. Insul.* **2004**, *11*, 481–490.

33. Van Laer, K.; Bogaerts, A. Fluid modelling of a packed bed dielectric barrier discharge plasma reactor. *Plasma Sources Sci. Technol.* **2016**, *25*, 015002.

34. Van Laer, K.; Bogaerts, A. Influence of gap size and dielectric constant of the packing material on the plasma behaviour in a packed bed DBD reactor: A fluid modelling study. *Plasma Process. Polym.* **2017**, *14*, doi:10.1002/ppap.201600129.

35. Van Laer, K.; Bogaerts, A. How bead size and dielectric constant affect the plasma behaviour in a packed bed plasma reactor: A modelling study. *Plasma Sources Sci. Technol.* **2017**, *26*, 085007.

36. Wang, W.; Kim, H.H.; Van Laer, K.; Bogaerts, A. Streamer propagation in a packed bed plasma reactor for plasma catalysis applications. *Chem. Eng. J.* **2018**, *334*, 2467–2479.

37. Kang, W.S.; Kim, H.H.; Teramoto, Y.; Ogata, A.; Lee, J.Y.; Kim, D.W.; Hur, M.; Song, Y.H. Surface streamer propagations on an alumina bead: Experimental observation and numerical modeling. *Plasma Sources Sci. Technol.* **2018**, *27*, 015018.

38. Hensel, K.; Martišovitš, V.; Machala, Z.; Janda, M.; Leštinský, M.; Tardiveau, P.; Mizuno, A. Electrical and optical properties of AC microdischarges in porous ceramics. *Plasma Process. Polym.* **2007**, *4*, 682–693.

39. Zhang, Y.; Wang, H.Y.; Zhang, Y.R.; Bogaerts, A. Formation of microdischarges inside a mesoporous catalyst in dielectric barrier discharge plasmas. *Plasma Sources Sci. Technol.* **2017**, *26*, 054002.

40. Zhang, Y.R.; Van Laer, K.; Neyts, E.C.; Bogaerts, A. Can plasma be formed in catalyst pores? A modeling investigation. *Appl. Catal. B Environ.* **2016**, *185*, 56–67.

41. Lieberman, M.A.; Lichtenberg, A.J. *Principles of Plasma Discharges and Materials Processing*; John Wiley & Sons: Hoboken, NJ, USA, 2005.

42. Furman, M.; Pivi, M. Probabilistic model for the simulation of secondary electron emission. *Phys. Rev. Spec. Top.* **2002**, *5*, 124404.

43. Phelps, A.; Petrovic, Z.L. Cold-cathode discharges and breakdown in argon: Surface and gas phase production of secondary electrons. *Plasma Sources Sci. Technol.* **1999**, *8*, R21.

44. Pancheshnyi, S.; Biagi, S.; Bordage, M.; Hagelaar, G.; Morgan, W.; Phelps, A.; Pitchford, L. The LXCat project: Electron scattering cross sections and swarm parameters for low temperature plasma modeling. *Chem. Phys.* **2012**, *398*, 148–153.

45. LXCat Website. Biagi-v8.9 Database. Available online: www.lxcat.net (accessed on 11 June 2015).

46. Kim, H.H.; Prieto, G.; Takashima, K.; Katsura, S.; Mizuno, A. Performance evaluation of discharge plasma process for gaseous pollutant removal. *J. Electrost.* **2002**, *55*, 25–41.

47. Song, H.K.; Choi, J.W.; Yue, S.H.; Lee, H.; Na, B.K. Synthesis gas production via dielectric barrier discharge over Ni/γ-Al$_2$O$_3$ catalyst. *Catal. Today* **2004**, *89*, 27–33.

48. Van Durme, J.; Dewulf, J.; Sysmans, W.; Leys, C.; Van Langenhove, H. Efficient toluene abatement in indoor air by a plasma catalytic hybrid system. *Appl. Catal. B Environ.* **2007**, *74*, 161–169.

49. Kim, H.H.; Teramoto, Y.; Sano, T.; Negishi, N.; Ogata, A. Effects of Si/Al ratio on the interaction of nonthermal plasma and Ag/HY catalysts. *Appl. Catal. B Environ.* **2015**, *166*, 9–17.

50. Tskhakaya, D.; Matyash, K.; Schneider, R.; Taccogna, F. The Particle-In-Cell Method. *Contrib. Plasma Phys.* **2007**, *47*, 563–594.

51. Nieter, C.; Cary, J.R. VORPAL: A versatile plasma simulation code. *J. Comput. Phys.* **2004**, *196*, 448–473.

52. Birdsall, C.K. Particle-in-cell charged-particle simulations, plus Monte Carlo collisions with neutral atoms, PIC-MCC. *IEEE Trans. Plasma Sci.* **1991**, *19*, 65–85.

53. Nanbu, K.; Mitsui, K.; Kondo, S. Self-consistent particle modelling of dc magnetron discharges of an O$_2$/Ar mixture. *J. Phys. D Appl. Phys.* **2000**, *33*, 2274.

MDPI

Article

Isotope Labelling for Reaction Mechanism Analysis in DBD Plasma Processes

Paula Navascués [1] , Jose M. Obrero-Pérez [1], José Cotrino [1,2], Agustín R. González-Elipe [1] and Ana Gómez-Ramírez [1,2,*]

1   Laboratory of Nanotechnology on Surfaces, Instituto de Ciencia de los Materiales de Sevilla (CSIC-Universidad de Sevilla), Avda. Américo Vespucio 49, 41092 Sevilla, Spain; paula.navascues@icmse.csic.es (P.N.); jmanuel.obrero@icmse.csic.es (J.M.O.-P.); cotrino@us.es (J.C.); arge@icmse.csic.es (A.R.G.-E.)
2   Departamento de Física Atómica, Molecular y Nuclear, Universidad de Sevilla, Avda. Reina Mercedes, 41012 Sevilla, Spain
*   Correspondence: anamgr@us.es; Tel.: +34-954-48-95-00 (ext. 909248)

Received: 27 November 2018; Accepted: 28 December 2018; Published: 4 January 2019

Abstract: Dielectric barrier discharge (DBD) plasmas and plasma catalysis are becoming an alternative procedure to activate various gas phase reactions. A low-temperature and normal operating pressure are the main advantages of these processes, but a limited energy efficiency and little selectivity control hinder their practical implementation. In this work, we propose the use of isotope labelling to retrieve information about the intermediate reactions that may intervene during the DBD processes contributing to a decrease in their energy efficiency. The results are shown for the wet reforming reaction of methane, using D_2O instead of H_2O as reactant, and for the ammonia synthesis, using $NH_3/D_2/N_2$ mixtures. In the two cases, it was found that a significant amount of outlet gas molecules, either reactants or products, have deuterium in their structure (e.g., HD for hydrogen, CD_xH_y for methane, or ND_xH_y for ammonia). From the analysis of the evolution of the labelled molecules as a function of power, useful information has been obtained about the exchange events of H by D atoms (or vice versa) between the plasma intermediate species. An evaluation of the number of these events revealed a significant progression with the plasma power, a tendency that is recognized to be detrimental for the energy efficiency of reactant to product transformation. The labelling technique is proposed as a useful approach for the analysis of plasma reaction mechanisms.

Keywords: dielectric barrier discharge (DBD); isotope labelling; methane reforming; ammonia synthesis; plasma catalysis

---

1. Introduction

Plasma and plasma catalysis with dielectric barrier discharge (DBD) reactors have been widely utilized for a large variety of chemical processes, including the reforming of hydrocarbons [1–3], the abatement of contaminants [4–6], or the synthesis of ammonia [7–9]. There are two major shortcomings when dealing with DBD plasma reactions. The first one refers to the energy efficiency, which, in general, is still much lower than that required for the current chemical or catalytic procedures. The second refers to the selectivity, which is still an unsolved challenge when trying to favor the formation of a particular product in detriment to others [10]. These limitations stem from the same nature of the plasma processes where kinetics control the reaction pathways and thermodynamics is a secondary player in determining the final reaction outputs. In addition, although much attention has been paid to the influence of electrical operating conditions (voltage, frequencies, etc.), there is still limited knowledge about the influence of other working parameters, such as the residence time of the reactants, internal structure of the reactors, and therefore the distribution of gases within the discharge, and so on. In recent works on the synthesis of ammonia [9,11] and on the reforming of hydrocarbons [3,12], we have demonstrated how the reaction performance is affected by these parameters. In particular, using deuterated water for the wet reforming reaction of methane, we were able to prove that not only direct reactions transforming the reactants (methane and water) intro products (hydrogen and carbon Monoxide) take place in the plasma, but also a panoply of intermediate processes that, consuming energy, do not lead to the formation of new product molecules [3]. To our knowledge, this is a first attempt in the literature to characterize DBD gas synthesis mechanisms, using a methodology that, while widely used in conventional or enzymatic catalysis for the same purpose [13–17], has only been incipiently used to study the plasma removal of pollutants [18–20].

In the present paper, we want to further explore the use of the isotope labelling technique to unravel the reaction mechanisms in the DBD plasma reactions. We studied two reactions, the synthesis of ammonia using hydrogen and nitrogen as reactants, and the wet reforming of methane to yield CO plus hydrogen. In the first case, we carried out experiments with a ternary mixture of ammonia/hydrogen/nitrogen, which, in the normal operating conditions of our reactor, represents the outlet mixture obtained during ammonia synthesis (i.e., including the ammonia formed and the unreacted hydrogen and nitrogen). We show that treating a mixture of $N_2/D_2/NH_3$ (i.e., where H_2 has been substituted by D_2) does not significantly alter the ammonia content in the outlet as compared to the inlet gas mixture, but substantially modifies the distribution of D atoms between the ammonia and hydrogen molecules, in a proportion that depends on the plasma power. Using the same methodology, we revisited the wet reforming reaction studied in our previous work [3] using mixtures of CH_4 plus D_2O (instead of conventional H_2O) as reactants, and where we studied the distribution of D in the outlet gas molecules as a function of the applied power. A careful analysis of the distribution of deuterium isotopes in the different outlet molecules provides useful information about both the occurrence of completely inefficient secondary processes (i.e., processes that do not contribute to the formation of the desired product compounds) and their relative importance, depending on electrical operational conditions. From this study, we propose a general methodology for the use of labelling techniques, which may help to unravel the DBD mechanisms intervening during the plasma synthesis reactions.

2. Results and Discussion

Before analyzing the isotopic exchange processes that took place for the selected reactions, we will first discuss the methodological basis utilized for the analysis of isotope labelling using mass spectrometry (MS) and infrared spectroscopy (IR).

2.1. Analysis of Plasmas Induced Isotope Exchange Reactions

For the ammonia reaction, we reported a maximum nitrogen conversion of 7% according to Reaction (1) [9], where the nitrogen and hydrogen acting as reactants give rise to ammonia, as follows:

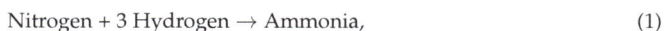

$$\text{Nitrogen} + 3\,\text{Hydrogen} \rightarrow \text{Ammonia,} \tag{1}$$

For the reported conditions of the maximum nitrogen conversion in the literature [11], for each 100 molecules of nitrogen and 100 of hydrogen in the inlet mixture, seven nitrogen molecules would transform into the ammonia. The outlet gas mixture would consist of 14 molecules of ammonia (product of Reaction (1)), 93 molecules of unreacted nitrogen, and 79 molecules of unreacted hydrogen. We attributed this relatively low reaction yield, in comparison with that attained in conventional catalytic processes [21] (yet the 7% found is one of the highest reported for the DBD synthesis of ammonia [7–9]), to the existence of back reactions leading to the formation of nitrogen and hydrogen from the formed ammonia (i.e., the inverse of Reaction (1)) or other intermediate processes, which result inefficient in rendering ammonia molecules. The isotope reaction experiment carried out in the present work does not pretend to increase the reaction yield, but instead evaluates the occurrence of intermediate processes that are neutral with respect to the formation of new ammonia molecules. For this purpose, we used a ternary gas mixture of nitrogen, hydrogen, and ammonia, approaching the composition of the outlet gas mixture of Reaction (1) reported in the literature [11], and ensuring that the amount of ammonia in the inlet and outlet mixtures remains invariable (i.e., to meet conditions under which there is no net production of ammonia). This does not mean that there are not intermediate reaction processes in the plasma, but that the formation of one ammonia molecule according to Reaction (1) is compensated with the decomposition of another one (i.e., inverse to Reaction (1)). A negligible formation of ammonia was indeed proved in the experiment, because the final concentration of N_2 detected by MS in the outlet gases remained constant after plasma activation (see below). As D_2 substituted H_2 in the inlet reaction mixture, the only source of H during the plasma activation of the ternary mixture was the NH_3 inlet gas feeding the reactor.

The typical MS and IR spectra of the outlet gases after the plasma activation of the $NH_3/D_2/N_2$ mixture for three power values is shown in Figure 1.

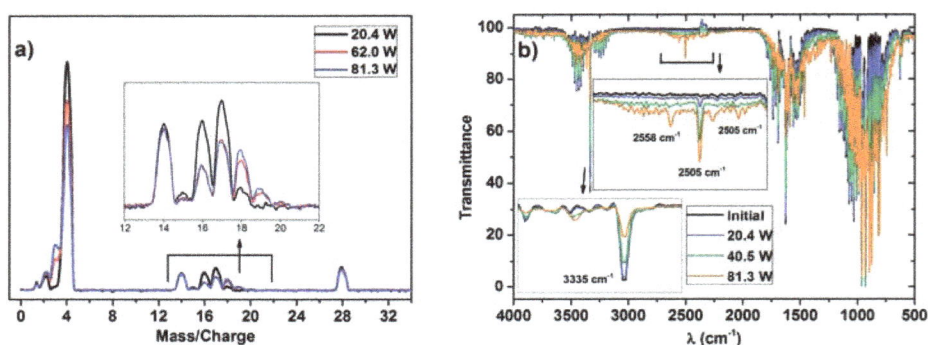

Figure 1. (**a**) Comparison of selected mass spectra recorded for the $NH_3/D_2/N_2$ mixture plasma treated at increasing powers in the dielectric barrier discharge (DBD) reactor; (**b**) infrared spectroscopy (IR) spectra of the outlet gases in the same experiments. The insets show an enlarge scale of the zones of bands attributed to differently labelled ammonia molecules (see text).

These MS spectra clearly show changes in the relative height of the peaks for the range of m/z values 0–4 (i.e., hydrogen–deuterium zone) and 13–20 (water and ammonia zones), which must be attributed to a progressively higher isotope exchange of H by D atoms at increasing powers. In concrete, the mass spectra reveal a clear increase of peaks at $m/z = 3$, attributable to HD, and others at $m/z = 18$ and 19, which must be attributed to NDH_2 and ND_2H (the fragmented ions are reported in Table 1). A minor contribution at $m/z = 20$, due to ND_3 molecules, could also be found in the spectra of the mixture activated with the maximum power. Meanwhile, a decrease in the intensity of the $m/z = 17$ peak indicates a parallel decrease in the NH_3 concentration in the outlet gases. In this set of experiments, no significant intensity change could be detected for the $m/z = 28$ (and $m/z = 14$) peak because of N_2, a behavior indicating that the concentration of nitrogen remains constant, and that, therefore, practically no new ammonia (i.e., including all of the labelled ND_xH_y molecules) is formed according to Reaction (1) during plasma activation.

A first account of the isotope labelling experiments for the wet reforming reaction was reported by the authors of [3], for a process that complied with the following stoichiometry:

$$\text{Methane + Water} \rightarrow \text{Carbon Monoxide + 3 Hydrogen,} \tag{2}$$

The outlet gases consisted of unreacted methane and water, carbon monoxide, and hydrogen. A conversion up to 50% of the initial methane flow was achieved under maximum operating power conditions (i.e., for 100 molecules of methane in the inlet mixture, there would be 50 molecules in of CO and 50 of methane in the outlet gas mixture). For the isotope labelling experiments, conventional water (i.e., H_2O) was substituted by deuterated water (D_2O). For this reaction, the MS reported in the literature [3] showed the appearance of m/z peaks, due to $H_2 + D_2 + HD$ (as hydrogen), $CH_4 + CH_3D + CH_2D_2$ (as methane), and $D_2O + DHO + H_2O$ (as water).

A quantitative evaluation of the percentages of the labelled molecules (i.e., incorporating D in their structure) in ammonia and hydrogen (Reaction (1)), and in hydrogen, methane, and water (Reaction (2)) are possible from the intensity of the MS peaks, taking into account the contributions to a particular peak of the molecular and molecular fragmented ions with this mass to charge (m/z) ratio. A summary of the different contributions to each particular m/z peak is reported in Table 1. The contributions of the doubly ionized species are disregarded in our analysis because of their very low probability. A quantitative evaluation of the percentage of labelled molecules can be made discounting the intensity of the residual masses always present in the MS analysis chamber (e.g., due to residual hydrogen, water, and hydrocarbons, an example of this can be seen in the Supporting Information, Figure S1) and also that of ionized molecular fragments with contributions that can be taken from fragmentation pattern libraries [22]. For example, to estimate the relative amount of NDH_2 in the outlet mixture, we proceed by assuming that $I (m/z = 18) = I (H_2O^+) + I (ND_2^+) + I (NDH_2^+)$, where $I (H_2O^+)$ is the intensity due to the residual water and is determined when measuring the spectrum of the initial mixture before switching on the plasma (see Figure S1), and $I (ND_2^+)$ is determined from the intensity of the ND_2H^+ ($m/z = 19$) and ND_3^+ ($m/z = 20$) peaks and their reported fragmentation patterns [21]. The results using this quantification procedure for the two investigated reactions will be presented in the next section.

Table 1. Contribution of molecular (in bold) and fragmented ions to the different peaks (m/z) in the mass spectrometry (MS) for Reactions (1) and (2). Hydrogen is common for the two reactions.

m/z Peak	Species		
	Hydrogen		
1	H^+		
2	D^+, $\mathbf{H_2^+}$		
3	HD^+		
4	$\mathbf{D_2^+}$		
	Ammonia [1]	**Methane** [2]	**Water** [2]
12	–	C^+	–
13	–	CH^+	–
14	N^+	CD^+, CH_2^+	–
15	NH^+	CDH^+, CH_3^+	–
16	NH_2^+, ND^+	$\mathbf{CH_4^+}$, CD_2^+, CDH_2^+	O^+
17	$\mathbf{NH_3^+}$, NDH^+	$\mathbf{CDH_3^+}$, CD_2H^+	OH^+
18	ND_2^+, $\mathbf{NDH_2^+}$	$\mathbf{CD_3H^+}$	$\mathbf{OH_2^+}$, OD^+
19	$\mathbf{ND_2H^+}$	$\mathbf{CD_4^+}$ [3]	ODH^+
20	$\mathbf{ND_3^+}$		OD_2^+

[1] Reaction (1). [2] Reaction (2). [3] Negligible intensity.

For the investigated ammonia reaction, the evaluation of the isotope exchange was confirmed by the data retrieved from the IR spectra. A rough evaluation of the series of spectra in Figure 1b indicates that the intensity profile of the vibrational/rotational band systems in the regions 3600–3100 cm^{-1}, 1800–1300 cm^{-1}, and 1250–700 cm^{-1} is characteristic of NH_3 (a typical spectrum of NH_3 gas is reported in the supported information, in Figure S2), and varies and becomes progressively shifted to lower wave numbers as the plasma power increases. In addition, a series of new little bands appear in the region of 2740–2400 cm^{-1}. This progressive shift and the appearance of new bands agree with the progressive formation of ND_2H and NDH_2 (and traces of ND_3 at the highest power), substituting the NH_3 molecules (note that the overall amount of ammonia remained invariable). This qualitative assessment of the spectral evolution and attribution of the bands coincides with the reported analysis of the IR spectra of NH_3, NH_2D, NHD_2, and ND_3 in the gas phase [23–25]. After a careful evaluation of these series of spectra, we could identify some specific bands that can be associated with NH_3 (3335 cm^{-1}), NH_2D (2505 cm^{-1}), NHD_2 (2558 cm^{-1}), and ND_3 (2420 cm^{-1}), which have been used to confirm the isotopic exchange deduced by the mass spectrometry analysis of the plasma activated mixture. However, as the extinction coefficient for each particular band is not easily accessible, the results will be semiquantitative and will be used just to confirm the tendencies deduced by MS.

2.2. Evaluation of Inefficient Reaction Events

The plasma induced reactions are triggered by the interaction of the plasma electrons with the gas molecules, giving rise to a series of activated intermediate species, radicals, and ions, which, through the intervention of a series of intermediate reactions, will eventually give rise to the product molecules of Reactions (1) and (2) detected in the outlet mixture. It is noteworthy that such intermediate reactions may involve not only "reactants", but also "products" molecules if they are present in the reaction medium. In the course of the DBD plasma processes, energy is wasted whenever these intermediate reactions do not give rise to "products" molecules. The use of labelled reactants tries to monitor the occurrence of the intermediate processes that are ineffective in producing "product" molecules. Some examples illustrating this type of inefficient intermediate reactions are as follows:

Reaction (1)

$$NH_3 + e^- \rightarrow NH^* + 2H^* + e^-$$

$$NH^* + D_2 \rightarrow NHD^* + D^*$$

$$NHD^* + D_2 \rightarrow \mathbf{NHD_2} + D^*$$

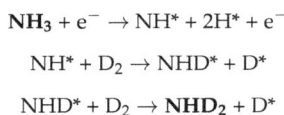

where one NH_3 molecule is being transformed into one NHD_2 molecule (or another indistinguishable NH_3 molecule when using H_2 instead of D_2) after three intermediate reactions, a set of processes that from the point of view of the reaction yield do not contribute to increasing the ammonia production (although it spends a considerable amount of the energy associated to plasma electrons). We must stress that in DBD process, particularly if they involve the use of catalysts, the intermediate reactions not involving electrons may take place either in the plasma phase or on the surface of the interelectrode pellets used to moderate the discharge (see reference [11], where we suggest this possibility for the synthesis of ammonia).

Reaction (2)

$$D_2O + e^- \rightarrow DO^* + D^* + e^-$$

$$\mathbf{CH_4} + D^* \rightarrow \mathbf{CH_3D} + H^*$$

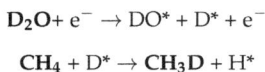

or alternatively

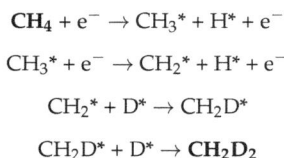

$$CH_4 + e^- \rightarrow CH_3^* + H^* + e^-$$

$$CH_3^* + e^- \rightarrow CH_2^* + H^* + e^-$$

$$CH_2^* + D^* \rightarrow CH_2D^*$$

$$CH_2D^* + D^* \rightarrow \mathbf{CH_2D_2}$$

or

$$CH_3^* + D^* \rightarrow \mathbf{CH_3D}$$

In this case, the energy of the electrons would be spent in dissociating the water or the methane used as reactants and, through a series of intermediate reactions, in creating intermediate species that react to form new CH_xD_y molecules without giving rise to hydrogen and CO as reaction products (Reaction (2)).

Some of the intermediate species quoted above in the examples of the intermediate reactions for Reactions (1) and (2) (e.g., H*, HO*, or NH*), have been effectively detected by optical emission spectroscopy during these reactions [9,11,26].

One of the purposes of the present study using labelled reactants is to develop a methodology to semi-quantitatively assess the occurrence of intermediate reactions. For this end, we will proceed in the following two steps: (i) determine the percentual distribution of the distinct labelled molecules of a given compound (e.g., the percentage of CH_4, CDH_3, and CD_2H_2) in the outlet gases for each experimental condition, (ii) approach the relative number of exchange processes that take place in each experiment. For step (i), we used the evaluation procedure based on the m/z peak intensities described in Section 2.1. For step (ii), we proceed as follows: we defined a relative number of reaction events (REs) for each compound as the sum of the percentage of a given isotopically marked molecule, multiplied by the number of exchanged isotope atoms, corrected by the flow ratio of this particular molecule in the outlet flow (note that in the case of the wet reforming Reaction (2), the total outlet flow increases with respect to the inlet). For example, the REs for methane (i.e., including all forms of isotopically labelled molecules) will be determined as follows:

$$RE_{methane} = [\%CDH_3 + \%CD_2H_2 \times 2 + \%CD_3H \times 3] \times [\text{partial flow of methane/total outlet flow}], \quad (3)$$

where, for example, $\%CD_3H_3$ is the percentage of this labelled molecule, referred to the total number of methane molecules. In this case, it is multiplied by three, because three H atoms in a parent CH_4 reactant molecule have been substituted by D. A similar argument holds for the other isotopically exchanged molecules.

Similarly, specific RE numbers can be defined for a specific isotopic molecular form. For example, for CD_3H we will have the following:

$$RE_{CD3H} = [\%CD_3Hx3] \times [\text{partial flow of } CD_3H/\text{total outlet flow}], \quad (4)$$

Similar definitions of REs can be applied to the other labelled molecules (e.g., CD_2H_2) detected for Reaction (2) and for the ammonia Reaction (1), although, in this case, the inlet and outlet flows are the same, as there is practically no net formation or decomposition of ammonia.

Similar relative numbers of the exchange reaction events can be determined for water (RE_{water}) and hydrogen ($RE_{hydrogen}$), and for their specific labelled molecules. For a given reaction, the total number of exchange reaction events (TRE) can be then calculated as the sum of all REs. We will make the assumption that TRE is as an indication of the number inefficient intermediate reactions occurring during the plasma process and, therefore, of the amount of energy wasted in the overall DBD plasma process. This assumption is somehow arbitrary and clearly underestimates the actual number of inefficient intermediate processes taking place to yield a given labelled molecule. This is so, because the RE and TRE numbers defined as in Reactions (3) or (4) do not take into account all of the possible intermediate processes contributing to isotope exchange reactions, see, for example, the intermediate reactions shown as the examples in "Reaction (2)" above. Moreover, these definitions of RE and TRE do not take into account the intermediate reactions involving either H_2 or H* atoms that are present in the system, and therefore do not give rise to isotopically exchanged molecules in the final isotopic molecular mixture.

In the following sections, we will determine the REs and TREs for Reactions (1) and (2) as a function of the power consumption, and discuss the evolution found in these parameters as a way to semi-quantitatively estimate the energy ratio wasted in the intermediate reactions, which are inefficient to render product molecules from the reactant molecules. These considerations rely on the assumption that elementary reaction rates are not significantly affected by the type of isotope bonded to the excited molecules. In reality, in conventional low pressure plasma, the rate of elementary reactions can be little affected by the type of isotope [27]. However, in atmospheric pressure, non-equilibrium plasmas excited with high AC voltages at relatively low frequency (i.e., conditions utilized in DBD discharges) energy are mainly used to induce very high electron temperatures and high vibrational temperatures [28], where the effect of the isotope mass will be negligible and therefore the above hypothesis is fully justified.

2.3. Ammonia Synthesis: Intermediate Exchange Reactions

Figure 2 shows the percentages of the isotopic labelled species produced as a function of the applied power. These percentages have been estimated by the analysis of the MS spectra in Figure 1a, recorded at increasingly higher powers. This plot clearly shows that the percentage of NDH_2 and ND_2H molecules relatively increases with the power. Simultaneously, the percentages of HD and, to a lesser extent, H_2 also increase with this parameter. A similar tendency was obtained by evaluating the evolution of the absorbance intensity of the specific IR bands attributed to the different ammonia species (see Figure S3 in the Supporting Information). This evaluation has only a semi-quantitative character, because the absorption coefficient of each band should be taken into account for quantification. As practically no change in the amount of ammonia (i.e., including all forms of labelled ammonia molecules) occurs during this series of experiments (as already mentioned, for the selected nitrogen/hydrogen/ammonia mixture, ammonia molecules are formed and decomposed at equivalent rates), we must assume that all of the energy applied to the reactor is used in inducing intermediate REs that are inefficient from the point of view of the ammonia synthesis.

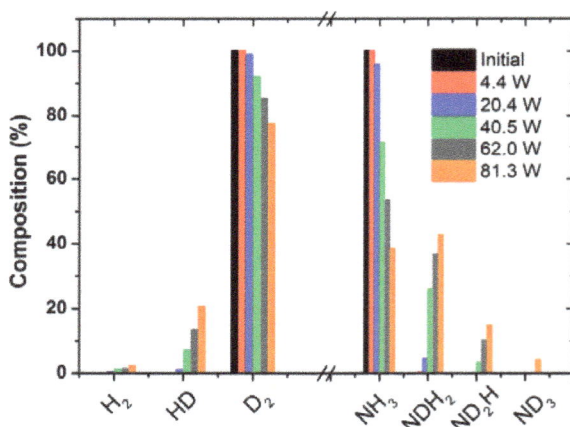

Figure 2. Percentage of isotopic labelled species detected for Reaction (1) as a function of plasma power.

Figure 3 shows the evolution of the individual REs and TRE calculated for Reaction (1). It is apparent in these plots that, for a power higher than 20 W, the RE values for the ammonia and hydrogen species increase with power in a rather lineal way. The final result is that the TRE values present a continuously increasing tendency when the applied power increases. A translation of this evidence for a real process aiming at the synthesis of ammonia from a binary mixture N_2/H_2 (Reaction (1)) is that although increasing the power could contribute to an increase in the ammonia yield by promoting the direct reaction between nitrogen and hydrogen, this increase would likely be damped by the progressive waste of energy involved in promoting an increasingly higher number of intermediate reaction events, as evidenced in Figure 3. An obvious conclusion of this evidence is that in order to increase the ammonia production and decrease the energy consumption, it would be required to change other reaction parameters and/or to modify the design and operational mode of the reactor.

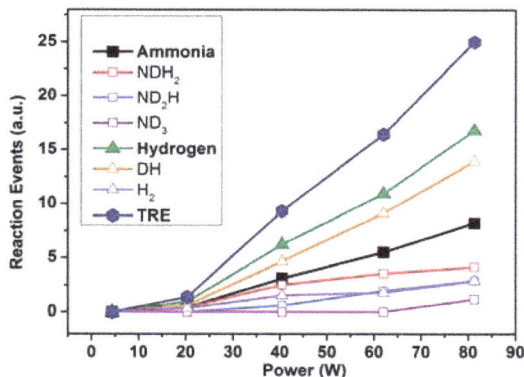

Figure 3. Evolution of reaction events (REs) and total number of exchange reaction events (TRE) for Reaction (1) as a function of plasma power.

2.4. Wet Reforming of Methane: Intermediate Exchange Reactions and Reaction Efficiency

The percentages of the different molecules detected by MS were reported in the literature [3], together with the percentage of methane transformed into carbon monoxide and hydrogen in each case (percentages of converted methane through Reaction (2) were 11%, 21%, and 50% for the three applied plasma powers). Figure 4 shows the REs for methane (summing their two labelled variants, also presented in this plot), water (sum of HDO and H_2O), and hydrogen (sum of HD and D_2), and the total value of TRE as a function of power consumption. The TRE depicts an increasing evolution with power that reveals a progressive increase in the number of inefficient events with this parameter. However, unlike the rather lineal evolution of the TRE reported for ammonia in Figure 3, the evolution of TRE in Figure 4 is characterized by a decreasing slope with power, indicating a certain saturation of the isotopic exchange probability. This tendency suggests that increasing the power may increase the reaction yield, as the number of REs does not progress lineally with it. In agreement with this, a reaction yield of 50% was found for the highest power applied in this experiment [3]. Another interesting feature deduced from Figure 4 is that, independently of the applied power, the number of events involved in the formation of CH_2D_2 is much higher than that rendering CH_3D. Tentatively, we associated this difference to the known relatively higher stability of CH_3^* radicals [29] and its smaller tendency to saturate the carbon bonds more than other reactive species such as CH_2^* (in other words, the intermediate reaction $CH_3^* + D^* \rightarrow CH_3D$ would have a much lower cross section probability than $CH_2^* + D^* \rightarrow CH_2D^*$). This isotope labelling result agrees with the current mechanistic models of the wet reforming reaction [30].

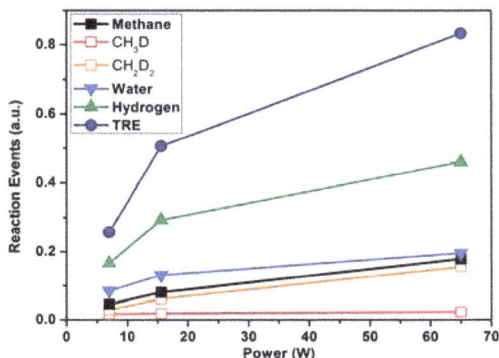

Figure 4. Evolution of REs and TRE for Reaction (2) as a function of plasma power.

3. Materials and Methods

Experiments have been carried out in a parallel plate DBD reactor made of stainless steel, which incorporates two electrodes of the same material, and 7.6 cm of diameter separated by a distance of 10 mm (i.e., gap between electrodes) in the case of the $NH_3/D_2/N_2$ mixture, and 3 mm for the wet reforming of methane. The bottom electrode was grounded while the top electrode was supplied with a high AC voltage at frequencies in the range of the kHz. The space separating the electrodes was filled with pellets of PZT (lead zirconate titanate). A more detailed description of the reactor and the electrical supply can be found in our previous works [3,9,11]. A scheme of the experimental set-up is reported in the Supporting Information (Figure S4).

The gases entered the reactor through the center of the bottom electrode thanks to a small grid with a 1 cm diameter communicating the gas inlet tubes and the discharge zone. The reactants were dosed by mass flow controllers and for D_2O with an automatic syringe. The D_2 and D_2O were supplied by Air Liquide (Alphagaz, Paris, France) and Sigma-Aldrich (St. Louis, MI, USA), respectively. The following mass flows were used for the experiments, as follows:

Ammonia reaction: NH_3 3.7 cm^3·min^{-1}, N_2 8.6 cm^3·min^{-1}, D_2 25.8 cm^3·min^{-1}

Wet reforming of methane: CH_4 4.8 cm^3·min^{-1}; D_2O 9 cm^3·min^{-1}

The reactor was kept at 130 °C for the wet reforming reaction in order to avoid any condensation of liquid water. For the $NH_3/D_2/N_2$ mixture, initially, at room temperature, the reactor walls naturally reached a temperature around 60 °C after two hours of operation (i.e., at the steady state, when an analysis of the outlet mixture was carried out). Under these operating conditions, and for a given experiment, we can neglect any significant influence of temperature changes on process efficiency, as the DBD processes reported were carried out at quite different values of this parameter [31].

The outlet gas mixture was analyzed with a mass spectrometer (Sensorlab, Prima Plus−Pfeiffer Vacuum, Asslar, Germany). Samples of the outlet gases were dosed into the mass spectrometer through a leak valve and a capillary tube to avoid any preferential enrichment of some species with respect to others (see Supporting Information, Figure S4). In the case that several molecules or fragments contribute to a given m/z peak, their particular contributions were estimated from the reported fragmentation patterns of the different molecular species [22].

In the case of the ammonia reaction, the mass spectrometry data were complemented by the infrared (IR) analysis of the outlet gases in an Agilent Cary 630 FTIR Spectrometer located in series with the outlet flow of the gases. The IR spectra were recorded in the region between 4000 and 500 cm^{-1}, with a resolution of 1 cm^{-1}.

In the case of the $NH_3/D_2/N_2$ mixture, the reactor was operated with a squared AC signal at a frequency of 5 kHz, and a width of 80 µs in the positive side and 120 µs in the negative (see Figure S5 in the Supporting Information). The maximum to minimum voltage difference was varied between 3.5 and 7.9 kV, rendering powers values of 4.4, 20.4, 40.5, 62.0, and 81.3 W, as determined from the corresponding Lissajous plots.

For the wet reforming reaction, the system was operated in the conventional AC mode using frequencies of 0.4, 1, and 5 kHz, and voltages of 2.9, 2.5, and 2.6 kV in the sinusoidal mode. The power consumptions were 7.0, 15.6, and 65.0 W.

The determination of the power consumption of the discharge was carried out by analyzing the Lissajous curves recorded with an oscilloscope, according to a procedure previously described [3,9,10]. The experiments were carried out at increasing currents, keeping (i) the voltage for the wet reforming of methane and (ii) the frequency (and the others mentioned characteristics of the pulse) for the $NH_3/D_2/N_2$ mixture approximately constant. The plot of the Lissajous curves obtained in the latter case for the different powers applied to the reactor is presented in Figure 5. Their shape is quite similar to that recorded when operating the reactor with a mixture $N_2 + H_2$, thus indicating that the two reaction mixtures behave similarly. Unlike the oval shape of the Lissajous curve obtained when operating the reactor in the conventional AC sinusoidal mode [12] used for the wet reforming reaction [3], the square shape of the Lissajous curves in Figure 5 is consistent with the fact that we are operating the reactor with a squared AC voltage in a packed bed reactor. Furthermore, the non-centered Lissajous figure is in agreement with the asymmetric squared signal.

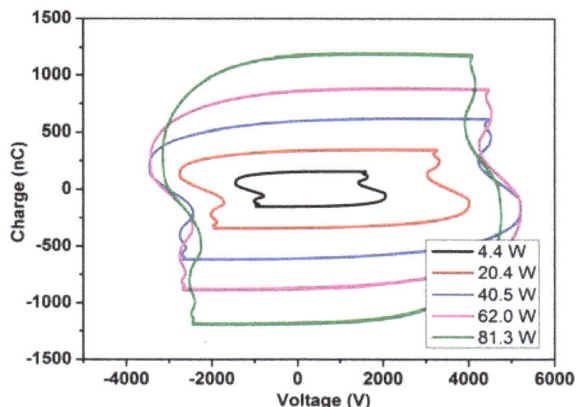

Figure 5. Set of Lissajous curves recorded when activating the $NH_3/D_2/N_2$ at increasing powers.

4. Conclusions

In this work, we present a systematic description of the isotope labelling technique to study plasma reactions. We show that elementary reactions, otherwise not accessible by the analysis of products, can be brought into the scene using this labelling technique, which provides a semi-quantitative procedure to estimate its occurrence as a function of applied power. Using MS data as a basis for the analysis, we introduced the concept of an inefficient reaction event and proposed a way of assessing the value of this parameter for each reactant and labelled molecule. The obtained values are quite relevant regarding the energy efficiency of the DBD reactions, and reveal that a considerable part of the energy is wasted in inducing events that do not give rise to the desired reaction products. This is firstly proved for a ternary gas mixture of ammonia, deuterium, and nitrogen, with which all of the energy was consumed in isotope exchange reaction events without any net production of new ammonia molecules. A similar analysis for the wet reforming of methane using deuterated water as a reactant yields a similar result, although the progression towards the formation of carbon monoxide and hydrogen as final products was more favorable that in the case of the ammonia reaction. The differences in the slopes of the plots representing the calculated numbers of the reaction events as a function of the applied power can be taken as a hint to predict the possibility of modifying the reaction yield, applying more power to the reactor. For the ternary mixture, the found linear evolution of the number of the total reaction events with applied power precludes that the reaction yield could increase with the applied power. This is not the case for the methane reforming reaction, and the methane conversion may increase up to 50% at the maximum power.

Supplementary Materials: The following are available online at http://www.mdpi.com/2073-4344/9/1/45/s1, Figures S1–S5. Figure S1: MS spectra (zone m/z from 13 to 20) taken for the ternary mixture $N_2+D_2+NH_3$ before (black line) and after (red line) switching on the plasma. The residual intensity at $m/z = 18$ is due to the residual water always present in the MS chambers. After application of plasma there is a change in the relative intensities of m/z peaks at 16, 17, 18 and 19, while the m/z peak at 14 remains constant. These changes are attributed to isotope exchange processes affecting to some of the initially detected NH_3 molecules that become transformed into NH_2D, NHD_2 and ND_3 (see the text); Figure S2: FTIR spectra recorded for binary N_2+D_2 (black line) and ternary $N_2+D_2+NH_3$ (red line) mixtures before plasma ignition. This analysis disregards the presence of water in the reactor chamber, since the zone 1300–2000 cm^{-1} only presents bands for the ternary mixture, which correspond to NH_3; Figure S3. Evolution of the intensity of the IR absorption bands attributed to different ammonia species during isotope labelling processes induced by DBD plasma; Figure S4: Scheme of the experimental set-up. A more detailed description of the reactor and the electrical supply can be found in our previous works. Outlet gas flow is represented with the point line; Figure S5: Squared AC curves for the different maximum to minimum voltage difference (3.5, 6.7, 8.7, 8.7 and 7.9 kV) and power (4.4, 20.4, 40.5, 62.0 and 81.3 W) in the case of the $NH_3/D_2/N_2$ mixture. Squared signals are expected to provide a higher efficiency that sinusoidal ones due to their higher Vrms value.

Author Contributions: The experimental part was conducted by P.N. and J.M.O.-P.; the bibliographic support and definition of the methodology of analysis was provided by J.C.; and the concept definition, discussion of results, and proposal of methodology and manuscript writing was done by A.G.-R. and A.R.G.-E.

Acknowledgments: The authors thank the European Regional Development Funds program (EU-FEDER) and the MINECO-AEI (201560E055 and MAT2016-79866-R and network MAT2015-69035-REDC) for financial support.

Conflicts of Interest: The authors declare no conflict of interest.

References

1. Tu, X.; Whitehead, J.C. Plasma-catalytic dry reforming of methane in an atmospheric dielectric barrier discharge: Understanding the synergistic effect at low temperature. *Appl. Catal. B Environ.* **2012**, *125*, 439–448. [CrossRef]

2. Wang, W.; Snoeckx, R.; Zhang, X.; Cha, M.S.; Bogaerts, A. Modeling plasma-based CO_2 and CH_4 conversion in mixtures with N_2, O_2, and H_2O: The bigger plasma chemistry picture. *J. Phys. Chem. C* **2018**, *122*, 8704–8723. [CrossRef]

3. Montoro-Damas, A.M.; Gómez-Ramírez, A.; Gonzalez-Elipe, A.R.; Cotrino, J. Isotope labelling to study molecular fragmentation during the dielectric barrier discharge wet reforming of methane. *J. Power Sources* **2016**, *325*, 501–505. [CrossRef]

4. Kim, H.-H. Nonthermal plasma processing for air-pollution control: A historical review, current issues, and future prospects. *Plasma Process. Polym.* **2004**, *1*, 91–110. [CrossRef]

5. Gómez-Ramírez, A.; Montoro-Damas, A.M.; Rodríguez, M.A.; González-Elipe, A.R.; Cotrino, J. Improving the pollutant removal efficiency of packed-bed plasma reactors incorporating ferroelectric components. *Chem. Eng. J.* **2017**, *314*, 311–319.

6. Dobslaw, D.; Schulz, A.; Helbich, S.; Dobslaw, C.; Engesser, K.-H. VOC removal and odor abatement by a low-cost plasma enhanced biotrickling filter process. *J. Environ. Chem. Eng.* **2017**, *5*, 5501–5511. [CrossRef]

7. Peng, P.; Chen, P.; Schiappacasse, C.; Zhou, N.; Anderson, E.; Chen, D.; Liu, J.; Cheng, Y.; Hatzenbeller, R.; Addy, M.; et al. A review on the non-thermal plasma-assisted ammonia synthesis technologies. *J. Clean. Prod.* **2018**, *177*, 597–609. [CrossRef]

8. Hong, J.; Aramesh, M.; Shimoni, O.; Seo, D.H.; Yick, S.; Greig, A.; Charles, C.; Prawer, S.; Murphy, A.B. Plasma catalytic synthesis of ammonia using functionalized-carbon coatings in an atmospheric-pressure non-equilibrium discharge. *Plasma Chem. Plasma Process.* **2016**, *36*, 917–940. [CrossRef]

9. Gómez-Ramírez, A.; Montoro-Damas, A.M.; Cotrino, J.; Lambert, R.M.; González-Elipe, A.R. About the enhancement of chemical yield during the atmospheric plasma synthesis of ammonia in a ferroelectric packed bed reactor. *Plasma Process. Polym.* **2017**, *14*, 1600081. [CrossRef]

10. Gómez-Ramírez, A.; Rico, V.J.; Cotrino, J.; González-Elipe, A.R.; Lambert, R.M. Low temperature production of formaldehyde from carbon dioxide and ethane by plasma-assisted catalysis in a ferroelectrically moderated dielectric barrier discharge reactor. *ACS Catal.* **2014**, *4*, 402–408. [CrossRef]

11. Gómez-Ramírez, A.; Cotrino, J.; Lambert, R.M.; González-Elipe, A.R. Efficient synthesis of ammonia from N_2 and H_2 alone in a ferroelectric packed-bed DBD reactor. *Plasma Sources Sci. Technol.* **2015**, *24*, 065011. [CrossRef]

12. Montoro-Damas, A.M.; Brey, J.J.; Rodríguez, M.A.; González-Elipe, A.R.; Cotrino, J. Plasma reforming of methane in a tunable ferroelectric packed-bed dielectric barrier discharge reactor. *J. Power Sources* **2015**, *296*, 268–275. [CrossRef]

13. Wei, J.; Iglesia, E. Isotopic and kinetic assessment of the mechanism of methane reforming and decomposition reactions on supported iridium catalysts. *Phys. Chem. Chem. Phys.* **2004**, *6*, 3754. [CrossRef]

14. Sprung, C.; Arstad, B.; Olsbye, U. Methane steam reforming over a $Ni/NiAl_2O_4$ model catalyst—Kinetics. *ChemCatChem* **2014**, *6*, 1969–1982. [CrossRef]

15. Luo, J.Z.; Yu, Z.L.; Ng, C.F.; Au, C.T. CO_2 /CH_4 Reforming over Ni–$La_2 O_3$/5A: An investigation on carbon deposition and reaction steps. *J. Catal.* **2000**, *194*, 198–210. [CrossRef]
16. Zhang, Z.; Verykios, X.E. Mechanistic aspects of carbon dioxide reforming of methane to synthesis gas over Ni catalysts. *Catal. Lett.* **1996**, *38*, 175–179. [CrossRef]
17. Luk, L.Y.P.; Ruiz-Pernía, J.J.; Adesina, A.S.; Loveridge, E.J.; Tuñón, I.; Moliner, V.; Allemann, R.K. Chemical ligation and isotope labeling to locate dynamic effects during catalysis by dihydrofolate reductase. *Angew. Chem. Int. Ed.* **2015**, *54*, 9016–9020. [CrossRef]
18. Daou, F.; Vincent, A.; Amouroux, J. Point and multipoint to plane barrier discharge process for removal of NO_x from engine exhaust gases: Understanding of the reactional mechanism by isotopic labeling. *Plasma Chem. Plasma Process.* **2003**, *23*, 309–325. [CrossRef]
19. Robert, S.; Francke, E.; Cavadias, S.; Gonnord, M.F.; Amouroux, J. Descomposition of acetaldehyde in air in a dielectric barrier discharge. *High Temp. Mater. Process. Int. Q. High-Technol. Plasma Process.* **2011**, *15*, 15–22. [CrossRef]
20. Vincent, A.; Daou, F.; Santirso, E.; Moscosa, M.; Amouroux, J. Experimental and simulation study of NO_x removal with a DBD wire-cylinder reactor. *High Temp. Mater. Process. Int. Q. High-Technol. Plasma Process.* **2003**, *7*, 267–275. [CrossRef]
21. Ertl, G. Reactions at surfaces: From atoms to complexity (Nobel Lecture). *Angew. Chem. Int. Ed.* **2008**, *47*, 3524–3535. [CrossRef] [PubMed]
22. S.E. Stein NIST Mass Spec Data Center. *"Mass Spectra" in NIST Chemistry WebBook*; NIST Standard Reference Database Number 69; Linstrom, P.J., Mallard, W.G., Eds.; National Institute of Standards and Technology: Gaithersburg, MD, USA, 2009.
23. Snels, M.; Hollenstein, H.; Quack, M. The NH and ND stretching fundamentals of $^{14}ND_2H$. *J. Chem. Phys.* **2003**, *119*, 7893–7902. [CrossRef]
24. Shimanouchi, T. *Tables of Molecular Vibrational Frequencies*; NSRDS-NBS: Washington, DC, USA, 1972; Volume 1.
25. Snels, M.; Fusina, L.; Hollenstein, H.; Quack, M. The v_1 and v_3 bands of ND_3. *Mol. Phys.* **2000**, *98*, 837–854. [CrossRef]
26. Wang, Y.-F.; Tsai, C.-H.; Chang, W.-Y.; Kuo, Y.-M. Methane steam reforming for producing hydrogen in an atmospheric-pressure microwave plasma reactor. *Int. J. Hydrog. Energy* **2010**, *35*, 135–140. [CrossRef]
27. Ivanov, M.V.; Schinke, R. Recombination of ozone via the chaperon mechanism. *J. Chem. Phys.* **2006**, *124*, 104303. [CrossRef]
28. Fridman, A. *Plasma Chemistry*; Cambridge University Press: Cambridge, UK, 2008.
29. Sugai, H.; Kojima, H.; Ishida, A.; Toyoda, H. Spatial distribution of CH_3 and CH_2 radicals in a methane rf discharge. *Appl. Phys. Lett.* **1990**, *56*, 2616–2618. [CrossRef]
30. Nozaki, T.; Fukui, W.; Okazaki, K. Reaction enhancement mechanism of the nonthermal discharge and catalyst hybrid reaction for methane reforming. *Energy Fuels* **2008**, *22*, 3600–3604. [CrossRef]
31. Harling, A.M.; Kim, H.-H.; Futamura, S.; Whitehead, J.C. Temperature dependence of plasma−catalysis using a nonthermal, atmospheric pressure packed bed; the destruction of benzene and toluene. *J. Phys. Chem. C* **2007**, *111*, 5090–5095. [CrossRef]

catalysts

MDPI

Article

Plasma Catalysis: Distinguishing between Thermal and Chemical Effects

Guido Giammaria [1], Gerard van Rooij [2] and Leon Lefferts [1,*]

[1]   Catalytic Processes and Materials Group, University of Twente, 7522 NB Enschede, The Netherlands;
       g.giammaria@utwente.nl
[2]   Nonequilibrium Fuel Conversion, DIFFER, 5612 AJ Eindhoven, The Netherlands; G.J.vanRooij@differ.nl
*   Correspondence: l.lefferts@utwente.nl; Tel.: +31-53-489-2922

Received: 21 January 2019; Accepted: 7 February 2019; Published: 16 February 2019

Abstract: The goal of this study is to develop a method to distinguish between plasma chemistry and thermal effects in a Dielectric Barrier Discharge nonequilibrium plasma containing a packed bed of porous particles. Decomposition of $CaCO_3$ in Ar plasma is used as a model reaction and $CaCO_3$ samples were prepared with different external surface area, via the particle size, as well as with different internal surface area, via pore morphology. Also, the effect of the CO_2 in gas phase on the formation of products during plasma enhanced decomposition is measured. The internal surface area is not exposed to plasma and relates to thermal effect only, whereas both plasma and thermal effects occur at the external surface area. Decomposition rates were in our case found to be influenced by internal surface changes only and thermal decomposition is concluded to dominate. This is further supported by the slow response in the CO_2 concentration at a timescale of typically 1 minute upon changes in discharge power. The thermal effect is estimated based on the kinetics of the $CaCO_3$ decomposition, resulting in a temperature increase within 80 °C for plasma power from 0 to 6 W. In contrast, CO_2 dissociation to CO and O_2 is controlled by plasma chemistry as this reaction is thermodynamically impossible without plasma, in agreement with fast response within a few seconds of the CO concentration when changing plasma power. CO forms exclusively via consecutive dissociation of CO_2 in the gas phase and not directly from $CaCO_3$. In ongoing work, this methodology is used to distinguish between thermal effects and plasma–chemical effects in more reactive plasma, containing, e.g., H_2.

Keywords: nonequilibrium plasma; plasma catalysis; gas temperature; calcium carbonate decomposition

1. Introduction

Plasma catalysis is receiving more and more attention in the last few years, since the specific interactions between plasma and catalyst surface may lead to synergistic effects [1–4]. One of the earliest plasma catalytic applications is the abatement of volatile organic compounds (VOC) [5,6], while in the last decade research has been focused more on CO_2 conversion [7–9], conversion of hydrocarbons via reforming, and coupling [10–12], as well as activation of N_2 [13,14]. Reforming of hydrocarbons is an example of an endothermic reaction, where plasma catalysis holds promise because of activation of hydrocarbons at low temperature, but also because electrical energy would be used to generate the required heat. Methane coupling and CO_2 dissociation are examples of thermodynamically hill-up reactions, which are clearly more challenging.

Nonequilibrium plasma, e.g., microwave of Dielectric Barrier Discharge (DBD) plasma, is especially attractive because it operates at relatively low temperatures [15–17]. Consequently, catalyst sintering is prevented. Moreover, low temperatures are a necessity to enable catalysis in the first place, facilitating initial adsorption that decreases entropy. Starting and stopping plasma reactors is much faster than usual thermal reactors, which is an advantage when fast capacity changes are required, e.g., in connection with intermittent energy supply and storage.

DBD plasma is frequently used for studying plasma catalytic conversion. The high AC voltages applied at relatively low frequency (50 to 10^5 Hz) produces a nonequilibrium plasma with very high electron temperatures (1–10 eV equal to 10^4–10^5 K), rather high vibrational temperatures (10^3 K), and rather low rotational and translational temperatures in the plasma zone, typically in the order or smaller than 100 K [6,15–17]. The fact that energy would be directed directly to bond breaking, without the need to heat up the gas mixture completely, is very attractive as heat exchangers to recover the heat would become redundant. The presence of a dielectric between the two electrodes prevents the formation of hot plasma in a single spark, forming instead several microfilaments, resulting in a more uniform plasma. The low gas temperature allows the application of a catalyst directly in the plasma generation zone without fast deactivation, maximizing the interaction between active species and the catalytic phase. Furthermore, DBD plasma can be generated at atmospheric pressure, which is interesting from the application point of view. Very promising results were presented in the last years on several topics, such as CO_2 conversion [9,18–27] and CH_4 reforming [12,13,28–34], in terms of high conversion and selectivity. However, the main issue of DBD remains the low energy efficiency achieved, i.e., the ratio between chemical energy stored in the produced molecules and electrical energy applied, which rarely surpasses 10%. This is explained by dissociative excitation by electron impact, involving a large activation barrier, dominating over vibrational excitation [7,35–37]. A more suitable technique for vibrational excitation is microwave plasma, which uses GHz frequencies [38–40]. However, the temperature increase is more pronounced in microwave plasma, limiting the opportunities for plasma catalysis.

Interaction between plasma and catalyst can proceed in many ways [1–4]. Obviously, the plasma will introduce new chemical species including activated species, radicals, and ions, which may all adsorb on the catalyst opening new reaction pathways and influencing the products distribution. Plasma can also induce photocatalytic effects by UV irradiation, impingement of charged particles and thermal fluctuations. The surface and subsurface of a catalyst can be modified by plasma via poisoning, implantation, sputtering, and etching. The presence of a catalyst influences the plasma by changing the electrical field distribution, but also modifying the free volume and the residence time in the plasma zone. Plasma also affects the temperature of the system, obviously influencing reaction rates of chemical reactions.

Unfortunately, it is not possible to measure the temperature in a DBD plasma catalytic reactor directly. Application of a thermocouple inside the plasma is not possible due to the high electric fields present. Nevertheless, thermocouples were used a few millimeters outside the plasma zone inside the reactor tube, or outside the reactor tube just alongside the plasma zone [41–44]. Furthermore, many attempts have been done to measure temperatures indirectly, for instance by emission spectroscopy of UV–Vis radiation probing electronic transitions in nitrogen and hydroxyl groups [45–48], by UV absorption spectroscopy [47], or by infrared emission [23,44]. Unfortunately, these methods have serious limitation depending on the reactor material properties as well as the packed bed properties.

This study proposes a method to distinguish between thermal effect and plasma chemistry effects in fixed bed DBD plasma reactors. The decomposition of calcium carbonate is used as a model system for packed beds containing porous particles. It is well known that thermal decomposition results in formation of exclusively CaO and CO_2 [49], whereas formation of CO would indicate that plasma chemistry is involved. A pure thermal effect is likely when using an Ar plasma, since no chemistry is expected between activated argon species and $CaCO_3$. The choice for $CaCO_3$ as model system in combination with a DBD reactor is inspired by its relevance for CO_2 separation. The calcium looping cycle consists of carbonation of CaO for capturing followed by calcination of $CaCO_3$ in order to recycle calcium oxide and to produce pure CO_2 [49]. Bottleneck is the calcination reaction that requires high temperatures in order to achieve high CO_2 concentrations in the outlet, i.e., at least 950 °C to achieve 1 bar CO_2 [50]. Such temperatures result in sintering, decreasing the CO_2 capture capacity when calcium oxide is recycled multiple times [51–53]. Using a DBD plasma during the calcium carbonate decomposition might circumvent the need for such high temperatures, and in addition CO_2 will be converted by plasma into CO, converting electrical energy into chemical energy and producing an added-value product.

The method to distinguish between thermal and plasma chemistry is based on the fact that plasma cannot exist in the pores inside particles if they are smaller than a few micrometers, as can be understood from Paschen's Law, which is generally accepted [54,55]. It was recently reported in a theoretical study that penetration of plasma in pores is possible to some extent [56]; however, we will discuss that under our conditions the plasma is limited to the interparticle volume and the external surface area is exposed under the conditions applied. The internal surface area, caused by the presence of small pores in the material is not exposed to the plasma, but would be influenced by any thermal effect. The theory is explained in detail in the section Methods in Appendix B and shown in Figure A13. The method then consists of two approaches: first, the effects of both the internal surface area as well as external surface area will be explored, and second, the dynamics of the performance on changing plasma power will be evaluated. Decomposition rate and eventual further reactions of the carbon dioxide product will be assessed. Argon plasma is used as a reference for the method to be developed as to distinguish between thermal effects and plasma–chemical effects.

2. Results

X-ray fluorescence (XRF) measurements confirmed the purity of CaO (99.12%) containing some minor impurities, i.e., SiO_2 (0.16%), MgO (0.12%), and Al_2O_3 (0.095%).

Table 1 in the Materials and Methods Section presents the surface area and particles sizes of the five prepared samples. The surface areas reported for samples A and B are well reproducible, but it should be noted that systematic errors may be larger, given the relatively low value of the surface areas. The surface area of batch III (samples C–E) is below the detection limit of the N_2 physisorption equipment, from which we deduce that the surface area is below 0.5 m²/g. In any case, the total surface of the samples increases in the order C<B<A. Remarkably, the surface areas of the parent oxides are much higher, confirming the theory that formation of a carbonate layer induces closure of small pores, due to the lower density of $CaCO_3$ (2.71 g cm^{-3}) compared to CaO (3.35 g cm^{-3}). However, the order in surface area remains the same, reassuring that the surface area of the samples is in the order A>B>C>D>E. Figure 1 shows the pore size distribution measured with mercury porosimetry for the carbonated samples synthesized from calcium ascorbate (sample B, Figure 2a) and from calcium carbonate (sample C, Figure 2b). Sintering at 900 °C for 24 h (sample C) causes formation of large pores of typically 400 nm, compared to sample A which was treated with CO_2 only at 630 °C. Remarkably, the pore volume of the sintered sample is ~20%, much larger than sample B (~5%), indicating that smaller pores collapsed favoring enlargement of the bigger pores.

Table 1. Characteristics of the samples prepared from three different precursors.

Sample Code	Precursor (Batch #)	Carbonation Time at 630 °C (h)	Sintering Time at 900 °C (h)	CaO S.S.A. (m² g⁻¹)	CaCO₃ S.S.A. (m² g⁻¹)	Particles Diameter (μm)
A	Ca Gluconate (I)	5	0	46.2	1.7 ± 0.1	250–300
B	Ca Ascorbate (II)	4	0	23.2	0.8 ± 0.1	250–300
C	CaCO₃ (III)	5	24	10.1	<0.5	250–300
D	CaCO₃ (III)	5	24	10.1	<0.5	100–125
E	CaCO₃ (III)	5	24	10.1	<0.5	38–45

(a)   (b)

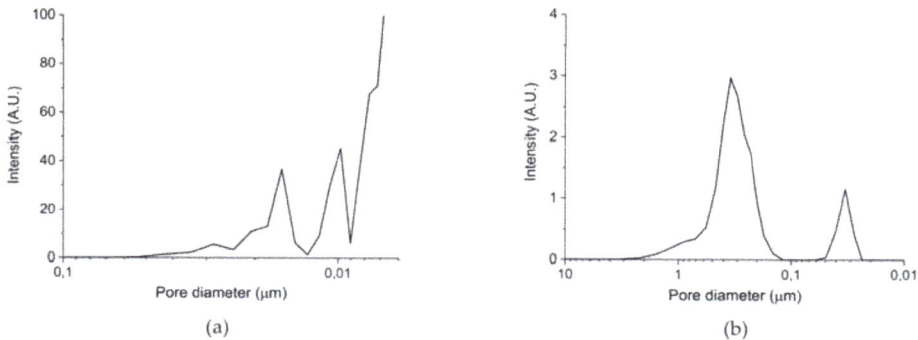

Figure 1. (**a**) Pore size distribution for sample B batch II and (**b**) pore size distribution for sample C batch III according Hg porosimetry.

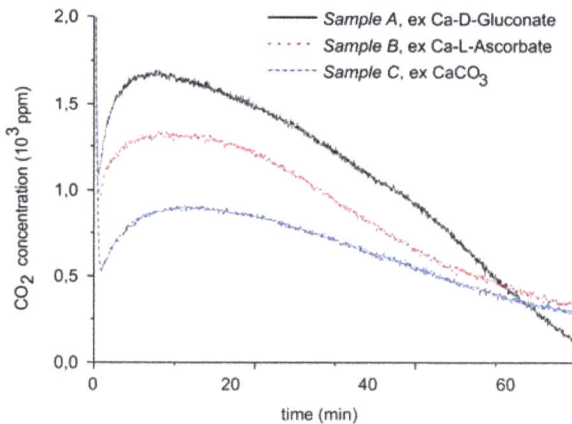

Figure 2. CO_2 concentration monitored by mass spectrometry during decomposition of carbonated sample A (solid line), sample B (dotted line), and sample C (dashed line). The temperature is 630 °C, flow rate is 30 mL min^{-1}, and the gas is pure Ar.

Figure 2 presents the results of isothermal decomposition at 630 °C of the carbonated samples A–C, showing that the CO_2 concentration generated via decomposition is within 10% constant during typically 20 min, after an induction time of a few minutes. In general, this is observed when decomposition is limited to max 50% of $CaCO_3$ present initially. The amounts of CO_2 desorbed from the samples A, B, and C are 4, 3.8, and 2.8 mg, respectively, equivalent with 0.66, 0.61, and 0.39 g_{CO2}/g_{CaO}, respectively. The amounts of CO_2 decrease in the same order as surface area. It can be estimated that the thickness of the $CaCO_3$ layer is in the order of 30 nm for all three sample. This is consistent with the observation that the maximum CO_2 concentration during decomposition experiments, between 900 to 1700 ppm, which is significantly lower than the 7000 ppm thermodynamic equilibrium CO_2 concentration at 630 °C [50]. Consequently, the CO_2 concentration is determined by kinetics, instead of thermodynamics, and thus also by the surface area.

Figure 3 presents the effect of plasma power, measured with a Lissajous plot (see Appendix B), on the decomposition of carbonated sample C. No plasma was applied during the first two minutes and CO_2 is the only product observed, while in the presence of plasma CO and O_2 are also produced, next to CO_2. Every two minutes a different voltage is applied, and the power is measured after ca. 1 minute. Changing the plasma power causes a fast response of the CO concentration in the order of seconds, while the CO_2 concentration needs typically a minute to stabilize. The O_2 concentration shows a delay; this effect is not understood at this time, but it may be speculated that interaction with CaO is responsible. Figure 3 shows that steady-state decomposition was achieved for the two lower plasma power values. The highest power setting of 9.6 W caused exhaustion so that the product concentrations are likely to be underestimated. Additional experiments were performed at constant maximum power as shown in Figure 4. Two experiments were performed at 2.1 W and demonstrate a reproducibility within ±5%, even though plasma was applied with a 1.5 min delay in the second experiment (Figure 4b). Figure 4c is performed at 4.4 W, showing a higher decomposition rate and shorter steady state duration. Therefore, the power was not further increased for this sample and in general the maximum power was limited to 5.1 W.

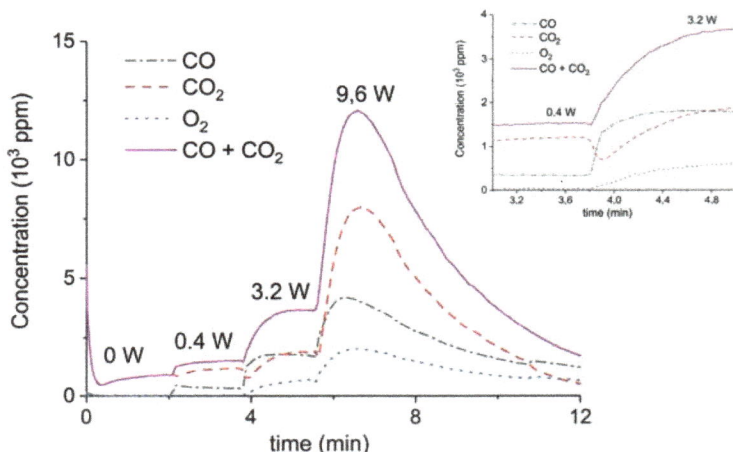

Figure 3. Concentration of the products of $CaCO_3$ decomposition (sample C) as function of time at 630 °C in pure argon, flow rate of 30 mL/min, plasma power is changed every 2 min with values of 0, 0.4, 3.2, and 9.6 W.

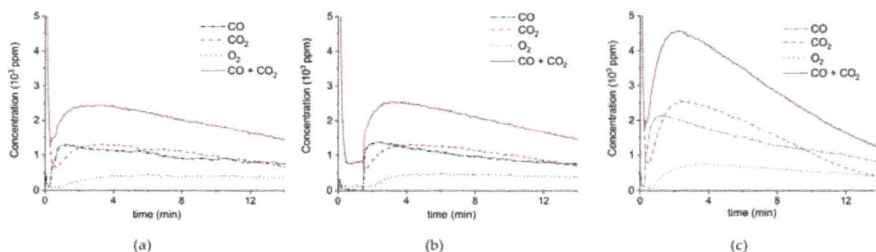

Figure 4. Concentration of the products of $CaCO_3$ decomposition (sample C) as function of time at 630 °C in pure argon, flow rate of 30 mL/min, plasma power is 2.1 W (**a,b**) and 4.4 W (**c**). Plasma is turned on at the beginning of decomposition (a,c) or after 1.5 min (b).

Figure 5 shows three typical results on the effect of the specific surface area on the decomposition at 2.1 W plasma power, by comparing the samples A (ex calcium gluconate, Figure 6a), B (ex calcium ascorbate, Figure 6b), and C (ex calcium carbonate Figure 6c), keeping particle size constant (250–300 μm). The total decomposition rate, as calculated based on the sum of the rates of formation of CO and CO_2, seems to increase with increasing specific surface area, as also observed in the absence of plasma. During the decomposition of sample B, the power has been turned off after 9 min, resulting in a rapid decrease in the CO and CO + CO_2 concentrations. Figure 6 presents all data on the decomposition rate measured on the three samples when changing the plasma power, showing that the rate of decomposition at the same power is significantly lower for the sample with the lowest specific surface area. The difference between sample A and B is not larger than experimental scatter, although the data suggest a slightly higher rate for sample A.

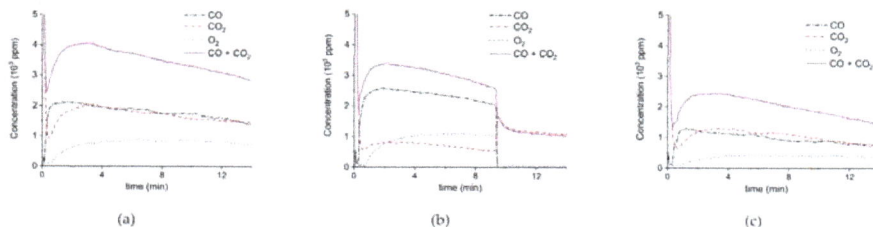

(a) (b) (c)

Figure 5. Concentration of the products of decomposition of sample A (**a**), sample B (**b**), and sample C (**c**), with a temperature of 630 °C in pure argon and flow rate of 30 mL/min; plasma power is 2.1 ± 0.1 W.

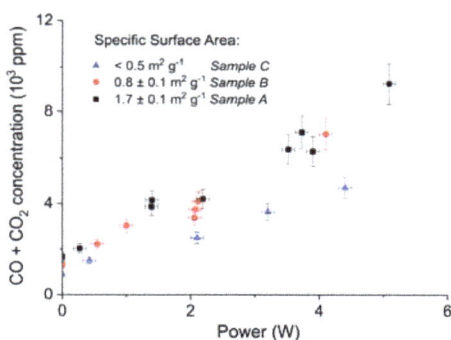

Figure 6. Sum of CO_2 and CO concentrations plotted as function of power obtained during decomposition of $CaCO_3$ samples A, B, and C with different surface areas; all the experiments performed at 630 °C, in pure argon, and flow rate 30 mL/min.

Figure 7 shows the influence of particles size on $CaCO_3$ decomposition by comparing sample C (250–300 µm, Figure 7a), D (100–125 µm, Figure 7b), and E (38–45 µm, Figure 7c) at plasma powers varying between 1.3 and 2.1 W. Figure 5d shows details on the response time after switching on the plasma for the three samples. The time to reach steady state is 50, 40, and 10 seconds for samples C, D, and E, respectively. These times are in reasonable agreement with the Fourier times of $CaCO_3$ particles of these sizes. The different powers do not allow direct comparison of the decomposition rates. Instead, the effect of plasma power on the decomposition rate for samples C, D, and E is presented in Figure 8, showing that the particle size has no effect on the decomposition rate within experimental error. All the experiments addressed in Figures 6 and 8 are shown in detail in Figures A1–A6 in Appendix A.

Figure 9 shows the results of a series of plasma enhanced decomposition experiments in the presence CO_2 in the feed gas, measured on sample C. All experiments were done with one sample by performing 22 carbonation and decomposition cycles. The stability of the sample was verified by repeating decomposition measurements in the absence of plasma, demonstrating invariable results within 5% as shown in Figure A15 in Appendix A.

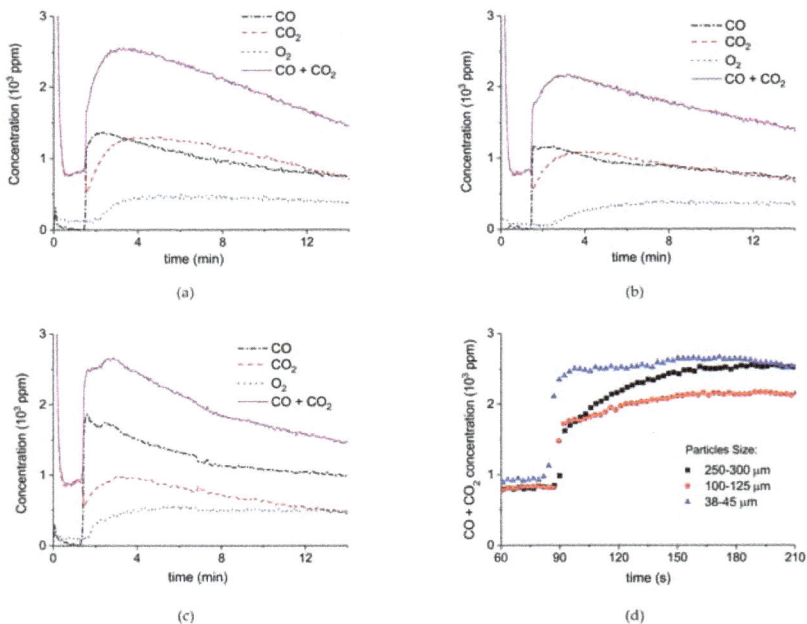

Figure 7. Concentration of the products of decomposition of sample C (**a**), sample D (**b**), and sample E (**c**) as function of time with a temperature of 630 °C in pure argon and flow rate of 30 mL/min; plasma power is 2.1 W for sample C, 1.3 W for sample D, and 1.7 W for sample E. (**d**) Initial response of CO + CO_2 concentration after activating the plasma for all three samples.

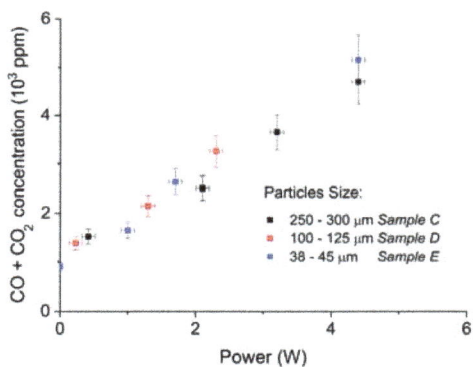

Figure 8. Sum of CO_2 and CO concentrations as function of power during decomposition of $CaCO_3$ samples with different particles size (samples C, D, and E); all the experiments were performed at 630 °C, in pure argon, and at a flow rate 30 mL/min.

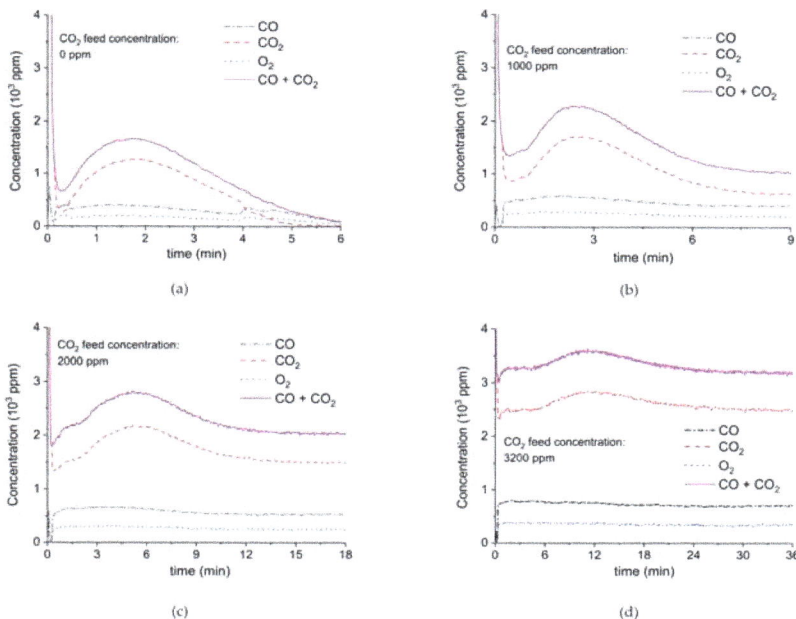

Figure 9. CO + CO$_2$ (solid line), CO$_2$ (dashed line), CO (dash-dotted line), and O$_2$ concentrations monitored by mass spectrometry during CaCO$_3$ decomposition in different CO$_2$ feed concentrations in Ar and different plasma power. The amount of CaO is 6 mg, temperature is 630 °C, and the flow rate is 90 mL min^{-1}. The fed CO$_2$ concentration is 0 ppm (**a**), 1000 ppm (**b**), 2000 ppm (**c**), and 3200 ppm (**d**). The plasma power is 1.5 W (**a**), 1.3 W (**b**), 1.3 W (**c**), and 1.5 W (**d**).

Figure 9 shows the effect of CO$_2$ in the feed on plasma-enhanced decomposition at similar plasma powers, i.e., between 1.3 and 1.5 W. The reaction rate substantially decreases in presence of extra CO$_2$, as can be estimated based on the sum of the concentrations of CO and CO$_2$ minus the CO$_2$ concentration at the inlet. Furthermore, the CO concentration increases with increasing CO$_2$ concentration.

In order to correct for the differences in power, additional experiments at other plasma powers were performed as shown in Figures A6–A9 in Appendix A and the results allowed interpolation of the CO and CO$_2$ values to 1.5 W plasma power, resulting in Figure 10. As the CO$_2$ concentration varies along the axis of the reactor, an averaged value is calculated according to Equation (1):

$$[CO_2]_{ave} = \frac{[CO_2]_{in} + [CO_2]_{out}}{2} \tag{1}$$

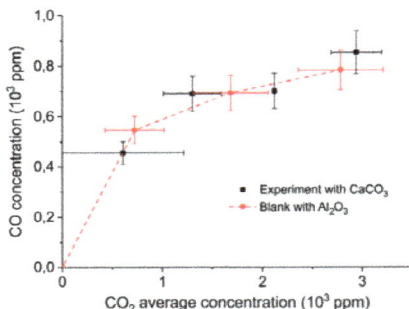

Figure 10. CO outlet concentration vs. CO_2 average concentration obtained in a packed bed of 10 mg $CaCO_3$ and 90 mg Al_2O_3 at 630 °C with different CO_2 concentrations in Ar and a flow rate of 90 mL/min (squares); all results interpolated to 1.5 W plasma power. The same experiment was also performed with 100 mg Al_2O_3 only (circles). Error margins of the CO concentration are provided, as well as the window of the CO_2 concentrations (note these are not error margins).

The spread in the CO_2 concentration provides the window of concentrations occurring along the axis of the reactor; note this is not an error margin. Figure 11 confirms that the CO concentration indeed increases with increasing CO_2 concentration in the gas phase and this is observed for both $CaCO_3$ as well as the blank experiment with α-Al_2O_3 only, which is shown in Figure A10 in Appendix A.

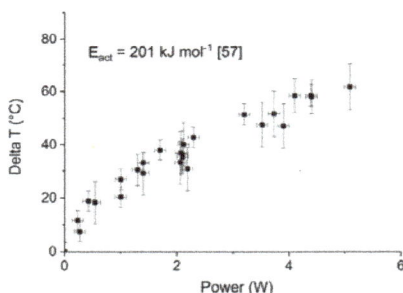

Figure 11. Average increase of temperature in the plasma zone as function of power during the $CaCO_3$ decomposition previously shown.

3. Discussion

Scheme 1 presents a simple reaction scheme describing $CaCO_3$ decomposition, both in the absence and presence of plasma. Thermal decomposition without plasma involves exclusively R_1 (solid arrow), whereas in the presence of plasma both R_2 and R_3 might contribute. The observed enhancement of decomposition by plasma might be caused by plasma chemistry according R_2 and/or by enhancing R_1 via increasing temperature, as will be discussed below. Furthermore, we will discuss whether R_2 or R_3 is responsible for the formation of CO.

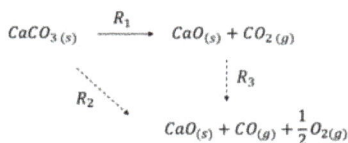

$$CaCO_{3\,(s)} \xrightarrow{\ R_1\ } CaO_{(s)} + CO_{2\,(g)}$$

$$\Big\downarrow R_3$$

$$\overset{R_2}{\searrow}$$

$$CaO_{(s)} + CO_{(g)} + \tfrac{1}{2}O_{2(g)}$$

Scheme 1. $CaCO_3$ decomposition in Ar plasma.

3.1. Formation of CO

Figure 10 shows that the CO formation rate is not affected by the presence of $CaCO_3$ in the plasma at low CO_2 feed concentrations. It is enhanced by feeding additional CO_2. Hence, we conclude that CO formation occurs only in the gas phase (R_3) and not on the $CaCO_3$ surface (R_2). The trend in Figure 11 also indicates that the order of CO formation in the CO_2 concentration is clearly smaller than one. This is qualitatively in line with the results of Ramakers et al. [25] and Butterworth et al. [27], reporting that Ar can enhance the activation of CO_2, which is attributed to the fact that the electron density is enhanced because Ar is much easier ionized than CO_2. On the other hand, one may also speculate that higher CO_2 concentration increases the probability of recombination of CO and O, quenching the reaction.

The formation of CO responds extremely fast to switching the plasma on, as can be seen in Figures 3–5 and Figure 7, much faster than the CO_2 response as will be discussed below. This is in line with the conclusion that CO formation is plasma controlled. In cases when CO_2 concentration increase slowly in time (Figures 4 and 5), it can be seen that the CO concentration follows, which is in line with the conclusion that the consecutive pathway R_3 is dominant over R_2.

3.2. Thermal Effect or Plasma Chemistry?

The decomposition rate at a fixed power depends on the total surface area, as can be observed from Figure 6. On the other hand, the decomposition rate remains constant within experimental error when varying the particles size and consequently the external surface area, as shown in Figure 8. Therefore, the external surface area has no influence on the rate of decomposition at any power. According to the Paschen law, typically plasma cannot exist in pores smaller than 6 μm, implying that any plasma chemistry on the surface of $CaCO_3$ can exclusively contribute at the external surface of the particles. Recent work [56] revealed that penetration into relatively large pores is possible; however, the pores in $CaCO_3$ are much smaller, the electrical field is two orders of magnitude lower compared to [56], and the calculated penetration depth is limited to 5 μm, which is much smaller than the particle size used. Therefore, it seems reasonable to assume that penetration of plasma in this study is negligible. As the external surface area has no significant effect, we conclude that R_2 does not contribute. The enhancing effect of plasma power on the decomposition rate, as well as the observation that increasing the specific surface area increases the rate of decomposition (both in Figure 6), both suggest that plasma induced temperature increase is responsible for the increase in the decomposition rate. In fact, this is a result that can be expected operating with an Ar plasma because a chemical reaction between Ar ions and $CaCO_3$ would not be expected. Work on DBD plasmas containing H_2 and H_2O is ongoing in which plasma induced chemical reactions are much more likely, in which we will use the methodology developed in this work.

Figure 11 shows the apparent temperature increase, as estimated based on the temperature that would be required to account for the increase in decomposition rate, based on the kinetics of the decomposition reaction [57]. Remarkably, all observations converge to a single line independent of both surface area and particle size. The order of magnitude of the temperature increase is quite similar to results reported in literature. Typical temperatures estimated in DBD plasmas range up to typically 200 K [17,23,41–48]. It should be noted that determination of the temperature is cumbersome, e.g., the temperature of the exiting gas provides only a minimum value because of rapid heat exchange between small reactors and environment, whereas infrared cameras and UV–Vis spectroscopy measurements have limited accuracy. Nevertheless, the order of magnitude agrees well with our observations. In short, although experimental details vary, a temperature increase of 50 °C due to 4 W plasma power input is concluded.

The temperature regulation of the oven played an important role in our study. It stabilized the temperature at a few millimeters outside the low voltage electrode at 630 °C. Therefore, any power input from the plasma will result in a decrease of the electrical power to the oven. Hence, the temperature effect of the plasma is actually larger than estimated above, in contrast to all experiments performed at room temperature without any kind of temperature control.

The conclusion that thermal effects are dominant is further supported by the fact that the typical response time of the decomposition rate is in the order of one minute for large particles and somewhat faster for small particles (Figures 4 and 8d). The order of magnitude agrees well with the Fourier time of $CaCO_3$ particles of the sizes used. In any case, the response times are longer than response times observed for CO formation, as discussed above, which is in line with the conclusion that decomposition is thermally controlled, whereas CO_2 dissociation is obviously plasma controlled.

The conclusion is also reinforced by the fact that the sum of CO_2 and CO concentrations obtained by decomposition of sample C at 630 °C and 3.2 W plasma power corresponds to the CO_2 concentration obtained by decomposition at 680 °C (i.e., increasing the temperature of 50 °C as calculated in Figure 11) without plasma of the same sample, as observed by comparing Figures A3a and A11 in Appendix A.

The proposed method is validated, since it enables to distinguish between thermal and plasma chemistry effect on the decomposition rate. This method is applicable to other systems in which a plasma is in contact with a fixed bed of porous particles, including supported catalysts. It should be noted that in general the increase in gas temperature in a DBD plasma cannot be neglected, as done frequently in many studies. Second, only the external surface of catalysts particles interacts with plasma: plasma–catalyst synergy is therefore maximal for nonporous catalytic materials. Further work is currently ongoing to apply this method for $CaCO_3$ decomposition in more reactive plasma, e.g., H_2 plasma, as well as for plasma catalytic conversions.

4. Materials and Methods

4.1. Plasma Reactor

Figure 12 shows a schematic representation of the equipment used to measure plasma enhanced decomposition of $CaCO_3$. The fixed bed reactor is fed with either pure Ar, or a mixture of Ar containing 5% CO_2. The temperature of the oven is controlled with a Euro-Therm controller with an accuracy of ± 0.5 °C between room temperature and 1000 °C. The isothermal zone is 8 cm long at 900 °C, defined as the part of the reactor with less than ± 1 °C temperature variation. A Quadrupole Mass Spectrometer (MS) measures the composition of the gas downstream of the reactor. The MS signal for CO_2 (44 m/e) is calibrated between 0.16% and 5% CO_2, resulting in a linear relationship as shown in Figure A14. The reactor is a 4 mm inner- and 6 mm outer diameter quartz tube. The inner electrode is a stainless-steel rod of 1 mm diameter placed coaxially in the center of the reactor section. The outer electrode is a 1 cm long stainless-steel tube with 6 mm inner diameter, enclosing a plasma zone of 0.035 cm^3 in volume. The amount of $CaCO_3$ sample was limited to 10.5 ± 0.3 mg in order to prevent CO_2 concentrations approaching thermodynamic equilibrium, thus minimizing reabsorption of CO_2. The 10 mg sample was mixed with 90 mg α-Al_2O_3, filling the plasma zone completely and preventing any bypassing. An AC voltage of up to 10 kV peak to peak was applied to the inner electrode with a frequency of 23.5 kHz using a PMV 500–4000 power supply, while the outer electrode is connected to the ground via a probe capacitor of capacity 4 nF. The power of the plasma was calculated using the Lissajous method by measuring the voltage on the inner electrode with a Tektronix P6015A high voltage probe and on the outer electrode with a TT–HV 250 voltage probe, as described in the literature [58]. A sample of Lissajous plot is shown in Figure A12 in Appendix B.

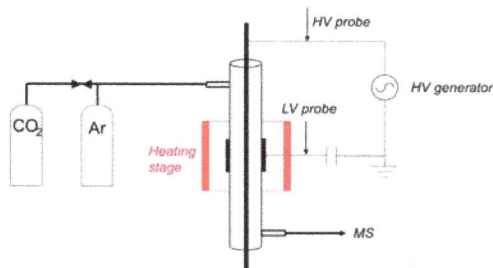

Figure 12. Schematic of the setup to study decomposition of $CaCO_3$ in Ar plasma. The generator can provide up to 30 kV peak to peak with a frequency range of 23.5 to 66 kHz. The plasma zone is 1 cm long and the reactor is a quartz tube with 6 mm outer diameter and 4 mm inner diameter. The inner electrode is a stainless-steel rod of 1 mm diameter.

4.2. Calcium Oxides Preparation

Three different precursors have been used to synthesize calcium oxide, respectively (batch I) calcium L-ascorbate-di-hydrate (99%, Sigma-Aldrich, St. Louis, MO, USA), (batch II) calcium D-gluconate-monohydrate (99%, Alfa Aesar, Haverhill, MA, USA), and (batch III) calcium carbonate (99%, Sigma-Aldrich, St. Louis, MO, USA). The precursors were calcined in 20% O_2 in N_2 at atmospheric pressure, heating the sample to 900 °C (heating rate 15 °C/min), and keeping the temperature at 900 °C for 3 h. The calcined products were pelletized (pressure 160 bar), crushed and sieved in different particle size range: 250–300 μm, 100–125 μm, and 38–45 μm.

4.3. Carbonation

Five $CaCO_3$ samples have been produced via carbonation of CaO. The oxide synthesized from calcium ascorbate (batch II) has been treated in situ with 5% CO_2 in Ar at 630 °C for 4 h (heating rate 15 °C/min), the other two batches I and III were treated in a calcination oven with 20% CO_2 in N_2 at 630 °C for 5 h (heating rate 15 °C/min). The oxide synthesized from calcium carbonate (III) was consecutively sintered in pure CO_2 at 900 °C for 24 h (heating rate 15 °C/min), as summarized in Table 1. The resulting three samples (A, B, and C) were crushed and sieved, obtaining particles sizes in the range between 250 and 300 μm. The material made from $CaCO_3$ was also obtained with smaller particles, between 100 and 125 μm (sample D) and between 38 and 45 μm (sample E), respectively.

4.4. Characterization

The specific surface area, pore volume, and pore size distribution of the samples were measured both in CaO form as well as in $CaCO_3$ form, after carbonation. The samples were first degassed at 300 °C in vacuum for 3 h. The surface area was calculated based on the BET isotherm for N_2 adsorption at −196 °C in a Tristar 3000 analyzer (Micromeritics, Norcross, GA, USA). The pore size distribution was measured by Hg porosimetry. The chemical composition was determined with X-ray fluorescence analysis in a S8 Tiger (Bruker, Billerica, MA, USA).

4.5. Experimental Procedure

The carbonated samples (ex situ) were heated up in 5% CO_2 in Ar to the temperature at which decomposition is to be measured, in order to prevent any premature decomposition. The decomposition reaction is initiated by switching the gas composition from 5% CO_2 to pure Ar, at a constant flow rate of 30 mL/min. Isothermal decomposition experiments have been performed at different plasma powers by varying the applied voltage. The plasma power was varied during the experiment in case of low decomposition rates, allowing observations of steady state CO_2 concentrations for each plasma power. In case $CaCO_3$ is exhausted too fast, only one single power was applied. The rate of decomposition is calculated based on the sum of CO_2 and CO concentrations in the exit of the reactor as measured with MS.

Sample C (Table 1, ex calcium carbonate, 250–300 μm) was measured by performing 20 carbonation-decomposition cycles. The sample was recarbonated, after a decomposition experiment, in the reactor (in situ) by CO_2 absorption at 630 °C in 5% CO_2 in Ar for 30 min in a constant flow of 90 mL/min. The carbonated samples were decomposed using a constant power plasma in the presence of a relatively low CO_2 concentration in the feed, varied between 0 to 3200 ppm. During these 20 cycles, blank experiments were done every few cycles by decomposing in the absence of plasma, in order to ensure that the sample did not change in the course of the experiments.

Blank experiments were performed with 100mg of α-Al_2O_3 with particles size of 250 to 300 μm in the absence of any $CaCO_3$, operating with low CO_2 concentrations in the feed, i.e., 1000, 2000, and 3200 ppm. The plasma power was varied between 0 and 10 W and the responses of the CO and CO_2 concentrations were measured with MS.

5. Conclusions

The effect of argon plasma on calcium carbonate decomposition was herein assessed by means of a comparative method which allowed us to distinguish between thermal effects and plasma chemistry, based on reaction rates and dynamics. It represents a systematic method to distinguish between thermal effects versus plasma chemistry effect in fixed DBD plasma applications. Application of a DBD Ar plasma causes two effects when decomposing $CaCO_3$.

First, the rate of $CaCO_3$ decomposition increases. We conclude that this effect is purely a thermal effect, based on the fact that the rate of decomposition is enhanced when the total surface area is increased, whereas the external surface area has no influence. If the contact of plasma with $CaCO_3$ would dominate, the opposite would be expected. Furthermore, the dynamics of $CaCO_3$ decomposition follow the dynamics of heat transfer in $CaCO_3$ particles.

Second, plasma induces formation of CO. We conclude that this occurs via decomposition of CO_2 in the gas phase, based on the observation that the rate of CO formation is ruled by the CO_2 concentration as well as the observation that dynamic changes are very fast, as expected for plasma effect.

Author Contributions: Conceptualization, G.G., L.L., and G. van R.; Methodology, G.G. and L.L.; Validation, G.G., L.L., and G. van R.; Formal Analysis, G.G. and L.L.; Investigation, G.G. and L.L.; Resources, L.L.; Data Curation, G.G. and L.L.; Writing—Original Draft Preparation, G.G.; Writing—Review and Editing, G.G. and L.L.; Supervision, L.L. Project Administration, G.G. and L.L.; Funding Acquisition, L.L.

Funding: This work was supported by Netherlands Organization for Scientific Research (NWO).

Acknowledgments: We acknowledge Bert Geerdink, Karin Altena-Schildkamp, and Tom Velthuizen for Technical assistance; Tom Butterworth and Floran Peeters from DIFFER for the fruitful discussions and help in understanding; Tesfaye Belete and Micheal Gleeson from DIFFER also for their fruitful discussions and critical feedback; and Vera Meyner, the technical staff of the department of Chemistry of University of Antwerp, and Frank Morssinkhof for the continuous help in material characterization.

Conflicts of Interest: The authors declare no conflict0s of interest.

Appendix A.

Appendix A.1. Decomposition in Argon Plasma of All the Samples

Figures A1–A5 show the results of all the $CaCO_3$ decomposition experiments on samples A–E, respectively, in presence of plasma and without recycling. The results with power higher than 6 W and further than the second step of power have been rejected due to exhaustion.

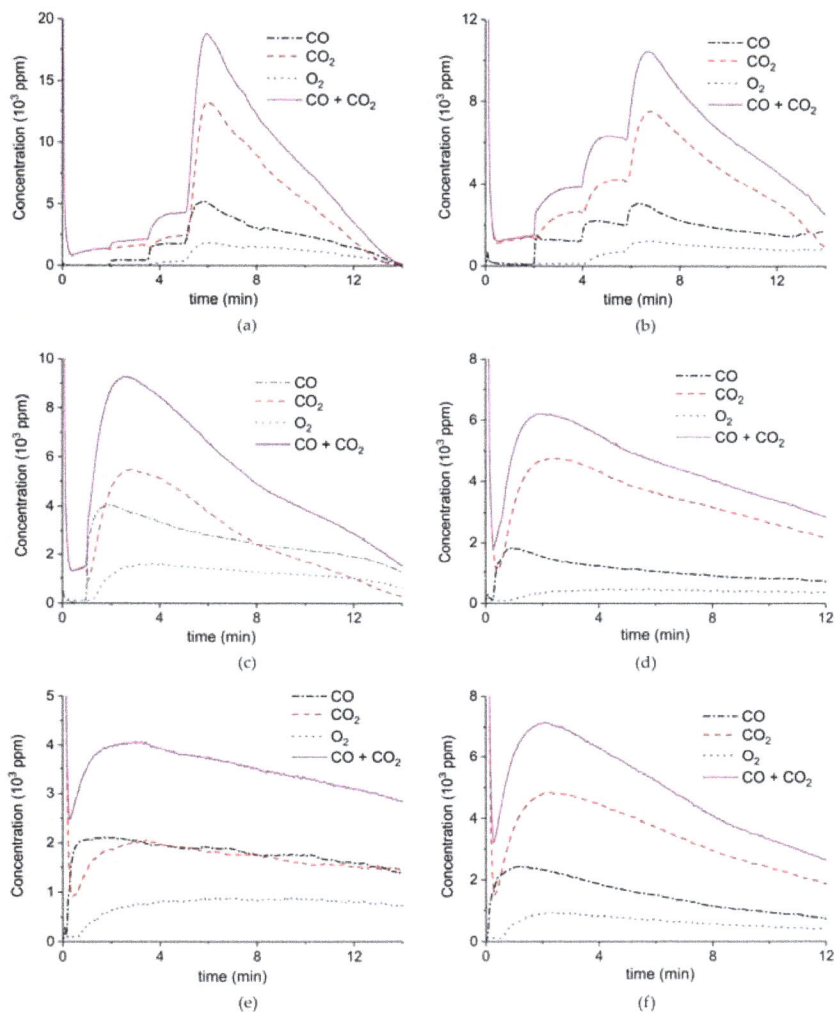

Figure A1. Concentration of the products of $CaCO_3$ decomposition (sample A) as function of time at 630 °C in pure argon and flow rate of 30 mL/min. Plasma power is 0.3, 1.4, and 8.4 W (**a**); 1.4, 3.5, and 6.4 W (**b**); 5.1 W (**c**); 3.9 W (**d**); 2.2 W (**e**); and 3.7 W (**f**). Plasma is turned on after 2 min and changed every 2 min (a,b), after 1.5 min (c), or at the beginning of the decomposition (d–f).

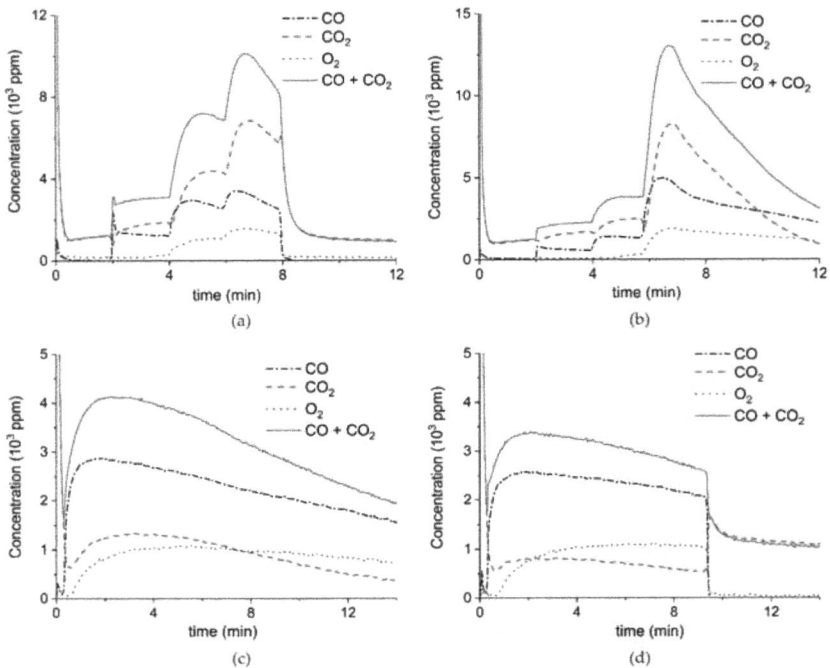

Figure A2. Concentration of the products of CaCO$_3$ decomposition (sample B) as function of time at 630 °C in pure argon and flow rate of 30 mL/min. Plasma power is 1, 4.1, and 5.8 W (**a**); 0.5, 2.1, and 8.5 W (**b**); and 2.1 W (**c,d**). Plasma is turned on after 2 min and changed every 2 min (a,b) or at the beginning of the decomposition (c,d).

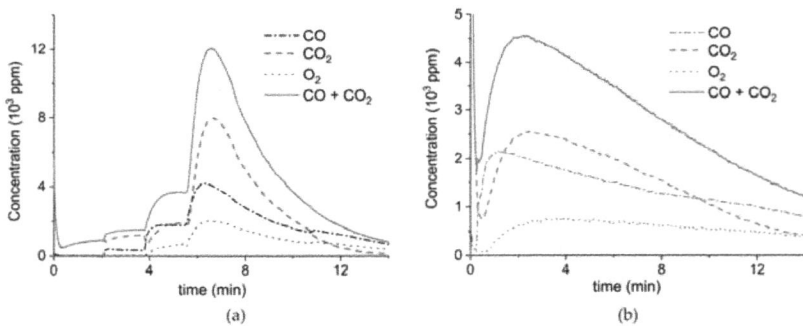

Figure A3. *Cont.*

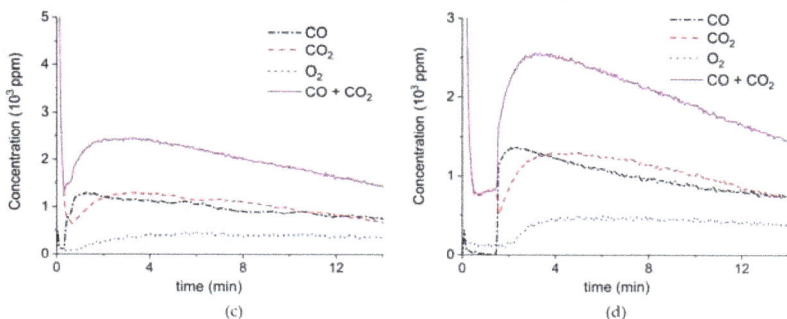

Figure A3. Concentration of the products of CaCO$_3$ decomposition (sample C) as function of time at 630 °C in pure argon and flow rate of 30 mL/min. Plasma power is 0.4, 3.2, and 9.6 W (**a**); 4.4 W (**b**); and 2.1 W (**c,d**). Plasma is turned on after 2 min and changed every 2 min (**a**) or at the beginning of the decomposition (**b,c**) or after 1.5 min (**d**).

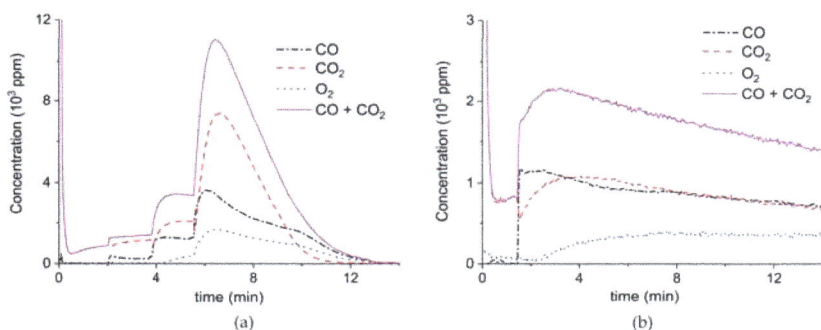

Figure A4. Concentration of the products of CaCO$_3$ decomposition (sample D) as function of time at 630 °C in pure argon and flow rate of 30 mL/min. Plasma power is 0.2, 2.3, and 8.1 W (**a**) and 1.3 W (**b**). Plasma is turned on after 2 min and changed every 2 min (**a**) or after 1.5 min (**b**).

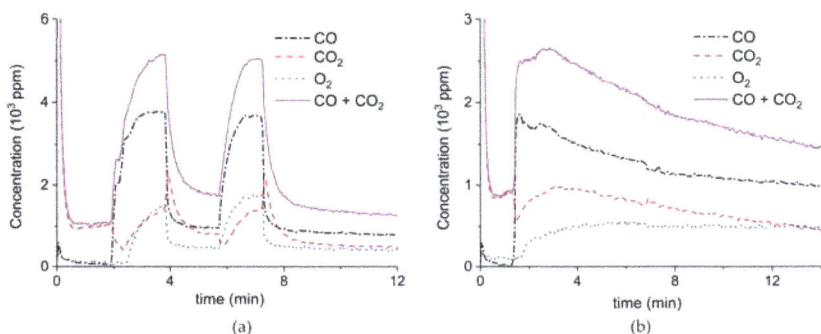

Figure A5. Concentration of the products of CaCO$_3$ decomposition (sample E) as function of time at 630 °C in pure argon and flow rate of 30 mL/min. Plasma power is 4.4, 1, 4.9, and 1 W (**a**) and 1.7 W (**b**). Plasma is turned on after 2 min and changed every 2 min (**a**) or after 1.5 min (**b**).

Figures A6–A9 show the results of all the CaCO$_3$ decomposition experiments on sample C in presence of plasma when the sample is recycled 20 times, including experiments with extra CO$_2$ added.

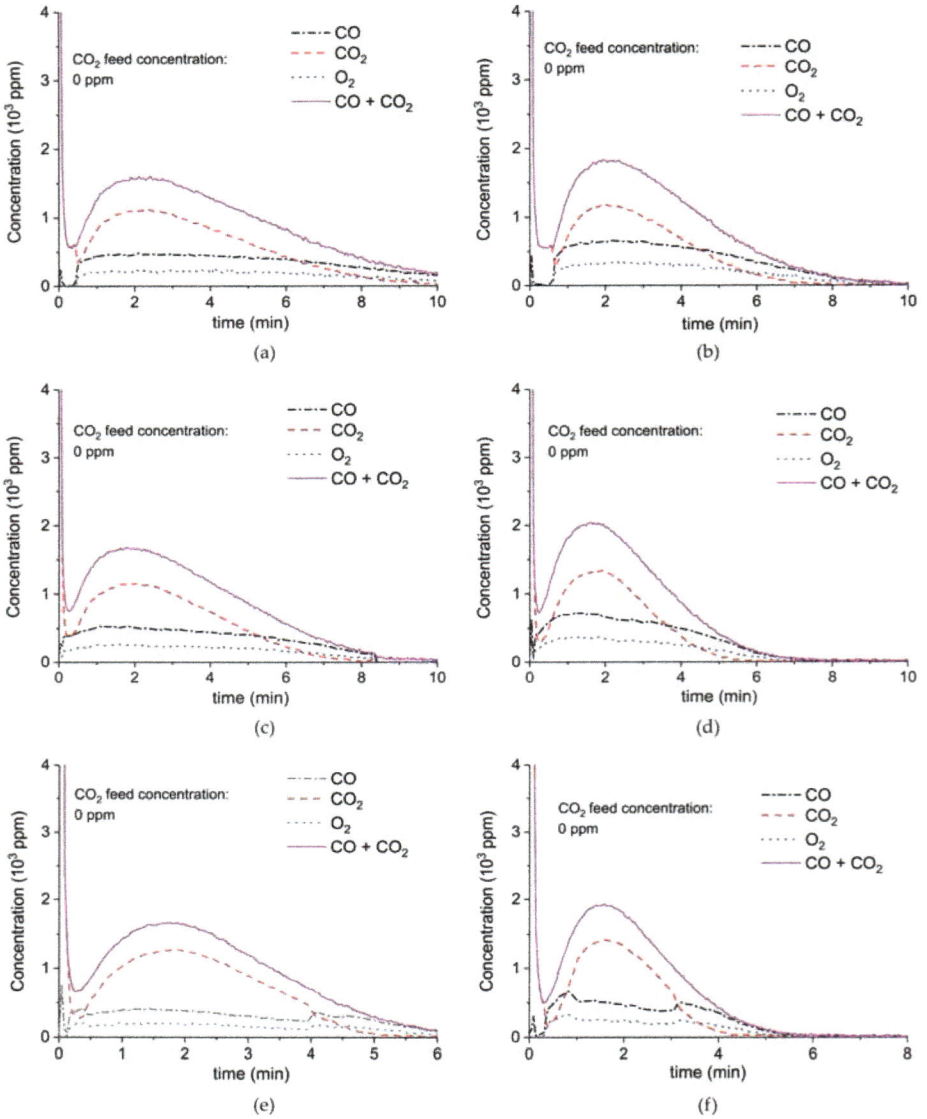

Figure A6. *Cont.*

(g)

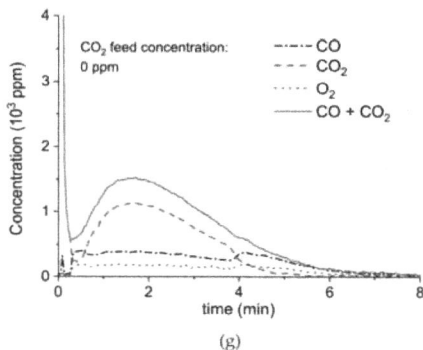

Figure A6. Concentration of the products of CaCO₃ decomposition (sample C) as function of time at 630 °C, flow rate of 90 mL/min. The fed CO₂ concentration is 0 ppm and the plasma power is 1.5 W (**a**), 2.1 W (**b**), 1.7 W (**c**), 2.2 W (**d**), 1.5 W (**e**), 2 W (**f**), and 1.1 W (**g**). The plasma is applied at the beginning of the reaction. The decomposition cycles are in order #2, #3, #7, #8, #10, #11, #19.

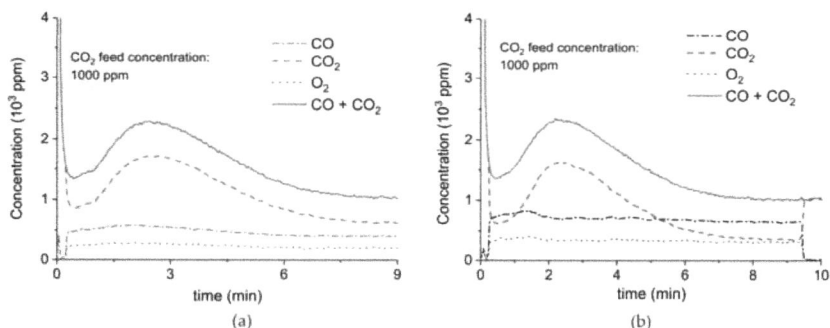

(a)

(b)

Figure A7. Concentration of the products of CaCO₃ decomposition (sample C) as function of time at 630 °C, flow rate of 90 mL/min. The fed CO₂ concentration is 1000 ppm and the plasma power is 1.3 W (**a**) and 1.5 W (**b**). The plasma is applied at the beginning of the reaction. The decomposition cycles are in order #20, #21.

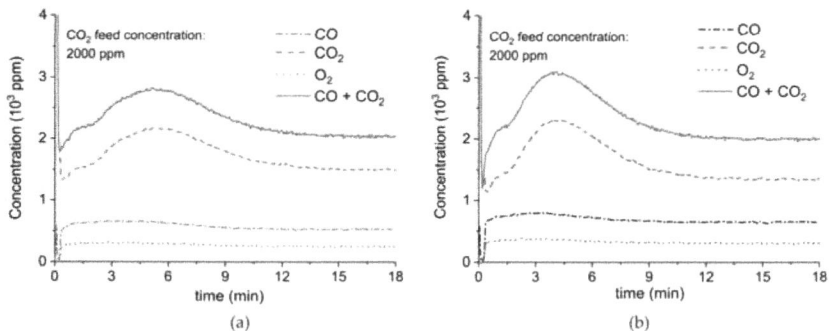

(a)

(b)

Figure A8. Concentration of the products of CaCO₃ decomposition (sample C) as function of time at 630 °C, flow rate of 90 mL/min. The fed CO₂ concentration is 2000 ppm and the plasma power is 1.4 W (**a**) and 1.9 W (**b**). The plasma is applied at the beginning of the reaction. The decomposition cycles are in order #13, #14.

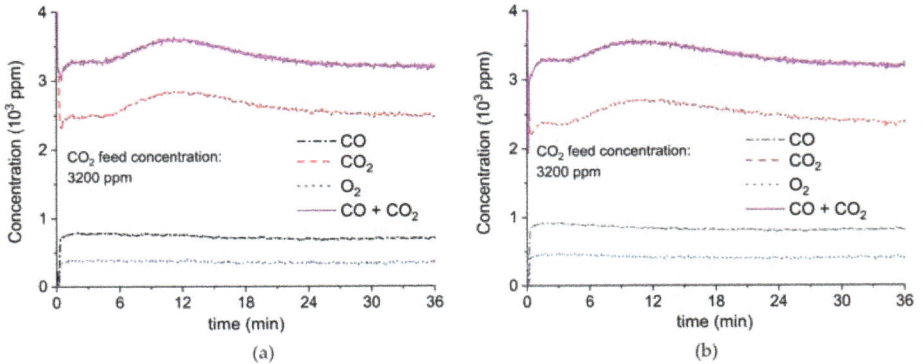

Figure A9. Concentration of the products of CaCO₃ decomposition (sample C) as function of time at 630 °C, flow rate of 90 mL/min. The fed CO_2 concentration is 3200 ppm and the plasma power is 1.4 W (**a**) and 1.5 W (**b**). The plasma is applied at the beginning of the reaction. The decomposition cycles are in order #16, #17.

Appendix A.2. CO_2 Dissociation with Only α-Al₂O₃

Figure A10 shows the result of CO_2 dissociation experiments on a sample consisting of 100 mg α-Al₂O₃ with particle size of 250–300 μm. The applied gas composition is respectively 1000 ppm (Figure A10a), 2000 ppm (Figure A10b), and 3200 ppm (Figure A10c) in Argon. The CO concentrations at 1.5 W plasma power are obtained by interpolation and showed in Figure 11.

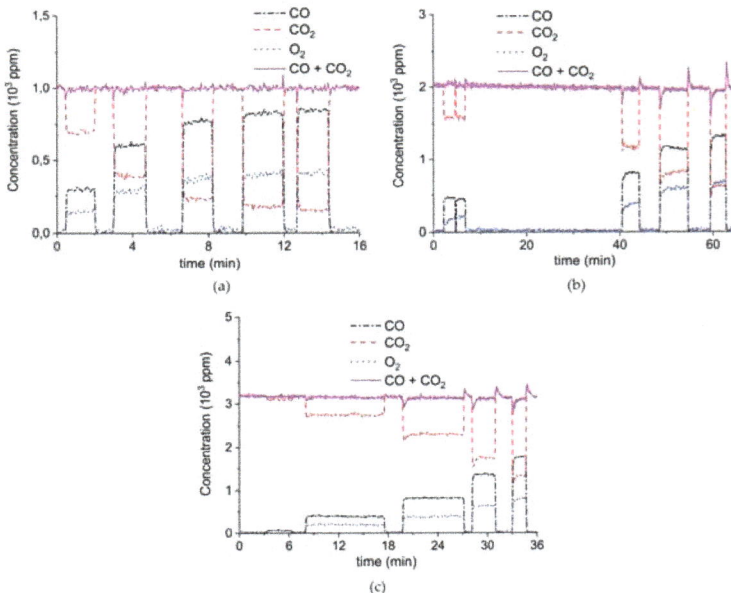

Figure A10. Concentration of the products of CO_2 dissociation on 100mg α-Al₂O₃, particle size of 250–300 μm, as function of time at 630 °C, flow rate of 90 mL/min. The fed CO_2 concentration is 1000ppm (**a**), 2000ppm (**b**), and 3200ppm (**c**); the plasma powers are 0.6, 1.6, 3.1, 4.8, and 7.2 W (a); 0.7, 1.8, 4.1, and 5.6 W (b); and 0.5, 1.5, 3.1, and 4.9 W (c). The plasma is applied for a few minutes and then turned off.

Figure A11 shows the result of a $CaCO_3$ decomposition experiment at 680 °C without plasma on sample C. It is observed that the CO_2 concentration at the steady state corresponds within the error margins to the sum of CO and CO_2 concentrations during decomposition at 630 °C with 3.2 W plasma power, shown in Figure A3a. The comparison supports the estimation of a 50 °C increase for 3.2 W, as shown in Figure 11.

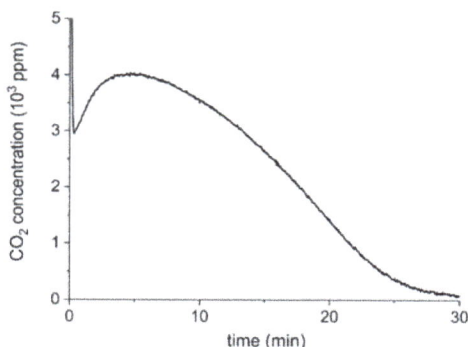

Figure A11. Concentration of CO_2 during CaCO3 decomposition (sample C) as function of time at 680 °C and a flow rate of 30 mL/min.

Appendix B.

Appendix B.1. Determination of Plasma Power

The Lissajous plot measured during decomposition of sample C in Ar with 3200 ppm extra CO_2 and plasma power 1.5 W, indicating plasma charge as function of voltage for one cycle, is shown in Figure A12. The hysteresis has a shape of parallelogram, indicating that a plasma is generated inside the reactor. The slopes of the two sides indicate respectively a cell capacity of 2.8 pF with plasma off and an effective capacity of 8 pF corresponding to the dielectric capacity and indicating that the plasma occupies the whole empty volume within the electrodes. The power of the plasma is calculated based on the area of the parallelogram:

$$P = \frac{1}{T} \cdot \int_0^T VdQ \qquad (2)$$

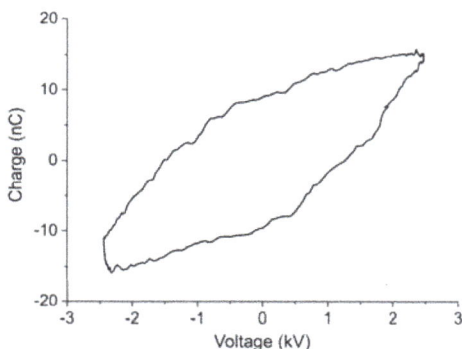

Figure A12. Lissajous plot of 1.5 W plasma applied on the $CaCO_3$ and Al_2O_3 mixture with 3200 ppm of CO_2 in Ar.

Appendix B.2. Method

The assumptions made in order to evaluate the argon plasma effect on calcium carbonate decomposition are the following:

- According to the Paschen's Law the electron filaments and active species cannot exist inside the pores of the carbonate particles, this is explained in the following paragraph.
- Temperature gradients between particles outer surface and inner core are minimized within much smaller times than the duration of the experiment, i.e., within 2 min, according to the Fourier time calculated in the specific case.

Figure A13 shows the voltage needed to generate a plasma as function of the distance between the charges at an argon pressure of 1 atm, where the parameters *A*, *B*, and γ are determined experimentally [59]. The physical mechanism can be explained for the 2 branches of the curve:

- $d_{gap} > 6\mu m$: In order to generate a discharge (that is produced by an avalanche of ionizations of Argon atoms by high energy electrons), we need a minimum electric field that is able to accelerate electrons above the ionization energy of Ar atoms in the mean free path of electrons at that pressure (ca. 0.5 μm). Since the electric field is inversely proportional to d_{gap} at constant voltage, V_b increases linearly with d_{gap}.
- $d_{gap} < 6\mu m$: the gap distance is comparable with the mean free path, so the probability of collisions that produce ionizations decreases. In order to have enough collisions to produce a discharge, the voltage has to be increased.

It follows that plasma cannot form in the pores of $CaCO_3$ particles within the plasma zone, since their diameter is below 2 μm.

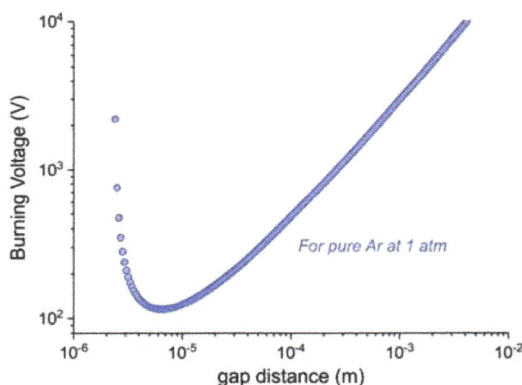

Figure A13. Burning voltage in function of the gap distance for argon at 1 atm, as calculated by [59].

In order to discriminate whether the gas temperature increment or the active species drives the $CaCO_3$ decomposition in presence of plasma, we compared the effect of plasma on the reaction rate on samples with different specific surface area and perform a similar comparison for samples with different particles size. By choosing different particles sizes or different specific surface areas, we can vary selectively the external surface, exposed to the active species and temperature increase, and the internal surface, exposed only to a temperature increase.

Appendix B.3. CO$_2$ and CO and Calibration

Figure A14 shows the CO_2 signal elaborated by MS (m/e = 44) for CO_2 concentrations up to 5% in Ar and in the inset CO_2 concentrations up to 1000 ppm are expanded. In this range the signal is linear with the CO_2 concentration, with a R^2 coefficient of 0.9996. The CO signal (m/e = 28) for the CO concentration of 5% in Ar is 1.15 ± 0.05 times larger than the CO_2 signal (m/e = 44) at the same CO_2 concentration.

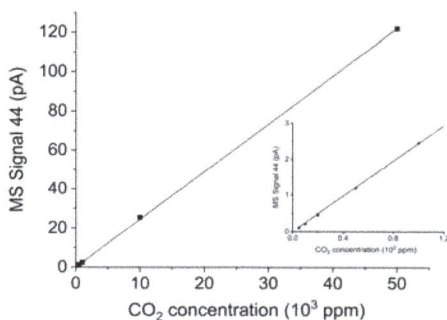

Figure A14. Mass spectrometer signal m/e = 44 for different CO_2 concentrations in Ar from 0 to 5%; inset: enhancement of the CO_2 concentration range of 0 to 1000 ppm.

Appendix B.4. Sample Stability during Absorption–Desorption Cycles

Figure A15 shows the aging of the sample synthesized from calcium carbonate with the largest particles size, within 22 absorption–desorption cycles: the concentration of CO_2 during the initial stage of decomposition does not show significant change during the 21 absorption–desorption cycles at 630°C in absence of extra CO_2 and plasma, as expected for a sample already sintered at 900 °C for 24 h. The experiments in presence of extra CO_2 and plasma were all performed from cycle 9 to cycle 22, where the sample shows the highest stability.

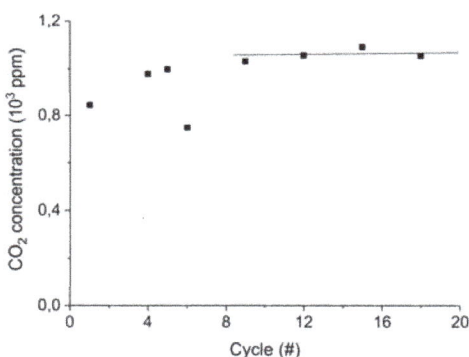

Figure A15. CO_2 concentration at the first stage of decomposition of 10 mg $CaCO_3$ for different cumulative number of cycles. The decompositions are performed at standard conditions, i.e., 630 °C without plasma in absence of CO_2 in the gas mixture.

Appendix B.5. Heat Transfer in the Packed Bed

Plasma is considered as a uniform heat source at temperature T_∞ around the particles which are assumed spherical and have an initial temperature T_0. We calculated the time needed to increase the temperature of the center of the particles of the 90% of the initial temperature difference $T_\infty-T_0$ as follows [60]

$$t = \frac{\rho C_p V}{hA} \ln\left(\frac{T_0 - T_\infty}{T - T_\infty}\right) \tag{3}$$

The physical constants in the equation are the following: ρ is the density of $CaCO_3$ (2710 kg/m^3); C_p is the heat capacity of $CaCO_3$ at 630 °C (1.26 J g^{-1} K^{-1}) [61]; V and A are the volume and section of the $CaCO_3$ particles, respectively, and h is the convective heat transfer coefficient of a dry gas in a packed bed reactor, approximated to 15 W m^{-2} K^{-1} [60]. As a result, the times needed to heat up the particles are respectively 4, 11, and 24 seconds for the smallest (38–45 μm), medium (100–125 μm), and largest (250–300 μm) particles, respectively. Equation (1) is an approximation for spherical particles without porosity, so we assume that more accurate calculation will give slightly longer times.

References

1. Neyts, E.C.; Ostrikov, K.; Sunkara, M.K.; Bogaerts, A. Plasma Catalysis: Synergistic Effects at the Nanoscale. *Chem. Rev.* **2015**, *115*, 13408–13446. [CrossRef] [PubMed]
2. Neyts, E.C.; Bogaerts, A. Understanding plasma catalysis through modelling and simulation—A review. *J. Phys. D Appl. Phys.* **2014**, *47*, 224010. [CrossRef]
3. Neyts, E.C.; Ostrikov, K.; Sunkara, M.K.; Bogaerts, A. Plasma-Surface Interactions in Plasma Catalysis. *Plasma Chem. Plasma Process.* **2014**, *36*, 13408–13446. [CrossRef]
4. Whitehead, J.C. Plasma-catalysis: The known knowns, the known unknowns and the unknown unknowns. *J. Phys. D Appl. Phys.* **2016**, *49*, 243001. [CrossRef]
5. Chen, H.L.; Lee, H.M.; Chen, S.H.; Chang, M.B.; Yu, S.J.; Li, S.N. Removal of volatile organic compounds by single-stage and two-stage plasma catalysis systems: A review of the performance enhancement mechanisms, current status, and suitable applications. *Environ. Sci. Technol.* **2009**, *43*, 2216–2227. [CrossRef] [PubMed]
6. Dobslaw, D.; Schulz, A.; Helbich, S.; Dobslaw, C.; Engesser, K.H. VOC removal and odor abatement by a low-cost plasma enhsnced biotrickling filter process. *J. Environ. Chem. Eng.* **2017**, *5*, 5501–5511. [CrossRef]
7. Kozàk, T.; Bogaerts, A. Splitting of CO_2 by vibrational excitation in non-equilibrium plasmas: A reaction kinetics model. *Plasma Sources Sci. Technol.* **2014**, *23*, 045004. [CrossRef]
8. Ashford, B.; Tu, X. Non-thermal plasma technology for the conversion of CO_2. *Curr. Opin. Green Sustain. Chem.* **2017**, *3*, 45–49. [CrossRef]
9. Michielsen, I.; Uytdenhouwen, Y.; Pype, J.; Michielsen, B.; Mertens, J.; Reniers, F.; Meynen, V.; Bogaerts, A. CO_2 dissociation in a packed bed DBD reactor: First steps towards a better understanding of plasma catalysis. *Chem. Eng. J.* **2017**, *326*, 477–488. [CrossRef]
10. Grofulović, M.; Silva, T.; Klarenaar, B.L.; Morillo-Candas, A.S.; Guaitella, O.; Engeln, R.; Pintassilgo, C.D.; Guerra, V. Kinetic study of CO_2 plasmas under non-equilibrium conditions. {I}. Relaxation of vibrational energy. *Plasma Sources Sci. Technol.* **2018**, *27*, 15019. [CrossRef]
11. Sobacchi, M.G.; Saveliev, A.V.; Fridman, A.A.; Kennedy, L.A.; Ahmed, S.; Krause, T. Experimental assessment of a combined plasma/catalytic system for hydrogen production via partial oxidation of hydrocarbon fuels. *Int. J. Hydrogen Energy* **2002**, *27*, 635–642. [CrossRef]
12. Tu, X.; Whitehead, J.C. Plasma-catalytic dry reforming of methane in an atmospheric dielectric barrier discharge: Understanding the synergistic effect at low temperature. *Appl. Catal. B Environ.* **2012**, *125*, 439–448. [CrossRef]
13. Chung, W.C.; Chang, M.B. Review of catalysis and plasma performance on dry reforming of CH_4 and possible synergistic effects. Renew. *Sustain. Energy Rev.* **2016**, *62*, 13–31. [CrossRef]
14. Patil, B.S.; Wang, Q.; Hessel, V.; Lang, J. Plasma N_2-fixation: 1900–2014. *Catal. Today* **2015**, *256*, 49–66. [CrossRef]
15. Fridman, A. *Plasma Chemistry*; Cambridge University Press: Cambridge, UK, 2008.
16. Tendero, C.; Tixier, C.; Tristant, P.; Desmaison, J.; Leprince, P. Atmospheric pressure plasmas: A review. *Spectrochim. Acta Part B At. Spectrosc.* **2006**, *61*, 2–30. [CrossRef]
17. Kim, H.H.; Teramoto, Y.; Negishi, N.; Ogata, A. A multidisciplinary approach to understand the interactions of nonthermal plasma and catalyst: A review. *Catal. Today* **2015**, *256*, 13–22. [CrossRef]
18. Aerts, R.; Somers, W.; Bogaerts, A. Carbon Dioxide Splitting in a Dielectric Barrier Discharge Plasma: A Combined Experimental and Computational Study. *ChemSusChem* **2015**, *8*, 702–716. [CrossRef]
19. Mei, D.; Zhu, X.; He, Y.L.; Yan, J.D.; Tu, X. Plasma-assisted conversion of CO_2 in a dielectric barrier discharge reactor: Understanding the effect of packing materials. *Plasma Sources Sci. Technol.* **2015**, *24*, 015001.

20. Nizio, M.; Albarazi, A.; Cavadias, S.; Amouroux, J.; Galvez, M.E.; da Costa, P. Hybrid plasma-catalytic methanation of CO_2 at low temperature over ceria zirconia supported Ni catalysts. *Int. J. Hydrogen Energy* **2016**, *41*, 11584–11592. [CrossRef]

21. Bogaerts, A.; Kozák, T.; van Laer, K.; Snoeckx, R. Plasma-based conversion of CO_2: Current status and future challenges. *Faraday Discuss.* **2015**, *183*, 217–232. [CrossRef]

22. van Laer, K.; Bogaerts, A. Improving the Conversion and Energy Efficiency of Carbon Dioxide Splitting in a Zirconia-Packed Dielectric Barrier Discharge Reactor. *Energy Technol.* **2015**, *3*, 1038–1044. [CrossRef]

23. Brehmer, F.; Welzel, S.; van de Sanden, M.C.M.; Engeln, R. CO and byproduct formation during CO_2 reduction in dielectric barrier discharges. *J. Appl. Phys.* **2014**, *116*, 123303. [CrossRef]

24. Uytdenhouwen, Y.; van Alphen, S.; Michielsen, I.; Meynen, V.; Cool, P.; Bogaerts, A. A packed-bed DBD micro plasma reactor for CO_2 dissociation: Does size matter? *Chem. Eng. J.* **2018**, *348*, 557–568. [CrossRef]

25. Ramakers, M.; Michielsen, I.; Aerts, R.; Meynen, V.; Bogaerts, A. Effect of argon or helium on the CO_2 conversion in a dielectric barrier discharge. *Plasma Process. Polym.* **2015**, *12*, 755–763. [CrossRef]

26. Zeng, Y.; Tu, X. Plasma-catalytic hydrogenation of CO_2 for the cogeneration of CO and CH_4 in a dielectric barrier discharge reactor: Effect of argon addition. *J. Phys. D Appl. Phys.* **2017**, *50*, 184004. [CrossRef]

27. Butterworth, T.; Elder, R.; Allen, R. Effects of particle size on CO_2 reduction and discharge characteristics in a packed bed plasma reactor. *Chem. Eng. J.* **2016**, *293*, 55–67. [CrossRef]

28. Kameshima, S.; Tamura, K.; Ishibashi, Y.; Nozaki, T. Pulsed dry methane reforming in plasma-enhanced catalytic reaction. *Catal. Today* **2015**, *256*, 67–75. [CrossRef]

29. Kameshima, S.; Tamura, K.; Mizukami, R.; Yamazaki, T.; Nozaki, T. Parametric analysis of plasma-assisted pulsed dry methane reforming over Ni/Al_2O_3 catalyst. *Plasma Process. Polym.* **2017**, *14*, 1600096. [CrossRef]

30. Ozkan, A.; Dufour, T.; Arnoult, G.; de Keyzer, P.; Bogaerts, A.; Reniers, F. CO_2-CH_4 conversion and syngas formation at atmospheric pressure using a multi-electrode dielectric barrier discharge. *J. CO2 Util.* **2015**, *9*, 78–81. [CrossRef]

31. Tu, X.; Gallon, H.J.; Twigg, M.V.; Gorry, P.A.; Whitehead, J.C. Dry reforming of methane over a Ni/Al_2O_3 catalyst in a coaxial dielectric barrier discharge reactor. *J. Phys. D Appl. Phys.* **2011**, *44*, 274001. [CrossRef]

32. Gallon, H.J.; Tu, X.; Whitehead, J.C. Effects of reactor packing materials on H_2 production by CO_2 reforming of CH_4 in a dielectric barrier discharge. *Plasma Process. Polym.* **2012**, *9*, 90–97. [CrossRef]

33. Zeng, Y.; Zhu, X.; Mei, D.; Ashford, B.; Tu, X. Plasma-catalytic dry reforming of methane over γ-Al_2O_3 supported metal catalysts. *Catal. Today* **2015**, *256*, 80–87. [CrossRef]

34. Zheng, X.; Tan, S.; Dong, L.; Li, S.; Chen, H. Silica-coated $LaNiO_3$ nanoparticles for non-thermal plasma assisted dry reforming of methane: Experimental and kinetic studies. *Chem. Eng. J.* **2015**, *265*, 147–156. [CrossRef]

35. Aerts, R.; Martens, T.; Bogaerts, A. Influence of vibrational states on CO_2 splitting by dielectric barrier discharges. *J. Phys. Chem. C* **2012**, *116*, 23257–23273. [CrossRef]

36. Bogaerts, A.; Wang, W.; Berthelot, A.; Guerra, V. Modeling plasma-based CO_2 conversion: Crucial role of the dissociation cross section. *Plasma Sources Sci. Technol.* **2016**, *25*, 55016. [CrossRef]

37. Pietanza, L.D.; Colonna, G.; D'Ammando, G.; Capitelli, M. Time-dependent coupling of electron energy distribution function, vibrational kinetics of the asymmetric mode of CO_2 and dissociation, ionization and electronic excitation kinetics under discharge and post-discharge conditions. *Plasma Phys. Control. Fusion* **2017**, *59*, 14035. [CrossRef]

38. Van Rooij, G.J.; van den Bekerom, D.C.; Den Harder, N.; Minea, T.; Berden, G.; Bongers, W.A.; Engeln, R.; Graswinckel, M.F.; Zoethout, E.; van de Sanden, M.C. Taming microwave plasma to beat thermodynamics in CO_2 dissociation. *Faraday Discuss.* **2015**, *183*, 233–248. [CrossRef] [PubMed]

39. Bongers, W.; Bouwmeester, H.; Wolf, B.; Peeters, F.; Welzel, S.; van den Bekerom, D.; den Harder, N.; Goede, A.; Graswinckel, M.; Groen, P.W.; et al. Plasma-driven dissociation of CO_2 for fuel synthesis. *Plasma Process. Polym.* **2017**, *14*, 1600126. [CrossRef]

40. Minea, T.; van den Bekerom, D.C.; Peeters, F.J.; Zoethout, E.; Graswinckel, M.F.; van de Sanden, M.C.; Cents, T.; Lefferts, L.; van Rooij, G.J. Non-oxidative methane coupling to C_2 hydrocarbons in a microwave plasma reactor. *Plasma Process. Polym.* **2018**, *15*, 1800087. [CrossRef]

41. Li, S.; Zheng, W.; Tang, Z.; Gu, F. Plasma heating and temperature difference between gas pellets in packed bed with dielectric barrier discharge under natural convection condition. *Heat Transf. Eng.* **2012**, *33*, 609–617. [CrossRef]

42. Jidenko, N.; Bourgeois, E.; Borra, J.P. Temperature profiles in filamentary dielectric barrier discharges at atmospheric pressure. *J. Phys. D Appl. Phys.* **2010**, *43*, 295203. [CrossRef]

43. Patil, B.S.; Cherkasov, N.; Lang, J.; Ibhadon, A.O.; Hessel, V.; Wang, Q. Low temperature plasma-catalytic NOx synthesis in a packed DBD reactor: Effect of support materials and supported active metal oxides. *Appl. Catal. B Environ.* **2016**, *194*, 123–133. [CrossRef]

44. Nozaki, T.; Hiroyuki, T.; Okazaki, K. Hydrogen enrichment of low-calorific fuels using barrier discharge enhanced Ni/γ-Al$_2$O$_3$ bed reactor: Thermal and nonthermal effect of nonequilibrium plasma. *Energy Fuels* **2006**, *20*, 339–345. [CrossRef]

45. Masoud, N.; Martus, K.; Figus, M.; Becker, K. Rotational and vibrational temperature measurements in a high-pressure cylindrical dielectric barrier discharge (C-DBD). *Contrib. Plasma Phys.* **2005**, *45*, 32–39. [CrossRef]

46. Rajasekaran, P.; Opländer, C.; Hoffmeister, D.; Bibinov, N.; Suschek, C.V.; Wandke, D.; Awakowicz, P. Characterization of dielectric barrier discharge (DBD) on mouse and histological evaluation of the plasma-treated tissue. *Plasma Process. Polym.* **2011**, *8*, 246–255. [CrossRef]

47. Du, Y.; Nayak, G.; Oinuma, G.; Ding, Y.; Peng, Z.; Bruggeman, P.J. Emission considering self-absorption of OH to simultaneously obtain the OH density and gas temperature: Validation, non-equilibrium effects and limitations. *Plasma Sources Sci. Technol.* **2017**, *26*, 095007. [CrossRef]

48. Florian, J.; Merbahi, N.; Wattieaux, G.; Plewa, J.M.; Yousfi, M. Comparative Studies of Double Dielectric Barrier Discharge and Microwave Argon Plasma Jets at Atmospheric Pressure for Biomedical Applications. *IEEE Trans. Plasma Sci.* **2015**, *43*, 3332–3338. [CrossRef]

49. Blamey, J.; Anthony, E.J.; Wang, J.; Fennell, P.S. The calcium looping cycle for large-scale CO$_2$ capture. *Prog. Energy Combust. Sci.* **2010**, *36*, 260–279. [CrossRef]

50. McBride, B.J.; Zehe, M.J.; Gordon, S. NASA Glenn Coefficients for Calculating Thermodynamic Properties of Individual Species. In *Technical Reports NASA*; NASA: Washington, DC, USA, 2002; Volume 291, p. 211556.

51. Fennell, P.S.; Pacciani, R.; Dennis, J.S.; Davidson, J.F.; Hayhurst, A.N. The effects of repeated cycles of calcination and carbonation on a variety of different limestones, as measured in a hot fluidized bed of sand. *Energy Fuels* **2007**, *21*, 2072–2081. [CrossRef]

52. Lysikov, A.I.; Salanov, A.N.; Okunev, A.G. Change of CO$_2$ carrying capacity of CaO in isothermal recarbonation-decomposition cycles. *Ind. Eng. Chem. Res.* **2007**, *46*, 4633–4638. [CrossRef]

53. Borgwardt, R.H. Sintering of nascent calcium oxide. *Chem. Eng. Sci.* **1989**, *44*, 53–60. [CrossRef]

54. Zhang, Y.R.; van Laer, K.; Neyts, E.C.; Bogaerts, A. Can plasma be formed in catalyst pores? A modeling investigation. *Appl. Catal. B Environ.* **2016**, *185*, 56–67. [CrossRef]

55. Hensel, K.; Katsura, S.; Mizuno, A. DC Microdischarges inside porous ceramics. *IEEE Trans. Plasma Sci.* **2005**, *33*, 574–575. [CrossRef]

56. Zhang, Q.Z.; Wang, W.Z.; Bogaerts, A. Importance of surface charging during plasma streamer propagation in catalyst pores. *Plasma Sources Sci. Technol.* **2018**, *27*, 065009. [CrossRef]

57. Giammaria, G.; Lefferts, L. Catalytic Effect of Water on Calcium Carbonate Decomposition. *Appl. Catal. B*, unpublished.

58. Peeters, F.J.J.; van de Sanden, M.C.M. The influence of partial surface discharging on the electrical characterization of DBDs. *Plasma Sources Sci. Technol.* **2015**, *24*, 15016. [CrossRef]

59. Ghassemi, M.; Mohseni, H.; Niayesh, K.; Shayegani, A.A. Dielectric Barrier Discharge (DBD) Dynamic Modeling for High Voltage Insulation. In Proceedings of the 2011 Electrical Insulation Conference (EIC), Annapolis, MD, USA, 5–8 June 2011.

60. Jess, A.; Wasserscheid, P. *Chemical Technology*; An Integrated Textbook; Wiley: Hoboken, NJ, USA, 2013.

61. Jacobs, G.K.; Kerrick, D.M.; Krupka, K.M. The High-Temperature Heat Capacity of Natural Calcite (CaCO$_3$). *Phys. Chem. Miner.* **1981**, *7*, 55–59. [CrossRef]

catalysts

MDPI

Article

DBD Plasma-ZrO$_2$ Catalytic Decomposition of CO$_2$ at Low Temperatures

Amin Zhou [1] , **Dong Chen [1]**, **Cunhua Ma [1,*]**, **Feng Yu [1,2,3]** and **Bin Dai [1,*]**

[1] School of Chemistry and Chemical Engineering, Shihezi University, Key Laboratory for Green Processing of Chemical Engineering of Xinjiang Bintuan, Shihezi 832003, China; m15699337928@163.com (A.Z.); liantian1986@sina.com (D.C.); yufeng05@mail.ipc.ac.cn (F.Y.)

[2] Engineering Research Center of Materials-Oriented Chemical Engineering of Xinjiang Production and Construction Corps, Shihezi 832003, China

[3] Key Laboratory of Materials-Oriented Chemical Engineering of Xinjiang Uygur Autonomous Region, Shihezi 832003, China

* Correspondence: mchua@shzu.edu.cn (C.M.); dbinly@126.com (B.D.);
Tel.: +86-(0)993-2058-176 (B.D.); Fax: +86-(0)993-2057-270 (B.D.)

Received: 27 April 2018; Accepted: 14 June 2018; Published: 23 June 2018

Abstract: This study describes the decomposition of CO$_2$ using Dielectric Barrier Discharge (DBD) plasma technology combined with the packing materials. A self-cooling coaxial cylinder DBD reactor that packed ZrO$_2$ pellets or glass beads with a grain size of 1–2 mm was designed to decompose CO$_2$. The control of the temperature of the reactor was achieved via passing the condensate water through the shell of the DBD reactor. Key factors, for instance discharge length, packing materials, beads size and discharge power, were investigated to evaluate the efficiency of CO$_2$ decomposition. The results indicated that packing materials exhibited a prominent effect on CO$_2$ decomposition, especially in the presence of ZrO$_2$ pellets. Most encouragingly, a maximum decomposition rate of 49.1% (2-mm particle sizes) and 52.1% (1-mm particle sizes) was obtained with packing ZrO$_2$ pellets and a 32.3% (2-mm particle sizes) and a 33.5% (1-mm particle sizes) decomposing rate with packing glass beads. In the meantime, CO selectivity was up to 95%. Furthermore, the energy efficiency was increased from 3.3%–7% before and after packing ZrO$_2$ pellets into the DBD reactor. It was concluded that the packing ZrO$_2$ simultaneously increases the key values, decomposition rate and energy efficiency, by a factor of two, which makes it very promising. The improved decomposition rate and energy efficiency can be attributed mainly to the stronger electric field and electron energy and the lower reaction temperature.

Keywords: self-cooling; dielectric barrier discharge; CO$_2$ decomposition; CO selectivity; packing materials

1. Introduction

The fast-growing consumption of fossil fuels has resulted in continually increasing emissions of carbon dioxide, which is identified as one of the major contributors to global warming. Therefore, the decrease of environmental pollution via CO$_2$ emissions has attracted worldwide attention. Different strategies are being developed to address the wasted CO$_2$ instead of releasing it into the atmosphere, such as: carbon capture and storage, transformation and utilization of carbon and CO$_2$ dissociation. Direct dissociation of CO$_2$ into other value-added fuels and chemicals provides a potential route for efficient utilization of CO$_2$ and reduction of CO$_2$ emissions [1]. Various progresses have been explored to convert CO$_2$ into other value-added chemicals, such as CO$_2$ reforming of CH$_4$ for hydrogen and CO$_2$ hydrogenation for the synthesis of methanol, methane, formaldehyde, dimethyl, etc. [2,3]. Additionally, direct decomposition of CO$_2$ into CO has also attracted great interest, which can not only relieve the pressure of economic growth, but also can achieve energy savings and emission reduction [4,5]. As a common feedstock for industry, CO is a widely-used chemical feedstock that can be used as a reactant to produce higher energy products. Not only can it be used for fuel synthesis, but also for the production of chemicals, such as organic acids, esters and other chemicals. Thus, the selective decomposition of CO_2 into CO is no doubt a promising candidate for clean energy and chemicals. However, due to the high structural stability of the CO_2 molecule, considerable energy is needed for CO_2 activation and decomposition. The conventional thermal-chemical process for CO_2 decomposition has many different levels of limited scope. For example, the thermodynamic equilibrium calculation of CO_2 conversion shows that CO_2 begins to split into CO and O_2 near 2000 K, yet with a very low conversion rate (<1%). The decomposition of CO_2 can only be carried out at an extraordinarily high temperature (3000–3500 K), which consumes high energy and involves considerable economic cost [6]. Nowadays, Non-Thermal Plasma (NTP) is a newly-developed technology, as an attractive alternative, which has been successfully applied in many fields, for instance gas purification and energy conversion [7–12]. It has advantages such as a non-equilibrium character, a low energy cost and a unique ability to initiate chemical reactions at low temperatures [13,14]. As a kind of non-thermodynamic equilibrium plasma, its distinct non-equilibrium character means the gas temperature in the plasma can be close to room temperature, whilst the electrons are highly energetic with a typical mean energy of 1–10 eV. NTP can initiate a series of chemical processes, including ionization, initiation and dissociation [15]. Various types of plasma, like glow discharge, corona discharge [16–18], microwave discharge, radio frequency discharge [19] and gliding arc discharge have been explored for CO_2 decomposition [20]. Dielectric Barrier Discharge (DBD) plasma was also tested for CO_2 splitting into CO, and it can generate high energetic electrons (1–10 eV) and initiate the chemical reactions while keeping the background temperature under ambient conditions [6]. It has been applied in many fields, such as the removal of NO_x [21], as well as the preparation of catalysts [22] and other materials [23]. The decomposition and conversion of the stable CO_2 using DBD plasma is no exception, which has attracted increasing attention for its unique abilities.

Many relevant works have been reported for the direct decomposition of pure CO_2, CO_2 conversion or reacting with other gases, such as adding inert gases to dilute the pure CO_2 under the assistance of DBD plasma [24,25]. It was reported that nitrogen is more effective for CO_2 decomposition among argon, nitrogen and helium as diluents, but this will produce unwanted by-products inevitably, which is not preferable from the industrial application point of view [26]. Therefore, many previous works have focused on the dissociation or conversion of CO_2 without diluting. Further fundamental works reported that different packing dielectric materials can enhance the conversion of CO_2 and improve the energy efficiency of the plasma process [14]. Yu et al. investigated the conversion of CO_2 in a packed-bed DBD reactor using silica gel, quartz, α-Al_2O_3, γ-Al_2O_3 and $CaTiO_3$ as packing materials and proved that the introduction of dielectric materials into the plasma reactor resulted in an increased electric field, which then increased the electron energy and led to an expected higher CO_2 conversion [27]. Similarly, a series of $Ca_xSr_{(1-x)}TiO_3$ has also been used as a dielectric material for the splitting of CO_2 in a DBD reactor to prove the importance of the high permittivity dielectrics, which can increase the discharge power of plasma accompanied by dense and strong micro-discharges, thereby significantly enhancing the decomposition of CO_2 [28–30]. Yap et al. [31] reported the best conversion with an Alternating Current (AC) sinusoidal activation when using the glass beads as packing materials. In recent reports, CeO_2 was packed to understand the oxygen storage/release capacity on the improvement of CO_2 conversion in the packed DBD reactor [32]. Duan et al. [33] obtained a CO_2 conversion of 41.9% in a CaO-packed DBD micro-plasma reactor. Mei et al. [34] proved that in the coaxial dielectric barrier discharge reactor, the discharge power was the most important factor that affected the CO_2 conversion. Furthermore, they showed that $BaTiO_3$ pellets exhibited a better performance than TiO_2 pellets and glass beads due to the higher dielectric constant and the better synergistic effect of plasma-catalysis [6,34]. Moreover, the design of the reactor is also of importance for CO_2 conversion [35].

In this study, a self-cooling coaxial cylinder DBD reactor was applied to decompose pure CO_2 into CO and O_2 at atmospheric pressure and ambient temperature, which was the same as the previous apparatus of Zhou et al. [36]. The commercial ZrO_2 pellets and glass beads were used as packing materials to demonstrate the improvement of the decomposition rate and energy efficiency for CO_2 splitting. Condensate water can take away the heat that is introduced in the process of plasma discharge to maintain the temperature of the reactor. Key factors like packing materials and discharge power, discharge length and bead size were investigated to evaluate the efficiency of CO_2 decomposition. Interestingly, a high decomposition rate and energy efficiency could be obtained. The physical discharge characteristics of DBD plasma were also investigated to understand the interactions between the dielectric materials and DBD plasma in CO_2 decomposition.

2. Results and Discussion

2.1. Effect of Discharge Length on CO_2 Decomposition Rate and Energy Efficiency

In order to further be sure of the influence of DBD plasma on the decomposition of CO_2, the discharge length of the unpacked DBD reactor was changed to conduct the experiment. As Figure 1a shows, the CO_2 decomposition rate increased clearly with the increasing discharge length. This can be attributed to competing effects primarily. Firstly, increasing the discharge length from 100–200 mm significantly increased the residence time of CO_2 gas in the reactor, which positively increased the probability of CO_2 molecules colliding with highly energetic electrons and reactive species [34]. However, a longer discharge region will need increasing surface area of the DBD reactor, leading to higher energy loss due to heat dissipation [37], as shown in Figure 1b. In this study, increasing the discharge length significantly increased the residence time of CO_2 in the reaction, which plays a more dominant role in the decomposition of CO_2 compared to the negative effects (e.g., increased energy loss); therefore, 200 mm was chosen as the optimum discharge length to conduct the experiment.

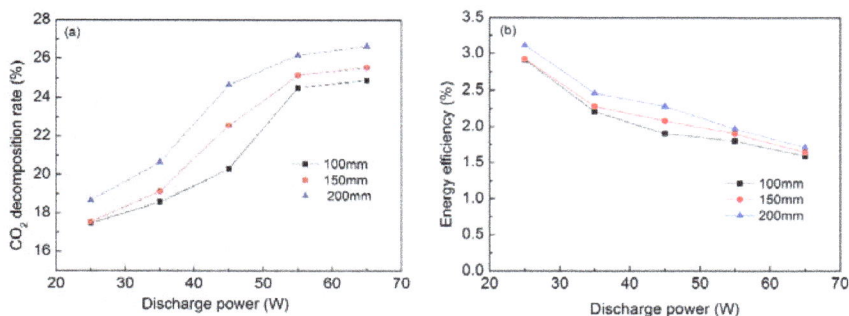

Figure 1. CO_2 decomposition rate (**a**) and energy efficiency (**b**) in the reactor without packing at different discharge powers (frequency: 12 kHz; feed flow rate: 20 mL/min; 20 °C condensate water).

2.2. Effect of Discharge Power and Beads Size on CO_2 Decomposition and Energy Efficiency

The CO_2 decomposition rate as a function of bead size at different discharge powers is illustrated in Figure 2. ZrO_2 and glass beads with a size of 1 mm and 2 mm were applied to investigate the influence on CO_2 decomposition. Figure 2a shows that the smaller bead size was more beneficial for CO_2 decomposition under other fixed conditions. When decreasing the beads size, more dielectric spheres would be needed to fill the reactor, which increased the discharge surface area, reinforced the surface discharge and hence, caused, a higher CO_2 decomposition rate. Similarly, Van Laer and Bogaerts [38] reported that a suitable range of bead size was needed to obtain a high CO_2 decomposition and energy efficiency.

Discharge power is also one of the key factors influencing CO_2 decomposition in the DBD plasma technique. According to associated reports, the discharge power determined whether there was sufficient energy for activating and decomposing the CO_2 molecule [6]. The CO_2 decomposition rate and energy efficiency at different discharge powers in the DBD reactor are shown in Figure 2a,b. There is a general observation that the CO_2 degradation rate increased with the increasing discharge power, but the energy efficiency was affected in the opposite manner. A higher discharge power meant more energy was injected into this system, therefore generating more chemically-active species and reactant molecules in the reaction, and enough energy would activate electron and reactant molecules, as well as increase the mutual collision opportunities between active species, so more chemical bonds would be broken and more active substances formed. However, the CO_2 decomposition rate tends to saturate when the discharge power rises above 55 W, so a suitable range of discharge power is necessary from the viewpoint of energy savings. Furthermore, the reason for the decreased energy efficiency with the increasing discharge power could be due to the lost energy, which was consumed as heat, and it can be evidenced by the increasing temperature of the condensate water. At the same discharge power of 55 W, the CO_2 decomposition rate in the unpacked reactor was 26.1%, but it reached 33.5% and 52.1% when the DBD reactor was packed with 1-mm glass beads and ZrO_2 pellets, respectively. Meanwhile, the energy efficiency only in ZrO_2 pellets packing was also improved by a factor of two.

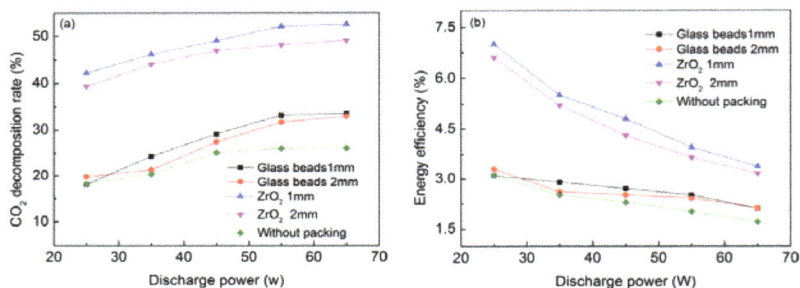

Figure 2. CO_2 decomposition rate (**a**) and energy efficiency (**b**) using different packing bead sizes at different discharge powers (frequency: 12 kHz; feed flow rate: 20 mL/min; 20 °C condensate water; discharge length: 200 mm).

Three reasons may account for this result. Firstly, filling materials in the discharge zone made a more stable, uniform and stronger discharge, and all these were favorable for higher CO_2 decomposition because it meant more CO_2 molecules were activated. Additionally, the intensity of the electric field between the contact points of the pellets to pellets or the pellets to the reactor wall could be enhanced because of the polarization of the dielectric materials. Though the electric field was enhanced by the increased discharge power regardless of the packing materials used [27], the presence of packing materials may have further strengthened the average electric field near contact points of the pellets, heightening the electric electron temperature, which facilitated electron collision [39,40], hence causing a higher CO_2 decomposition rate in the packed reactor. The morphology of ZrO_2 pellets may also be conducive to the transfer of electrons, thus accelerating the decomposition of CO_2. Because ZrO_2 is a kind of high performance structure ceramic material, it owns a higher dielectric constant (27) than glass beads (9), which is proven to play a significant role in CO_2 decomposition [39]. In addition, ZrO_2 is also a kind of basic oxide, and it plays an important role in CO_2 decomposition due to the acidity of CO_2. This was proved in Duan X's study: the base properties of the packing materials affected the chemisorption of CO_2 in the process of CO_2 decomposition [34]. That is advantageous to CO_2 decomposition.

Furthermore, ZrO_2 exhibits better reaction activity than glass beads possibly due to the fast oxygen ion migration rate of ZrO_2 [41,42], so the oxygen containing active substances was produced on the surface of ZrO_2 pellets, or oxygen ions could be quickly transferred, thus inhibiting the recombination of oxygen free radicals with the generated CO to promote the continuous decomposition of CO_2. In addition, the thermal conductivity of ZrO_2 is higher than that of glass. It may also contribute to the energy transfer in the reaction process and the activation of CO_2 molecules. Therefore, ZrO_2 enhanced the CO_2 decomposition compared to glass beads.

2.3. Effect of Packing Materials on Discharge Characteristics

It can be noticed from the V-Q curve of the CO_2 discharge that the characteristics vary in the reactor with and without packing. A typical filamentary discharge in the discharge with no packing can be observed, and this be confirmed by the numerous peaks in the discharge signal of Figure 3a. In contrast, as is shown in Figure 3b,c, packing ZrO_2 or glass beads into the discharge zone generates a typical packed-bed effect and leads to a transition in the discharge behavior from a filamentary discharge to a combination of surface discharge and filamentary discharge, because of the decrease of the spikes in the discharge signal.

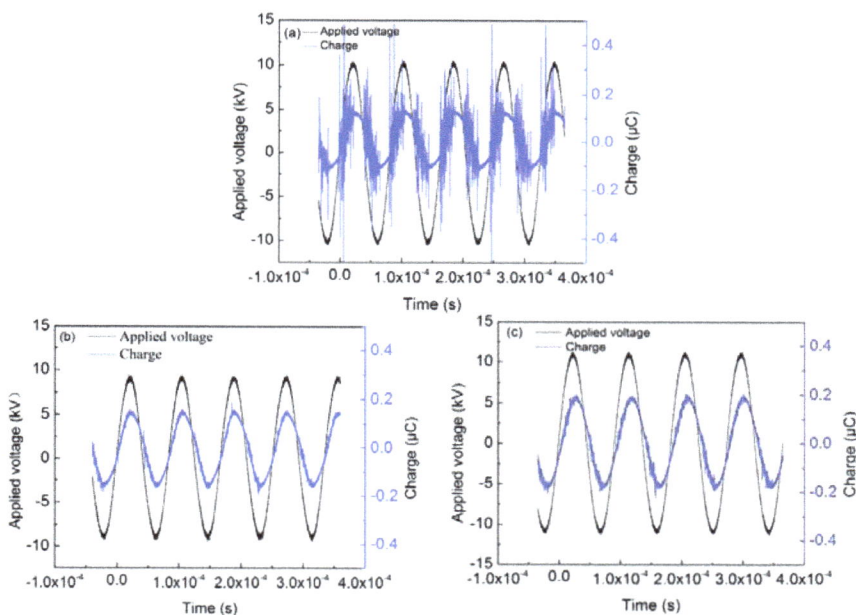

Figure 3. Discharge characteristics of CO_2 in the Dielectric Barrier Discharge (DBD) reactor with or without packing: (**a**) without packing; (**b**) glass beads; (**c**) ZrO_2 (discharge power: 55 W; feed flow rate: 20 mL/min; 20 °C condensate water; discharge length: 200 mm; beads size: 1 mm).

Figure 4 presents the Lissajous figure of CO_2 decomposition in the DBD reactor with or without packing materials, at the discharge power of 55 W. It shows the change of the discharge characteristic during the decomposition of CO_2. Compared with non-packing materials, the Lissajous figure changed from a parallelogram to an oval shape when ZrO_2 pellets or glass beads were filled in the DBD reactor. At the same discharge power, the applied voltage increased from 9.8 kV (pk-pk) without packing to 11.5 kV with the ZrO_2 packing and to 10.6 kV (pk-pk) with the glass bead packing. It was obviously observed that the discharge mode varied from filamentary discharge to a mode of filamentary discharge combined with surface discharge, and we could obtain a more stable and uniform discharge form. As a result, the introduction of packing materials into the DBD reactor had a promotional effect on the discharge characteristics. Yu et al. [27] also published a consistent report on the discharge characteristics of CO_2 conversion in the dielectric packed-bed plasma reactor, and the result was further proved by Mei et al. [6]. The introduction of packing materials caused the presence of a strong electrical field and thus led to a high electron energy near the contact points of the pellets, hence the improved CO_2 decomposition.

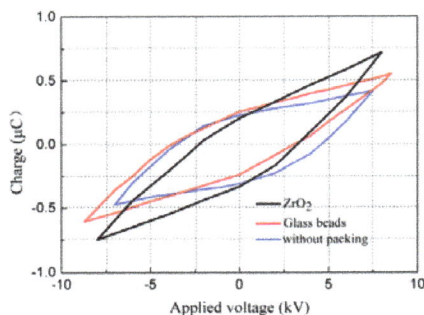

Figure 4. Lissajous figure of CO_2 decomposition in the DBD reactor with or without packing materials (discharge power: 55 W; feed flow rate: 20 mL/min; 20 °C condensate water; discharge length: 200 mm; beads size: 1 mm).

2.4. Effect of Packing Materials on CO Selectivity

The function of discharge power on CO selectivity is presented in the Figure 5a. The selectivity of CO throughout the reaction remained around 94–96%, and the difference of CO selectivity in different discharge power was not obvious in our experimental condition. It can be concluded that discharge power played a minor role in the CO selectivity. The CO_2 decomposition at the same discharge power and feed flow rate was also conducted to investigate the effect of packing materials on CO selectivity, as is shown in Figure 5b. The CO selectivities of the three different conditions were almost the same, and the CO selectivity remained at about 95% regardless of the packing materials. It can be concluded that packing materials played a minor role in the CO selectivity using self-cooling dielectric barrier discharge plasma.

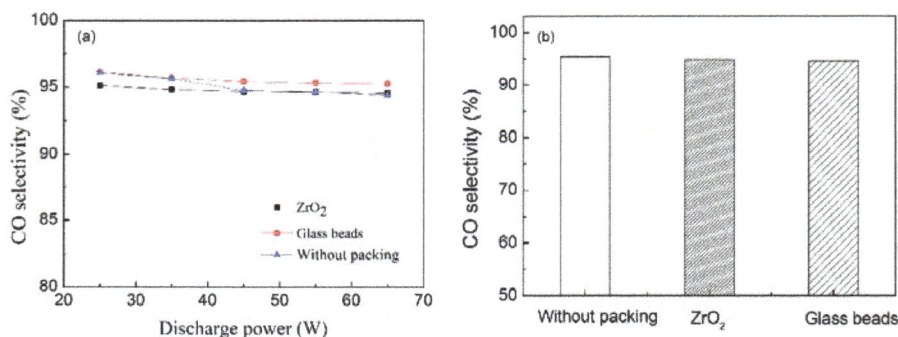

Figure 5. (**a**) CO selectivity at different discharge powers; (**b**) CO selectivity at a discharge power of 55 W (frequency: 12 kHz; feed flow rate: 20 mL/min; 20 °C condensate water; discharge length: 200 mm; bead size: 1 mm).

2.5. The Investigation of Carbon Deposition

After reaction, some black carbon deposition was clearly observed on the surface of the inner electrode and ZrO_2 pellets present in the discharge zone, suggesting that carbon was produced during the CO_2 decomposition in the DBD plasma combined with ZrO_2. Figure 6 is the SEM pictures of the surface of ZrO_2 pellets before and after the reaction; it shows that the surface of the ZrO_2 pellets was covered with carbon deposition. Because high energy electrons generated by the DBD plasma had a range of 1–10 eV [35], which was not enough for the dissociation of CO molecules, because of its high dissociation energy (11 eV), the carbon deposition may have come from the decomposition of CO_2 at temperatures higher than 20 °C of condensate water.

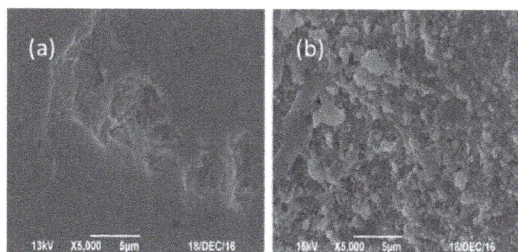

Figure 6. SEM images of carbon deposition on the surface of ZrO_2 pellets (**a**) before the reaction and (**b**) after the reaction.

2.6. Reaction of CO and O_2

To gain insight into the role of packing dielectric materials and the principles of the CO_2 decomposition, the oxidation of CO, which is the reverse reaction of CO_2 decomposition, was still investigated using ZrO_2 pellets and glass beads as the packing materials at different discharge powers, and the results are shown in Figure 7 CO conversion increased with the increasing discharge power. Apparently, the CO conversion without packing was higher than that of the packed DBD reactor. Though ZrO_2 exhibited a promotional effect on the CO_2 decomposition, it was not conducive for CO conversion at the same condition. In all cases, CO_2 was formed as the only product, and no carbon deposition was generated. This was in agreement with the above result that CO could not split into C in this condition due to its high dissociation energy of CO molecules (11.1 eV). Therefore, the DBD plasma conditions with electron energy in the range of 1–10 eV were optimum for the electron impact dissociation of CO_2 [27].

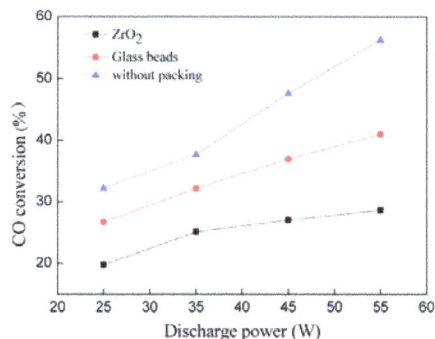

Figure 7. CO oxidation at different discharge powers (frequency: 12 kHz; feed flow rate: 20 mL/min; 20 °C condensate water; beads size: 1 mm).

2.7. Comparison of Obtained Values in Different Packed DBD Reactors

Table 1 shows the compared values of the CO_2 decomposition and energy efficiency using different packed DBD reactors. Though the DBD reactor made of quartz has been widely applied for CO_2 decomposition, the self-cooling DBD reactor made of Pyrex glass can also be utilized for CO_2 decomposition [37]. A relatively higher CO_2 decomposition rate, together with a stable CO selectivity, was obtained when using the same packing materials like ZrO_2 pellets. Fifty two-point-one percent (1-mm ZrO_2) of the CO_2 decomposition rate was an accurate value, which has been repeatedly verified.

Table 1. Obtained values in different packed DBD plasma reactors.

Packing Materials	ZrO_2	ZrO_2	$BaTiO_3$	$BaTiO_3$	$CaTiO_3$
Reactor	Quartz	Glass	Quartz	Quartz	Quartz
Decomposition rate (%)	42	52.1	28	38.3	20.5
Energy efficiency (%)	9.6	7.0	7.1	17	4.8
CO selectivity (%)	50	94–96	—	—	—
Power (w)	60	55	50	—	35.3
Reference	38	This work	6	35	27

3. Experimental Setup

Figure 8 shows the schematic diagram of the experimental setup, which is basically consistent with the previous apparatus in Zhou et al. [37]. It contains 3 parts: a DBD plasma reactor, a flow control device and a GC detection and analysis system. The experiments were performed using a self-cooling DBD reactor (BEIJING SYNTHWARE GLASS, Beijing, China) made of Pyrex, which consisted of a double coaxial cylinder, a High-Voltage (HV) electrode and a Low-Voltage (LV) electrode. The HV electrode was a stainless steel rod (2 mm in diameter), on which a 1 mm-thick copper wire was wrapped, and it was installed in the axis of the double coaxial glass cylinder and linked to an alternating current power supply. Condensate water used as a cooling agent was fed into the shell of the concentric cylinder DBD reactor. The LV electrode that grounded by a wire placed into condensate water. The HV and LV electrodes formed a cylindrical discharge space.

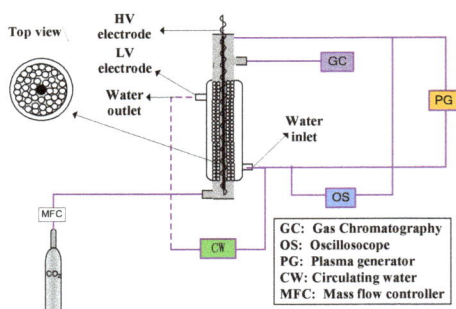

Figure 8. The schematic of the experimental setup.

During the course of the experiment, condensate water can take away the heat produced during plasma discharge, to keep the temperature of the reactor at the designed temperature. It has been proven that decreasing the temperature of the reactor was advantageous to CO_2 decomposition [43,44]. In this work, we controlled the temperature of the condensed water at 20 °C, while this value was optimized in Zhou et al. [37]. The flow rate of pure CO_2 was fixed at 20 mL/min. The applied voltage was measured by a high-voltage probe (Tektronix P6015A), and the current was recorded by a current monitor. The voltage on the external capacitor was measured to obtain the charge generated in the discharge. All of the electrical signals were sampled by a four-channel digital oscilloscope (TDS3054B). The discharge power was obtained by the area calculation of the *Q-U* Lissajous figure. The gas products were analyzed by gas chromatography (GC-2014C), which was equipped with a thermal conductivity detector (TCD), and H_2 was used as the carrier gas. For comparison, CO_2 decomposition was also carried out in the DBD reactor without packing materials.

In this study, ZrO_2 pellets and glass beads with a grain size of 1–2 mm were fully packed in the discharge area as dielectric packing materials, both of which were commercial reagents without any treatment (Shanghai Gongtao Ceramics Co., Ltd., Shanghai, China).

To evaluate the performance of the plasma process, the Specific Input Energy (SIE), CO_2 decomposing rate and selectivity towards CO were defined as follows:

$$CO_2 \text{ decomposition rate } (\%) = \frac{CO_2 \text{ decomposed}}{CO_2 \text{ input}} \times 100\%$$

$$CO \text{ selectivity } (\%) = \frac{CO \text{ formed}}{CO_2 \text{ decomposed}} \times 100\%$$

$$\text{Energy efficiency } (\eta) = \frac{\Delta H \times CO_2 \text{ decomposed} \times F}{60 \times V_m \times P} \times 100\%$$

where P is the discharge power, F is the feed flow rate (mL/min) of CO_2, ΔH is the enthalpy of reaction: $CO_2 \rightarrow CO + \frac{1}{2}O_2$, $\Delta H = 279.8$ kJ/mol and $V_m = 22.4$ L/mol, and V_m is the molar volume.

4. Conclusions

Decomposition of CO_2 into CO and O_2 has been carried out in a self-cooling DBD plasma reactor packed with ZrO_2 and glass beads at low temperatures and ambient pressure. It could be concluded that a longer discharge length and a smaller bead size were beneficial for CO_2 decomposition. The introduction of packing materials shifted the discharge mode from filamentary discharge to a combination of filamentary and surface discharge. In comparison with the CO_2 decomposition rate in the empty reactor and glass-packed reactor, which reached 26.1% and 33.5%, respectively, a CO_2 decomposition rate of 52.1% was obtained in the ZrO_2 (1 mm) packed DBD reactor at the same condition. ZrO_2 exhibited a superior promotional effect on the decomposition of CO_2 and energy efficiency by up to a factor of 1.9 and 2.1, respectively, compared with the result of the unpacked DBD reactor. Additionally, the main product of this work was CO, with slight carbon deposition that came from the dissociation of CO_2. The results indicated that the dielectric constant and particle morphology of the packing materials matter greatly in the decomposition of CO_2.

Author Contributions: A.Z. and B.D. designed and administered the experiments. A.Z. and B.D. performed the experiments. A.Z., D.C., C.M. and F.Y. collected and analyzed the data. All authors discussed the data and wrote the manuscript.

Acknowledgments: This work was financially supported by the National Natural Science Foundation of China (No. 21663022) and the High-level Talent Scientific Research Project in Shihezi University (No. RCZX201406).

Conflicts of Interest: The authors declare no conflict of interest.

References

1. Razali, N.; Lee, K.T.; Bhatia, S.; Mohamed, A.R. Heterogeneous catalysts for production of chemicals using carbon dioxide as raw material: A review. *Renew. Sustain. Energy Rev.* **2012**, *16*, 4951–4964. [CrossRef]
2. Centi, G.; Quadrelli, E.A.; Perathoner, S. Catalysis for CO_2 conversion: A key technology for rapid introduction of renewable energy in the value chain of chemical industries. *Energy Environ. Sci.* **2013**, *6*, 1711–1731. [CrossRef]
3. Lebouvier, A.; Iwarere, S.A.; Argenlieu, P.D.; Ramjugernath, D.; Fulcheri, L. Assessment of carbon dioxide dissociation as a new route for syngas production: A Comparative review and potential of plasma-Based technologies. *Energy Fuels* **2013**, *27*, 2712–2722. [CrossRef]

4. Ayzner, A.L.; Wanger, D.D.; Tassone, C.J.; Tolbert, S.H.; Schwartz, B.J. Room to improve conjugated polymer-based solar cells: Understanding how thermal annealing affects the fullerene component of a bulk heterojunction photovoltaic device. *J. Phys. Chem. C* **2008**, *112*, 18711–18716. [CrossRef]

5. Kim, S.S.; Sang, M.L.; Hong, S.C. A study on the reaction characteristics of CO_2 decomposition using iron oxides. *J. Ind. Eng. Chem.* **2012**, *18*, 860–864. [CrossRef]

6. Mei, D.; Zhu, X.; He, Y.L.; Yan, J.D.; Tu, X. Plasma-assisted conversion of CO_2 in a dielectric barrier discharge reactor: Understanding the effect of packing materials. *Plasma Sour. Sci. Technol.* **2015**, *24*, 015011. [CrossRef]

7. Aerts, R.; Tu, X.; De Bie, C.; Whitehead, J.C.; Bogaerts, A. An Investigation into the dominant reactions for ethylene destruction in non-thermal atmospheric plasmas. *Plasma Process. Polym.* **2012**, *9*, 994–1000. [CrossRef]

8. Harling, A.M.; Glover, D.J.; Whitehead, J.C.; Zhang, K. Novel method for enhancing the destruction of environmental pollutants by the combination of multiple plasma discharges. *Environ. Sci. Technol.* **2008**, *42*, 4546–4550. [CrossRef] [PubMed]

9. Liu, S.; Mei, D.; Shen, Z.; Tu, X. Nono-xidative conversion of methane in a dielectric barrier discharge reactor: Prediction of reaction performance based on neural model. *J. Phys. Chem. C* **2014**, *118*, 10686–10693. [CrossRef]

10. Snoeckx, R.; Aerts, R.; Tu, X.; Bogaerts, A. Plasma-based dry reforming: A computational study ranging from nanoseconds to seconds timescale. *J. Phys. Chem. A* **2013**, *117*, 4857–4970. [CrossRef]

11. Xin, T.; Whitehead, J.C. Plasma dry reforming of methane in an atmospheric pressure AC gliding arc discharge: Co-generation of Syngas and carbon nanomaterials. *Int. J. Hydrog. Energy* **2014**, *39*, 9658–9669.

12. Yu, L.; Tu, X.; Li, X.; Wang, Y.; Chi, Y.; Yan, J. Destruction of acenaphthene, fluorene, anthracene and pyrene by a dc gliding arc plasma reactor. *J. Hazard. Mater.* **2010**, *180*, 449–455. [CrossRef] [PubMed]

13. Tu, X.; Gallon, H.J.; Twigg, M.V.; Gorry, P.A.; Whitehead, J.C. Dry reforming of methane over a Ni/Al_2O_3 catalyst in a coaxial dielectric barrier discharge reactor. *J. Phys. D Appl. Phys.* **2011**, *44*, 274007. [CrossRef]

14. Tu, X.; Whitehead, J.C. Plasma-catalytic dry reforming of methane in an atmospheric dielectric barrier discharge: Understanding the synergistic effect at low temperature. *Appl. Catal. B Environ.* **2012**, *125*, 439–448. [CrossRef]

15. Wang, B.; Yan, W.; Ge, W.; Duan, X. kinetic model of the methane conversion into higher hydrocarbons with a dielectric barrier discharge microplasma reactor. *Chem. Eng. J.* **2013**, *234*, 354–360. [CrossRef]

16. Horvath, G.; Skalny, J.D.; Mason, N.J. FTIR study of decomposition of carbon doxide in DC corona discharges. *J. Phys. D Appl. Phys.* **2008**, *41*, 207–225. [CrossRef]

17. Mikoviny, T.; Kocan, M.; Matejcik, S.; Mason, N.J.; Skalny, J.D. Experimental study of negative corona discharge in pure carbon dioxide and its mixtures with oxygen. *J. Phys. D Appl. Phys.* **2003**, *37*, 64–73. [CrossRef]

18. Wen, Y.; Jiang, X. Decomposition of CO_2 Using Pulsed Corona Discharges Combined with Catalyst. *Plasma Chem. Plasma Process.* **2001**, *21*, 665–678. [CrossRef]

19. Kozak, T.; Bogaerts, A. Splitting of CO_2 by vibrational excitation in non-equilibrium plasmas: A reaction kinetics model. *Plasma Sour. Sci. Technol.* **2014**, *23*, 045004. [CrossRef]

20. Nunnally, T.; Gustol, K.; Rabinovich, A.; Fridman, A.; Gutsol, A.; Kemoun, A. Dissociation of CO_2 conversion in a low current gliding arc plasmatron. *J. Phys. D Appl. Phys.* **2011**, *44*, 274009. [CrossRef]

21. Zhao, D.; Yu, F.; Zhou, A.; Ma, C.; Dai, B. High-efficiency removal of NOx using dielectric barrier discharge nonthermal plasma with water as an outer electrode. *Plasma Sci. Technol.* **2018**, *20*, 014020. [CrossRef]

22. Zhang, M.; Li, P.; Zhu, M.; Tian, Z.; Dan, J.; Li, J.; Wang, Q. Ultralow-weight loading Ni catalyst supported on two-dimensional vermiculite for carbon monoxide methanation. *Chin. J. Chem. Eng.* **2017**. [CrossRef]

23. Wang, Y.; Yu, F.; Zhu, M.; Ma, C.; Zhao, D.; Wang, C.; Zhou, A.; Guo, X. N-doping of plasma exfoliated graphene oxide via dielectric barrier discharge plasma treatment for oxygen reduction reaction. *J. Mater. Chem. A* **2018**, *6*, 2011–2017. [CrossRef]

24. Indarto, A.; Yang, D.R.; Choi, J.W.; Lee, H.; Song, H.K. Gliding arc plasma processing of CO_2 conversion. *J. Hazard. Mater.* **2007**, *146*, 309–315. [CrossRef] [PubMed]

25. Nozaki, T.; Okazaki, K. Non-thermal plasma plasma catalysis of methane: Principles, energy efficiency, and applications. *Catal. Today* **2013**, *211*, 29–38. [CrossRef]

26. Ozkan, A.; Dufour, T.; Arnoult, G.; Keyzer, P.D.; Bogaerts, A.; Reniers, F. CO_2-CH_4 conversion and syngas formation at atmospheric pressure using a multi-electrode dielectric barrier discharge. *J. CO_2 Util.* **2015**, *9*, 74–81. [CrossRef]

27. Yu, Q.; Kong, M.; Liu, T.; Fei, J.; Zheng, X. Characterstics of the Decomposition of CO_2 in a Dielectric Packed-Bed Plasma Reactor. *Plasma Chem. Plasma Process.* **2012**, *32*, 153–163. [CrossRef]

28. Li, R.; Tang, Q.; Shu, Y.; Sato, T. Investigation of dielectric barrier discharge dependence on permittivity of barrier materials. *Appl. Phys. Lett.* **2007**, *90*, 131502. [CrossRef]

29. Li, R.; Tang, Q.; Yin, S.; Yamaguchi, Y.; Sato, T. Decomposition of carbon dioxide by the dielectric barrier discharge (DBD) Plasma Using $Ca_{0.7}Sr_{0.3}TiO_3$ barrier. *Chem. Lett.* **2004**, *33*, 412–413. [CrossRef]

30. Wang, S.; Zhang, Y.; Liu, X.; Wang, X. Enhancement of CO_2 conversion rate and conversion efficiency by homogeneous discharges. *Plasma Chem. Plasma Process.* **2012**, *32*, 979–989. [CrossRef]

31. Yap, D.; Tatibouet, J.M.; Batiot-Dupeyrat, C. Carbon dioxide dissociation to carbon monoxide by non-thermal plasma. *J CO_2 Util.* **2015**, *12*, 54–61. [CrossRef]

32. Ray, D.; Subrahmanyam, C. CO_2 decomposition in a packed dbd plasma reactor: Influence of packing materials. *RSC Adv.* **2016**, *6*, 39492–39499. [CrossRef]

33. Duan, X.; Hu, Z.; Li, Y.; Wang, B. Effect of dielectric packing materials on the decomposition of carbon dioxide using dbd microplasma reactor. *AIChE J.* **2015**, *61*, 898–903. [CrossRef]

34. Mei, D.; He, Y.; Liu, S.; Yan, J.; Tu, X. Optimization of CO_2 conversion in a cylindrical dielectric barrier discharge reactor using design of experiments. *Plasma Process. Polym.* **2016**, *13*, 544–556. [CrossRef]

35. Mei, D.; Zhu, X.; Wu, C.; Ashford, B.; Williams, P.T.; Tu, X. Plasma-photocatalytic conversion of CO_2 at low temperatures: Understanding the synergistic effect of plasma-catalysis. *Appl. Catal. B Environ.* **2016**, *182*, 525–532. [CrossRef]

36. Mei, D.; Tu, X. Conversion of CO_2 in a cylindrical dielectric barrier discharge reactor: Effects of plasma processing parameters and reactor design. *J. CO_2 Util.* **2017**, *19*, 68–78. [CrossRef]

37. Zhou, A.; Chen, D.; Dai, B.; Ma, C.; Li, P.; Yu, F. Direct decomposition of CO_2 using self-cooling dielectric barrier discharge plasma. *Greenh. Gases Sci. Technol.* **2017**, *7*, 721–730. [CrossRef]

38. Van Laer, K.; Bogaerts, A. Improving the conversion and energy efficiency of carbon dioxide splitting in a zirconia-packed dielectric barrier discharge reactor. *Energy Technol.* **2015**, *3*, 1038–1044. [CrossRef]

39. Ozkan, A.; Dufour, T.; Silva, T.; Britun, N.; Snyders, R.; Bogaerts, A.; Reniers, F. The influence of power and frequency on the filamentary behavior of a flowing DBD-application to the splitting of CO_2. *Plasma Sour. Sci. Technol.* **2016**, *25*, 025013. [CrossRef]

40. Dou, B.; Feng, B.; Wang, C.; Jia, Q.; Li, J. Discharge characteristics and abatement of volatile organic compounds using plasma reactor packed with ceramic Raschig rings. *J. Electrost.* **2013**, *71*, 939–944. [CrossRef]

41. Jiang, Q.; Chen, Z.; Tong, J.; Yang, M.; Jiang, Z.; Li, C. Direct thermolysis of CO_2 into CO and O_2. *Chem. Commun.* **2017**, *53*, 1188–1191. [CrossRef] [PubMed]

42. Mahato, N.; Banerjee, A.; Gupta, A.; Omar, S.; Balani, K. Progress in material selection for solid oxide fuel cell technology: A review. *Prog. Mater. Sci.* **2015**, *72*, 141–337. [CrossRef]

43. Zhang, K.; Zhang, G.; Liu, X.; Phan, A.N.; Luo, K. A study on CO_2 decomposition to CO and O_2 by the combination of catalysis and dielectric barrier discharges at low temperatures and ambient pressure. *Ind. Eng. Chem. Res.* **2017**, *56*, 3204–3216. [CrossRef]

44. Ozkan, A.; Bogaerts, A.; Reniers, F. Routes to increase the conversion and the energy efficiency in the splitting of CO_2 by a dielectric barrier discharge. *J. Phys. D Appl. Phys.* **2017**, *50*, 084004. [CrossRef]

catalysts

MDPI

Article

Altering Conversion and Product Selectivity of Dry Reforming of Methane in a Dielectric Barrier Discharge by Changing the Dielectric Packing Material

Inne Michielsen *, Yannick Uytdenhouwen, Annemie Bogaerts and Vera Meynen

Department of Chemistry, University of Antwerp, Universiteitsplein 1, B-2610 Wilrijk, Belgium;
Yannick.uytdenhouwen@uantwerpen.be (Y.U.); Annemie.Bogaerts@uantwerpen.be (A.B.);
vera.meynen@uantwerpen.be (V.M.)
* Correspondence: inne.michielsen@uantwerpen.be; Tel.: +32-3265-23-60

Received: 11 December 2018; Accepted: 2 January 2019; Published: 7 January 2019

Abstract: We studied the influence of dense, spherical packing materials, with different chemical compositions, on the dry reforming of methane (DRM) in a dielectric barrier discharge (DBD) reactor. Although not catalytically activated, a vast effect on the conversion and product selectivity could already be observed, an influence which is often neglected when catalytically activated plasma packing materials are being studied. The α-Al_2O_3 packing material of 2.0–2.24 mm size yields the highest total conversion (28%), as well as CO_2 (23%) and CH_4 (33%) conversion and a high product fraction towards CO (~70%) and ethane (~14%), together with an enhanced CO/H_2 ratio of 9 in a 4.5 mm gap DBD at 60 W and 23 kHz. γ-Al_2O_3 is only slightly less active in total conversion (22%) but is even more selective in products formed than α-Al_2O_3. $BaTiO_3$ produces substantially more oxygenated products than the other packing materials but is the least selective in product fractions and has a clear negative impact on CO_2 conversion upon addition of CH_4. Interestingly, when comparing to pure CO_2 splitting and when evaluating differences in products formed, significantly different trends are obtained for the packing materials, indicating a complex impact of the presence of CH_4 and the specific nature of the packing materials on the DRM process.

Keywords: dry reforming of methane; dielectric barrier discharge; packing materials; plasma catalysis

1. Introduction

An increasing energy and resource demand from a growing population and the impact it has on the environment, necessitate enhancing the share of renewable energy and replacing (part of the) fossil fuels, by recycling waste streams. These challenges have given the incentive for new methodologies that allow converting (two) greenhouse gasses (CO_2 and CH_4) into value added chemicals (like syngas, basic chemicals) and fuels [1,2].

Although syngas (CO and H_2) can be obtained in a two-step process, where hydrogen is added to CO—originating from CO_2 splitting into CO and O_2—it is much more efficient to produce it directly through dry reforming of methane (DRM) in a one-step process, with the possibility of directly forming higher hydrocarbons [3,4]. A recent study suggests that, together with the Fischer-Tropsch process, DRM is economically the most promising method for CO_2 conversion [5]. Indeed, when executing thermal DRM, a conversion of 100% can be reached (accompanied by an energy efficiency of 60%) or a maximum energy efficiency of 70% (of the thermal thermodynamic optimum for syngas formation), coinciding with a conversion of 83% [1]. These high values for conversion and energy efficiency are a definite advantage of thermal DRM but require a high temperature (900–1200 K) and a catalyst.

Moreover, this reaction is prone to coking, deactivating the catalyst and therefore decreasing the conversion and energy efficiency [6–10].

To avoid the need for severe reaction conditions and thus to provide a possible energy efficient alternative, plasma technology can be used for DRM. Although multiple different types of plasma reactors are being studied for their performance in DRM, the dielectric barrier discharge (DBD) is the most applied [1]. It operates at or near room temperature (far away from local thermal equilibrium) and the highest conversion and corresponding energy efficiency reported up to now (for a non-packed DBD reactor) are 60% and 12.8%, respectively [1]. A DBD reactor has the advantages of operating under mild reaction conditions, being easy to scale-up (evidenced via industrial ozone generation) and having the possibility to operate with renewable energy (e.g., it is highly intermittent) [3,11]. Moreover, in contrast to thermal dry reforming, plasma-based DRM can directly yield higher hydrocarbons, apart from syngas [1,12]. To steer the selectivity of the products formed and to further increase the conversion [1], a (catalytic) packing can be added to the DBD reactor. When combining plasma with a catalyst, another advantage compared to thermal DRM arises: plasma also reduces the susceptibility of the catalyst to coking [13].

For this reason, an increasing amount of research is being conducted towards CO_2 splitting and DRM in a DBD reactor, both with and without (catalytic) packing materials, as summarised in Tables 1 and 2, respectively. Even though there is a growing number of literature reports, the current literature indicates that still a substantial amount of work lies ahead to unravel all aspects of DRM in a packed bed DBD. Indeed, some important observations can be made when comparing literature, that underline the need for further research. For example, some papers contradict each other with respect to the influence of frequency on CO_2 splitting in a non-packed reactor, reporting either a rise or a drop or no influence of frequency on the results (see details in Table 1) [14–16]. Similar discrepancies have been observed for DRM in packed bed DBD plasma reactors. For instance, adding a packing has been reported to increase the conversion [17–24], while several other report a decrease in conversion of both CO_2 and CH_4 [24–26] and still other papers show only an effect on one of the two reacting gasses [12,23,27,28] (see Table 2). Moreover, also vast differences in selectivity are being described, even for similar packing materials, as detailed in Table 2. Also, it is clear that even for the non-packed reactor, both the process conditions and the reactor design already affect the selectivity tremendously [12,17,29].

Furthermore, very often catalytically activated packing materials are being introduced and discussed, without evaluating the impact of the non-activated packing on the DRM process [17–20,23–25,30,31], even though the latter can be expected to have an influence on the conversion and selectivity as well [12,13,32]. Indeed, in those papers where the packing materials are being studied with and without catalytic activation, an influence of the packing material itself can be observed [21,26,27,29,32]. For instance, Wang et al. reported the formation of liquid products and a significant influence on the selectivity, when comparing different catalytic activations with non-activated packing. Unfortunately, they only compared to one type of packing material [12]. Krawczyk et al. [21] and Sentek et al. [26] indicated only minor alterations in selectivity and conversion when adding a catalytic element on a certain packing, whereas different packing materials with the same active elements yielded major changes, suggesting that the packing itself could be responsible for the selectivity and conversion and not the catalytic element. Other research [23,29] shows a larger influence of the catalytic element on the conversion and selectivity. Unfortunately, these studies are limited to specific packing materials and do not allow to compare the impact of different non-activated packing materials, which would be necessary to elucidate a synergic combination of packing material and catalytic active site. However, a large influence on (and possible control over) the conversion and selectivity is in principle possible, depending on the packing, the catalytic element, the reactor and the operating conditions. The packing material on itself is not necessarily catalytically active (although it can have some catalytic activity or promoting effect as well) but it is typically used as support and/or influences the plasma characteristics in a physical way (e.g., through changes in electric field, discharge, sorption processes). Most used catalytic elements for catalytic activation of the packing are Ni, Co, Fe, Mn, Cu, Ag and Pd [1].

Also for other plasma-assisted processes, such as the abatement of diluted VOC's, numerous studies have shown that physical size and material properties of the packing materials in a DBD reactor play a role to convert chemicals [33–40]. Even more, VOC decomposition is mainly influenced by the adsorption process, rather than by the discharge characteristics [13,41].

Our previous work [32] suggested a large effect of the reactor setup and reactor/bead size combination on the impact of the packing material on conversion of pure CO_2. Thus, comparing results obtained in different reactor setups should be done with care. Therefore, the results obtained in different literature reports cannot be easily compared to one another and no general conclusion towards the impact of the packing material itself on DRM can be drawn. This points towards an important gap in the knowledge required to achieve a maximal synergy between the packing, the active catalytic element and the plasma, ultimately yielding higher conversions and a better selectivity towards the desired components in plasma-based DRM.

The aim of this work is thus to provide better insights in the influence of five different (dense, spherical) packing materials, with different chemistry and size, on the conversion and product fractions of DRM in a DBD reactor. Additionally, γ-alumina is evaluated as porous packing material. Although it is not possible to distinguish between catalytic effects of the packing material itself and physical effects caused by inserting the packing, it is important to note that there is no explicit catalytic activation of the packing materials in the present study, that is, we have not introduced an active element, such as applied in many literature reports (e.g., Cu, Fe, Ni, Co, Pd, Ag and so forth. See Table 2 and [1]) on or in the packing materials. Furthermore, the impact of these packing materials in DRM is being compared to the insights gained for pure CO_2 splitting, providing surprising and valuable information on the influence of adding CH_4. To our knowledge, such a detailed comparison has not yet been carried out before in literature.

Table 1. Summary of a selection of literature for CO$_2$ splitting in a non-packed and a packed DBD reactor.

Study	Reactor	Operating Conditions	Implementing Packing and/or Catalysts				Conclusion			Highest Conversion		Ref.
	Gap (mm)	Power (Watt)	Flow (mL/min)	Reactor Volume (cm³)	Frequency	SEI (kJ/L)	Packing/Catalyst	Shape	Packing Size			
CO$_2$ splitting/ non-packed	1.5	70	150	21.9	5–65 kHz	28				No influence of frequency	10%	[42]
	2	100–200	50–500	13.56	10–90 kHz	12–240				Conversion ≅ when flow rate ↓, T$_{gas}$ ≅, P ≅ Best frequency depends on power	30%	[15]
	2	10–97	50–2000	15.1	16.2–28.6 kHz	0.3–116				Conversion ≅ when flow rate ↓, barrier thickness ↓, frequency ↓ and power ≅	35%	[16]
	4	21.6–35.3	40	30.17	13 kHz	32.4–53				Conversion ≅ when power ≅ and discharge length ≅ Cokes: small on inner electrode	13%	[43]
	4.5, 3.5, 2.5 or 2	60	50	17.67	26.5 kHz	72				Conversion ≅ when flow rate ↓	12%	[32]
CO$_2$ splitting/ packed							Silica gel	Beads	20–40 mesh	silica gel < α-Al$_2$O$_3$ < quartz ≈ γ-Al$_2$O$_3$ < CaTiO$_3$ Cokes: limited on inner electrode	14%	[13]
	4	21.6–35.3	40	30.17	13 kHz	32.4–53	Quartz	Pellets with rigid edges	20–40 mesh	silica gel < α-Al$_2$O$_3$ < quartz ≈ γ-Al$_2$O$_3$ < CaTiO$_3$ Cokes: limited on inner electrode	16%	
							γ-Al$_2$O$_3$	Beads	20–40 mesh	silica gel < α-Al$_2$O$_3$ < quartz ≈ γ-Al$_2$O$_3$ < CaTiO$_3$ Cokes: limited on inner electrode	16%	

Table 1. *Cont.*

Study	Reactor Gap (mm)	Operating Conditions Power (Watt)	Implementing Packing and/or Catalysts Flow (mL/min)	Reactor Volume (cm³)	Frequency	SEI (kJ/L)	Conclusion Packing/Catalyst	Shape	Highest Conversion Packing Size		Ref.
							α-Al$_2$O$_3$	Beads	20–40 mesh	silica gel < α-Al$_2$O$_3$ < quartz ≈ γ-Al$_2$O$_3$ < CaTiO$_3$ Cokes: limited on inner electrode 15%	
							CaTiO$_3$	Beads	20–40 mesh	silica gel < α-Al$_2$O$_3$ < quartz ≈ γ-Al$_2$O$_3$ < CaTiO$_3$ Cokes: limited on inner electrode 20.5%	
	3	20–50	50	10.1	9 kHz	24–60	Glass	Beads	1 mm	Glass < BaTiO$_3$ 22% (16% without packing)	[44]
							BaTiO$_3$	Beads	1 mm	Glass < BaTiO$_3$ 28% (16% without packing)	
	4.5, 3.5, 2.5 or 2	60	50	17.67	26.5 kHz	72	Glass wool	Beads	1.25–2.24 mm	Conversion ≅ when # contact points ≅, void space volumes ≅ and bead/gap size ratio ≅. Impact of the packing material (chemistry and physical), also influenced by setup 10%	[32]
							Quartz wool	Beads	1.25–2.24 mm	Conversion ≅ when # contact points ≅, void space volumes ≅ and bead/gap size ratio ≅. Impact of the packing material (chemistry and physical), also influenced by setup 10%	

Table 1. *Cont.*

Study	Reactor	Operating Conditions	Implementing Packing and/or Catalysts				Conclusion				Highest Conversion	Ref.
	Gap (mm)	Power (Watt)	Flow (mL/min)	Reactor Volume (cm³)	Frequency	SEI (kJ/L)	Packing/Catalyst	Shape	Packing Size			
							SiO_2	Beads	1.25–2.24 mm	Conversion ≅ when # contact points ≅, void space volumes ≅ and bead/gap size ratio ≅. Impact of the packing material (chemistry and physical), also influenced by setup	16%	
							ZrO_2	Beads	1.25–2.24 mm	Conversion ≅ when # contact points ≅, void space volumes ≅ and bead/gap size ratio ≅. Impact of the packing material (chemistry and physical), also influenced by setup	19%	
							α-Al_2O_3	Beads	1.25–2.24 mm	Conversion ≅ when # contact points ≅, void space volumes ≅ and bead/gap size ratio ≅. Impact of the packing material (chemistry and physical), also influenced by setup	17%	
							$BaTiO_3$	Beads	1.25–2.24 mm	Conversion ≅ when # contact points ≅, void space volumes ≅ and bead/gap size ratio ≅. Impact of the packing material (chemistry and physical), also influenced by setup	26%	

117

Table 2. Summary of a selection of literature for DRM in a non-packed and a packed DBD reactor.

Study	Reactor	Operating Conditions							Implementing Packing and/or Catalysts			Selectivity	Conclusion	Highest Conversion	Ref.
	Gap (mm)	Power (Watt)	Flow (mL/min)	Reactor volume (cm³)	Frequency	Ratio CO_2/CH_4	T	SEI (kJ/L)	Packing/ Catalyst	Shape	Packing size	Highest achieved selectivity per component			
DRM/ non-packed	3	30–60	25–100	11.4	30–40 kHz	1	/	18–144				45% CO, 29% H_2, 5% C_2H_2/C_2H_4, 22% C_2H_6, 2% C_3H_6, 12% C_3H_8 (estimation)	Conversion ≅ when flow rate ↓ and P ≅	50.4% CH_4, 30.5% CO_2	[30]
	1	25–75	30–75	4.4	30 kHz	0.66–3	/	20–150				76% CO, 57% H_2	Conversion ≅ when flow rate ↓ P ≅. CH_4 conversion ↓, CO_2 conversion ≈ when ratio ↓	59.7% CH_4, 36.9% CO_2	[17]
	1	80–130	10–40	4.7	20 kHz	0.25–1	/	120–780				73.8% CO, 65.9% H_2, 18.0% C_2, 10.2% C_3, 6.2% C_4	Conversion ≅ when flow rate ↓ P ≅. CH_4 conversion ↓, CO_2 conversion ≈ when ratio ↓	64% CH_4, 34% CO_2	[29]
	3	10	40	2.12	12 kHz	4	/	15				20% CO, 34% H_2, <1% C_2H_2, <1% C_2H_4, 12% C_2H_6, 1% C_3H_6, <1% C_4H_{10}, 11.9% methanol, 11.9% ethanol, 33.7% acetic acid, 1.6% acetone, 0% HCHO	Impact depends on catalyst, both ≅ and ↓ conversion and differs from pure packing	18% CH_4, 15% CO_2	[12]
DRM/ packed	3	30–60	25–100	11.4	30–40 kHz	1	/	18–144	10 wt% Ni@γ-Al_2O_3	Pellets	0.5–1.7 mm	55% CO, 33% H_2, 10% C_2H_2/C_2H_4, 47% C_2H_6, 2% C_3H_6, 25% C_3H_8 (estimation)	Conversion ≅ when pellet size ≅, quartz wool is best, impact packing on selectivity	40.2% CH_4, 30.5% CO_2	[30]
									12% Ni/γ-Al_2O_3	?	?	43% CO, 53% H_2	12% Cu–12% Ni/γ-Al_2O_3 performs best, Ni content influences CO selectivity	30% CH_4, 24% CO_2	[30]

Table 2. Cont.

Study	Reactor	Operating Conditions							Implementing Packing and/or Catalysts			Selectivity	Conclusion	Highest Conversion	Ref.
	Gap (mm)	Power (Watt)	Flow (mL/min)	Reactor volume (cm³)	Frequency	Ratio CO_2/CH_4	T	SEI (kJ/L)	Packing/Catalyst	Shape	Packing size	Highest achieved selectivity per component			
DRM/packed	1	25–75	30–75	4.4	30 kHz	1	450 °C	20–150	12% Cu/γ-Al$_2$O$_3$	?	?	50% CO, 31% H$_2$	12% Cu-12% Ni/γ-Al$_2$O$_3$ performs best, Ni content influences CO selectivity	7% CH$_4$, 5% CO$_2$	[1?]
									1%Cu-12% Ni/γ-Al$_2$O$_3$	?	?	45% CO, 51% H$_2$	12% Cu-12% Ni/γ-Al$_2$O$_3$ performs best, Ni content influences CO selectivity	33% CH$_4$, 25% CO$_2$	
									5%Cu-12% Ni/γ-Al$_2$O$_3$	?	?	47% CO, 54% H$_2$	12% Cu-12% Ni/γ-Al$_2$O$_3$ performs best, Ni content influences CO selectivity	37% CH$_4$, 24% CO$_2$	
									12%Cu-12% Ni/γ-Al$_2$O$_3$	?	?	75% CO, 56% H$_2$	12% Cu-12% Ni/γ-Al$_2$O$_3$ performs best, Ni content influences CO selectivity	69% CH$_4$, 75% CO$_2$	
									16%Cu-12% Ni/γ-Al$_2$O$_3$	?	?	64% CO, 57% H$_2$	12% Cu-12% Ni/γ-Al$_2$O$_3$ performs best, Ni content influences CO selectivity	43% CH$_4$, 47% CO$_2$	
									5% Ni-12%Cu/γ-Al$_2$O$_3$	?	?	75% CO, 56% H$_2$	12% Cu-12% Ni/γ-Al$_2$O$_3$ performs best, Ni content influences CO selectivity	43% CH$_4$, 45% CO$_2$	
									16% Ni-12%Cu/γ-Al$_2$O$_3$	?	?	71% CO, 58% H$_2$	12% Cu-12% Ni/γ-Al$_2$O$_3$ performs best, Ni content influences CO selectivity	57% CH$_4$, 57% CO$_2$	
									20% Ni-12%Cu/γ-Al$_2$O$_3$	?	?	62% CO, 58% H$_2$	12% Cu-12% Ni/γ-Al$_2$O$_3$ performs best, Ni content influences CO selectivity	35% CH$_4$, 32% CO$_2$	

Table 2. *Cont.*

Study	Reactor		Operating Conditions						Implementing Packing and/or Catalysts			Selectivity	Conclusion	Highest Conversion	Ref.
	Gap (mm)	Power (Watt)	Flow (mL/min)	Reactor volume (cm³)	Frequency	Ratio CO_2/CH_4	T	SEI (kJ/L)	Packing/ Catalyst	Shape	Packing size	Highest achieved selectivity per component			
									γ-Al_2O_3	Crushed flakes	10–20 mesh	49.2% CO, 51% H_2, 9.7% C_2, 5.5% C_3, 3% C_4	Packing: CO_2 conversion $\cong$ CH_4 conversion ☑. After activation: conversion $\cong$, selectivity ☑ for H_2 and C_2	57.6% CH_4, 30.9% CO_2	
									2 wt% Ni @ γ-Al_2O_3	Crushed flakes	10–20 mesh	60.6% CO, 52.3% H_2, 9.8% C_2, 5.9% C_3, 3.2% C_4	Packing: CO_2 conversion $\cong$ CH_4 conversion ☑. After activation: conversion $\cong$, selectivity ☑ for H_2 and C_2	55.4% CH_4, 32.7% CO_2	
1	1	130	30	4.7	20 kHz	1	/	260	5 wt% Ni @ γ-Al_2O_3	Crushed flakes	10–20 mesh	60.9% CO, 51.9% H_2, 10.1% C_2, 5.9% C_3, 3.2% C_4	Packing: CO_2 conversion $\cong$ CH_4 conversion ☑. After activation: conversion $\cong$, selectivity ☑ for H_2 and C_2	55.7% CH_4, 33.5% CO_2	[20]
									7 wt% Ni @ γ-Al_2O_3	Crushed flakes	10–20 mesh	63.9% CO, 53.5% H_2, 10.6% C_2, 6.1% C_3, 3.6% C_4	Packing: CO_2 conversion $\cong$ CH_4 conversion ☑. After activation: conversion $\cong$, selectivity ☑ for H_2 and C_2	55.5% CH_4, 32.6% CO_2	
									10 wt% Ni @ γ-Al_2O_3	Crushed flakes	10–20 mesh	61.4% CO, 53% H_2, 10.6% C_2, 6.2% C_3, 3.4% C_4	Packing: CO_2 conversion $\cong$ CH_4 conversion ☑. After activation: conversion $\cong$, selectivity ☑ for H_2 and C_2	55.2% CH_4, 32.7% CO_2	

Table 2. Cont.

Study	Reactor	Operating Conditions							Implementing Packing and/or Catalysts			Selectivity	Conclusion	Highest Conversion	Ref.
	Gap (mm)	Power (Watt)	Flow (mL/min)	Reactor volume (cm³)	Frequency	Ratio CO₂/CH₄	T	SEI (kJ/L)	Packing/Catalyst	Shape	Packing size	Highest achieved selectivity per component			
	3	10	40	2.12	9 kHz	1	/	15	γ-Al$_2$O$_3$	?	?	23% CO, 55% H$_2$, <1% C$_2$H$_2$, <1% C$_2$H$_4$, 20% C$_2$H$_6$, 2% C$_3$H$_6$, <1% C$_4$H$_{10}$, 13% methanol, 9% ethanol, 20% acetic acid, 2% acetone, 0% HCHO	Impact depends on catalyst, both ≅ and ⬇ conversion and differs from pure packing	15% CH$_4$, 12.5% CO$_2$	[12]
									Cu/γ-Al$_2$O$_4$	?	?	14% CO, 35% H$_2$, <1% C$_2$H$_2$, <1% C$_2$H$_4$, 15% C$_2$H$_6$, 2% C$_3$H$_6$, <1% C$_4$H$_{10}$ 11% methanol, 11% ethanol, 42% acetic acid, 2% acetone, 0% HCHO	Impact depends on catalyst, both ≅ and ⬇ conversion and differs from pure packing	16% CH$_4$, 7.5% CO$_2$	
DRM/packed									Au/γ-Al$_2$O$_5$	?	?	20% CO, 42% H$_2$, <1% C$_2$H$_2$, <1% C$_2$H$_4$, 16% C$_2$H$_6$, 2% C$_3$H$_6$, <1% C$_4$H$_{10}$ 10% methanol, 10% ethanol, 30% acetic acid, 2% acetone, 5% HCHO	Impact depends on catalyst, both ≅ and ⬇ conversion and differs from pure packing	16% CH$_4$, 15% CO$_2$	
									Pt/γ-Al$_2$O$_6$	?	?	20% CO, 40% H$_2$, <1% C$_2$H$_2$, <1% C$_2$H$_4$, 17% C$_2$H$_6$, 2% C$_3$H$_6$, <1% C$_4$H$_{10}$ 10% methanol, 25% acetic acid, 9% ethanol, 2% acetone, 11% HCHO	Impact depends on catalyst, both ≅ and ⬇ conversion and differs from pure packing	17.5% CH$_4$, 13% CO$_2$	
	5.9	40	80	?	300 Hz	0.07–1	RT–600 °C	30	Glass	Beads	2 mm	70% CO, 19.5% H$_2$, 42.9% C$_2$, 15% C$_3$, 8.7% C$_4$	CH$_4$ concentration ≅ = C$_2$ ≅. Influence catalyst only > 200 °C. Effect glass = Al$_2$O$_3$	25% CH$_4$, 56.1% CO$_2$	[1]
									γ-Al$_2$O$_3$	Beads	2 mm	70% CO, 19.5% H$_2$, 42.9% C$_2$, 15% C$_3$, 8.7% C$_4$	CH$_4$ concentration ≅ = C$_2$ ≅. Influence catalyst only > 200 °C. Effect for CO$_2$: Al$_2$O$_3$	25% CH$_4$, 56.1% CO$_2$	
									La$_2$O$_3$/γ-Al$_2$O$_3$-Beads		2 mm	70% CO, 19.5% H$_2$, 42.9% C$_2$, 15% C$_3$, 8.7% C$_4$	CH$_4$ concentration ≅ = C$_2$ ≅. Influence catalyst only > 200 °C. for CO$_2$:	25% CH$_4$, 56.1% CO$_2$	

Table 2. *Cont.*

Study	Reactor	Operating Conditions							Implementing Packing and/or Catalysts			Selectivity	Conclusion	Highest Conversion	Ref.
	Gap (mm)	Power (Watt)	Flow (mL/min)	Reactor volume (cm³)	Frequency	Ratio CO_2/CH_4	T	SEI (kJ/L)	Packing/ Catalyst	Shape	Packing size	Highest achieved selectivity per component			
	2	40–240	40	?	5–20 kHz	1	/	60–360	Ni/γ-Al$_2$O$_3$	Nano-particles	100 nm	86% CO, 73% H$_2$	NiFe$_2$O$_4$#SiO$_2$ conversion and selectivity ≅, carbon deposit ✓	64.6% CH$_4$, 58% CO$_2$	[18]
									Ni-Fe/γ-Al$_2$O$_3$	Nano-particles	100 nm	87% CO, 74% H$_2$	NiFe$_2$O$_4$#SiO$_2$ conversion and selectivity ≅, carbon deposit ✓	68.7% CH$_4$, 60.5% CO$_2$	
									Ni-Fe/SiO$_2$	Nano-particles	100 nm	88% CO, 75% H$_2$	NiFe$_2$O$_4$#SiO$_2$ conversion and selectivity ≅, carbon deposit ✓	73.5% CH$_4$, 62.7% CO$_2$	
									NiFe$_2$O$_4$	Nano-particles	100 nm	89% CO, 77% H$_2$	NiFe$_2$O$_4$#SiO$_2$ conversion and selectivity ≅, carbon deposit ✓	77.4% CH$_4$, 67.1% CO$_2$	
									NiFe$_2$O$_4$#SiO$_2$	Nano-particles	100 nm	90% CO, 81% H$_2$	NiFe$_2$O$_4$#SiO$_2$ conversion and selectivity ≅, carbon deposit ✓	80% CH$_4$, 70.3% CO$_2$	
	2	150	40	?	5–100 kHz	1	/	225	Ni/SiO$_2$	?	?	87% CO, 73% H$_2$	Packing: conversion ≅, selectivity ≅	65% CH$_4$, 52% CO$_2$	[19]
									LaNiO$_3$/SiO$_2$	?	?	89% CO, 79% H$_2$	Packing: conversion ≅, selectivity ≅	82% CH$_4$, 69% CO$_2$	
									LaNiO$_3$	?	?	90% CO, 81% H$_2$	Packing: conversion ≅, selectivity ≅	84% CH$_4$, 72% CO$_2$	
									LaNiO$_3$@SiO$_2$	?	?	92% CO, 84% H$_2$	Packing: conversion ≅, selectivity ≅	88% CH$_4$, 78% CO$_2$	

Table 2. *Cont.*

Study	Reactor	Operating Conditions							Implementing Packing and/or Catalysts			Selectivity	Conclusion	Highest Conversion	Ref.
	Gap (mm)	Power (Watt)	Flow (mL/min)	Reactor volume (cm³)	Frequency	Ratio CO_2/CH_4	T	SEI (kJ/L)	Packing/ Catalyst	Shape	Packing size	Highest achieved selectivity per component			
DRM/ packed	4.5	50	50	?	30–40 kHz	1	/	60	Ni/Al_2O_3	Pellets	0.85–5 mm	25% CO, 45% H_2, 10% C_2, 5% C_3	non-packed: filamentary discharge, packed: combination of surface discharges microdischarges breakdown voltage and conversion ☑	18% CH_4, 13% CO_2	[5]
	3.5	1.4–4.8	40	27.2	50 Hz	0.5–2	/	2–7.2	Ni/Al_2O_3	Pellets	1 mm	35% CO, 56% H_2	Conversion ≅ with packing, Conversion ☑ when ratio ☑	52% CH_4, 43% CO_2, (38% CH_4, 23% CO_2 non-packed)	[20]
	3	19	16.7–33.3	?	6 kHz	1	130–340 °C	34–68	Al_2O_3	?	1–2 mm	19% CO, 24% H_2, 0.6% C_2H_2/C_2H_4, 10% C_2H_6, 0.3% C_3H_6, 6% C_3H_8, 1.3% CH_3OH	Conversion ≅ with packing	52% CH_4, 31% CO_2	[1]
									Fe/Al_2O_3	?	1–2 mm	14% CO, 21% H_2, 1.3% C_2H_2/C_2H_4, 9% C_2H_6, 0.3% C_3H_6, 5% C_3H_8, 1% CH_3OH	No effect of T or flow rate, Conversion ≅ with packing	46% CH_4, 20% CO_2	
									zeolite NaY	?	?	10% CO, 21% H_2, 1% C_2H_2/C_2H_4, 6% C_2H_6, 0.2% C_3H_6, 3% C_3H_8, 0% CH_3OH	No effect of T or flow rate, Conversion ≅ with packing	49% CH_4, 19% CO_2	
									zeolite Na ZSM-5	?	?	5% CO, 21% H_2, 0.1% C_2H_2/C_2H_4, 9% C_2H_6, 0% C_3H_6, 5% C_3H_8, 0% CH_3OH	Conversion ≅ with packing	65% CH_4, 40% CO_2	
	4	15–60	5–50	?	1–100 kHz	1	325–525 °C	18–720	$Ni/\gamma-Al_2O_3$	Grains	70–100 mesh	?	Conversion ≅ with packing (fluidized bed)	48% CH_4, 40% CO_2	[2]
	3	19	16.7–33.3	?	5.7–6 kHz	1–2	120–290 °C	34–68	Al_2O_3	?	1–2 mm	38% CO, 28% H_2, 11% C_2, 6% C_3, 4% C_4, 2% CH_3OH	Conversion ☑ with packing	55% CH_4, 31% CO_2	[3]
									Pd/Al_2O_3	?	1–2 mm	40% CO, 29% H_2, 15% C_2, 5% C_3, 3% C_4, 1% CH_3OH	Conversion ☑ with packing	51% CH_4, 28% CO_2	
									Ag/Al_2O_3	?	1–2 mm	38% CO, 29% H_2, 10% C_2, 5% C_3, 4% C_4, 2% CH_3OH	Conversion ☑ with packing	52% CH_4, 30% CO_2	

Table 2. Cont.

Study	Reactor Gap (mm)	Operating Conditions							Implementing Packing and/or Catalysts			Selectivity Highest achieved selectivity per component	Conclusion	Highest Conversion	Ref.
		Power (Watt)	Flow (mL/min)	Reactor volume (cm^3)	Frequency	Ratio CO$_2$/CH$_4$	T	SEI (kJ/L)	Packing/ Catalyst	Shape	Packing size				
	4.5	10–40	50	16.5	30–40 kHz	1	/	12–48	Quartz wool	?	?	28% CO, 22% H$_2$, 1% C$_2$H$_2$/C$_2$H$_4$, 7% C$_2$H$_6$, 0.5% C$_3$H$_6$, 4% C$_3$H$_8$ (estimation)	CH$_4$ conversion: quartz wool> no packing > Al$_2$O$_3$ > zeolite 3A	30% CH$_4$, 12% CO$_2$	[25]
									γ-Al$_2$O$_3$	pellets	500–850 µm	32% CO, 18% H$_2$, 2% C$_2$H$_2$/C$_2$H$_4$, 8% C$_2$H$_6$, 0.5% C$_3$H$_6$, 4% C$_3$H$_8$ (estimation)	CH$_4$ conversion: quartz wool> no packing > Al$_2$O$_3$ > zeolite 3A	23% CH$_4$, 8% CO$_2$	
									zeolite 3A	beads	2 mm	22% CO, 30% H$_2$, 19% C$_2$H$_2$/C$_2$H$_4$, 8% C$_2$H$_6$, 1% C$_3$H$_6$, 6% C$_3$H$_8$ (estimation)	CH$_4$ conversion: quartz wool> no packing > Al$_2$O$_3$ > zeolite 3A	7% CH$_4$, 3% CO$_2$	
									Ni/γ-Al$_2$O$_3$	?	?	37% CO, 33% H$_2$, 22% C$_2$H$_6$	Ni/γ-Al$_2$O$_3$ and Mn-Al$_2$O$_3$: CH$_4$ conversion ≅, yields CO and H$_2$ ≅	19% CH$_4$, 9% CO$_2$	[23]
									Co/γ-Al$_2$O$_4$	?	?	42% CO, 43% H$_2$, 30% C$_2$H$_6$	Ni/γ-Al$_2$O$_3$ and Mn/γ-Al$_2$O$_3$: CH$_4$ conversion ≅, yields CO and H$_2$ ≅	15% CH$_4$, 8% CO$_2$	
	2.5	7.5–15	25–200	11.6	50 Hz	0.11–9	/	2–36	Cu/γ-Al$_2$O$_5$	?	?	43% CO, 44% H$_2$, 30% C$_2$H$_6$	Ni/γ-Al$_2$O$_3$ and Mn/γ-Al$_2$O$_3$: CH$_4$ conversion ≅, yields CO and H$_2$ ≅	14% CH$_4$, 8% CO$_2$	
									Mn/γ-Al$_2$O$_6$	?	?	35% CO, 34% H$_2$, 24% C$_2$H$_6$	Ni/γ-Al$_2$O$_3$ and Mn/γ-Al$_2$O$_3$: CH$_4$ conversion ≅, yields CO and H$_2$ ≅	18% CH$_4$, 10% CO$_2$	
DRM/ packed									BaTiO$_3$	Beads	3 mm	50% CO, 56% H$_2$	BaTiO$_3$ size ✓ = conversions ✓	33% CH$_4$, 20% CO$_2$	[24]
	7.5	46–106	25–100	100	25 kHz	1	110 °C	28–254	Ni/SiO$_2$	Pellets	2–3 mm	56% CO, 54% H$_2$	Packing = conversions ✓	20% CH$_4$, 12% CO$_2$	
									NiFe/SiO$_2$	Pellets	2–3 mm	54% CO, 56% H$_2$	Packing = conversions ≅	28% CH$_4$, 15% CO$_2$	

2. Results

2.1. CO₂ Conversion in DRM and Comparison with CO₂ Splitting

The influence of four different packing materials (SiO_2, ZrO_2, α-Al_2O_3 and $BaTiO_3$) and three different sphere sizes (1.25–1.4; 1.6–1.8 and 2.0–2.24 mm diameter) on the CO_2, CH_4 and total conversion is displayed in Figures 1–3, respectively. Figure 1 shows the CO_2 conversion in DRM, compared to the conversion that we obtained before for pure CO_2 splitting [32], evidencing a clear impact of the presence of CH_4. Figure S1 in the Supplementary Materials shows all data on conversion (CO_{2-}, CH_{4-} and total conversion) combined in one graph, for comparison.

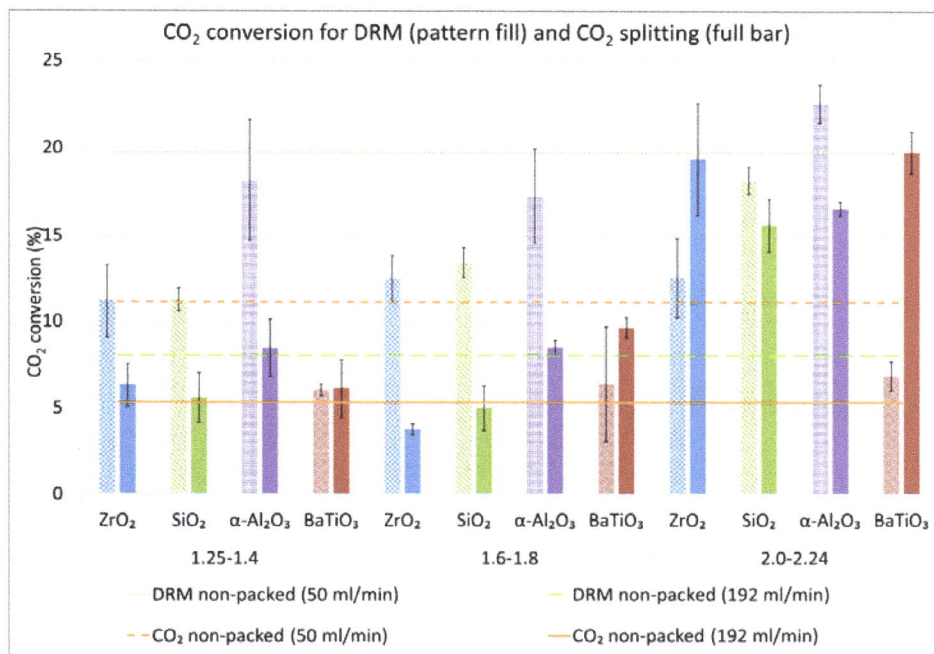

Figure 1. CO_2 conversion for different sphere sizes and materials, compared to the results for the non-packed reactor, at the same flow rate (50 mL/min) and at the same residence time (5.52 s; flow rate of 192 mL/min) for both DRM and pure CO_2 splitting. The bars with pattern fill show the results for DRM, whereas the full bars show the results for CO_2 splitting [32].

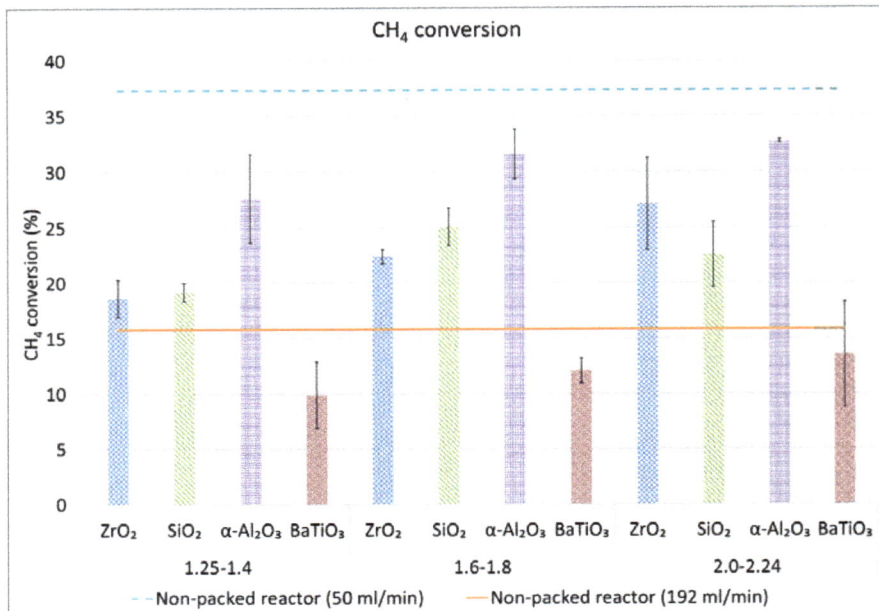

Figure 2. CH_4 conversion for different sphere sizes and materials, compared to the results for the non-packed reactor, at the same flow rate (50 mL/min) and at the same residence time (5.52 s; flow rate of 192 mL/min).

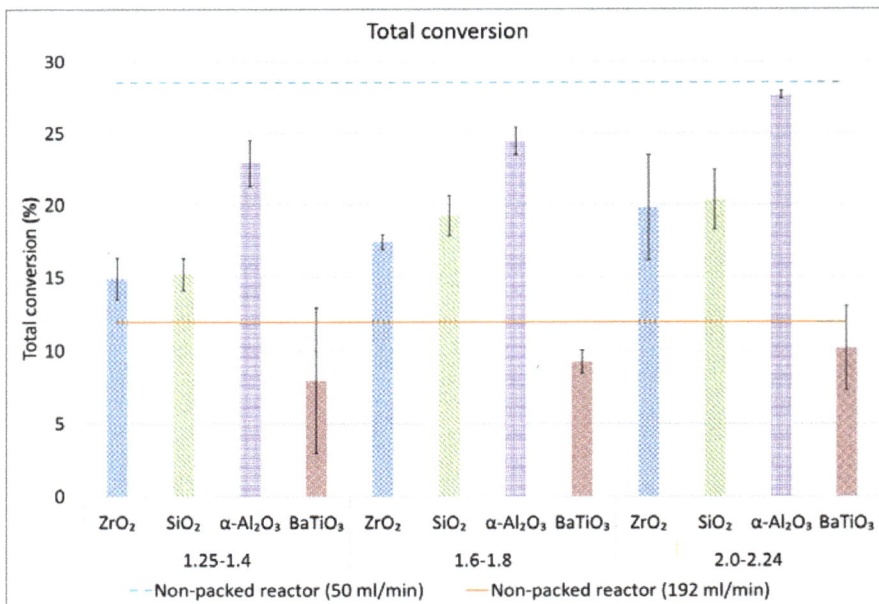

Figure 3. Total conversion for different sphere sizes and materials, compared to the results for the non-packed reactor, at the same flow rate (50 mL/min) and at the same residence time (5.52 s; flow rate of 192 mL/min).

The CO_2 conversion in DRM shows that, when comparing the packed bed reactor to the non-packed reactor, only the largest α-Al_2O_3 spheres achieve a higher CO_2 conversion than the non-packed reactor at the same flow rate. This indicates that only in this case, the positive influence of the packing compensates for the volume loss (and thus lower residence time) caused by introducing the packing. SiO_2, ZrO_2, α-Al_2O_3 (with the smaller bead sizes) and γ-Al_2O_3 do not reach this CO_2 conversion but still surmount the CO_2 conversion for the non-packed reactor at the same residence time. In the case of $BaTiO_3$, a negative effect of the packing is observed, even at the same residence time. Furthermore, a clear impact of the size of the packing materials can be observed, although the effect itself depends on the type of material. The order in which the materials perform is $BaTiO_3 < ZrO_2 < SiO_2 < \alpha$-$Al_2O_3$, although SiO_2 only performs better than ZrO_2 for the largest bead size. When looking at the effect of bead size, only SiO_2 and α-Al_2O_3 show a significantly increased conversion for the largest bead size, in comparison to the other bead sizes. In case of $BaTiO_3$ and ZrO_2, no significant impact of the bead size can be seen.

Interesting differences can be observed when comparing the CO_2 conversion in DRM with pure CO_2 splitting obtained in our previous experiments [32]. It is important to clarify that the total flow rate (and thus the residence time) is kept constant for CO_2 splitting and DRM but with DRM, the concentration of CO_2 is halved, as it has been 'diluted' with 50% CH_4. Diluting with another gas can influence the conversion, even when the diluting gas does not actively participate in the reactions [45]. Indeed, Ramakers et al. have shown that the absolute conversion increases (from 5% to 41%) with a decreasing percentage (from 100 to 5% in argon) of CO_2 [45]. In a 50/50 CO_2/Ar mixture, the rise in conversion of CO_2 is around a factor 1.6, compared to pure CO_2 splitting. Note, however, that the effective CO_2 conversion drops upon dilution with argon, because there is less CO_2 in the mixture. Our experiments clearly reveal that the absolute CO_2 conversion is also higher for DRM than for CO_2 splitting, with the exception of $BaTiO_3$ and 2.0–2.24 mm ZrO_2 packing. Indeed, in the non-packed reactor at 50 mL/min and 192 mL/min (straight lines in Figure 1), the conversion is (on average) a factor 1.8 and 1.5 higher in case of DRM, indicating that CH_4 aids the conversion of CO_2. This is confirmed by computer simulations for DRM in a non-packed DBD reactor, where the CO_2 conversion was largely determined by collision with CH_2 radicals [46], originating from CH_4 dissociation.

For DRM in the packed reactor, the CO_2 conversion is always higher when using SiO_2 and α-Al_2O_3 packing materials than for pure CO_2 splitting. However, the enhancement of the CO_2 conversion due to CH_4 depends on the size of the spherical packing material and is more significant for α-Al_2O_3 than for SiO_2, except for the bead size of 1.6–1.8 mm. For ZrO_2, a complex and striking behaviour depending on the bead size is observed: the conversion drops for DRM for the 2.0–2.4 mm bead size, while it is enhanced (even by a factor 3.3) for the 1.6–1.8 mm beads and to a lesser extent also for the 1.25–1.4 mm beads. Finally, CH_4 has a clearly negative effect in case of the 2.0–2.24 mm beads of $BaTiO_3$, while the conversion is (more or less) equal for CO_2 splitting and DRM for the other $BaTiO_3$ bead sizes. Last but not least, although $BaTiO_3$ in general performs best for CO_2 splitting, compared to the other packing materials, it yields the worst results for DRM.

2.2. CH_4 and Total Conversion

The first observation to be made from Figure 2 is that the CH_4 conversion is always higher than the CO_2 conversion, which is logical, since the dissociation energy of a C-H bond in CH_4 is 412 kJ/mol, while it is 743 kJ/mol for a C=O bond in CO_2 [47].

Comparing again to the non-packed reactor, it can be seen that in contrast to the CO_2 conversion, none of the packing materials allow a better conversion at the same flow rate. However, with the exception of $BaTiO_3$, all materials do perform better than the non-packed reactor at the same residence time. $BaTiO_3$ again performs worse than the non-packed reactor, even at the same residence time. The same trend is seen for the total conversion (Figure 3).

When comparing the results for the different bead sizes and materials, we can make the following observations: Similar to the CO_2 conversion, $BaTiO_3$ performs worst and α-Al_2O_3 performs best, for the four materials tested. Although the bead size had little impact on CO_2 conversion in case of ZrO_2, increasing the ZrO_2 bead size has a positive effect on the CH_4 conversion. On the other hand, the upward trend in conversion of CO_2 with increasing bead size of SiO_2 is much less pronounced for CH_4 conversion, showing even a slight drop for the largest SiO_2 bead size. Finally, also for α-Al_2O_3 the dependence of bead size is somewhat different for CH_4 and CO_2 conversion. In Table 3, we list the CH_4/CO_2 conversion ratios for all packing materials and sizes.

Table 3. Ratio of CH_4 conversion over CO_2 conversion and of the CO over H_2 product fraction, for the different sphere sizes and materials, as well as for the non-packed reactor.

		CH_4 Conversion/CO_2 Conversion	CO/H_2
	ZrO_2	1.7	5.5
	SiO_2	1.7	4.8
1.25–1.4 mm	α-Al_2O_3	1.5	9.5
	$BaTiO_3$	1.6	6.0
	ZrO_2	1.8	5.9
	SiO_2	1.9	4.7
1.6–1.8 mm	α-Al_2O_3	1.8	8.8
	$BaTiO_3$	1.9	6.3
	ZrO_2	2.2	6.4
	SiO_2	1.2	5.3
2.0–2.24 mm	α-Al_2O_3	1.5	9.0
	$BaTiO_3$	2.0	6.9
	γ-Al_2O_3	2.3	8.3
Non-packed reactor	50 mL/min	1.9	7.9
	192 mL/min	2.0	7.2

To interpret the above results, we compare to modelling results obtained by Snoeckx et al. [46], keeping in mind the differences between their work and this work (70 W and 35 kHz in a non-packed reactor, versus 62 W and 23.5 kHz in both non-packed and packed bed reactors, respectively). The conversion of both CO_2 and CH_4 as a function of residence time, as predicted by the model, is plotted in Figure 4. In our work, the residence time is kept constant at 5.52 s, for which the model predicts a CO_2 and CH_4 conversion of 4.6 and 9.2%, respectively. We obtained 8.1% and 15.8% conversion for CO_2 and CH_4, respectively, in the non-packed reactor, while the packed bed reactor (with 2.0–2.24 mm α-Al_2O_3) can reach 22.5% (CO_2) and 32.8% (CH_4) conversion. Note that our obtained values in the non-packed reactor are almost a factor 2 higher than the calculated values but it is not possible to make an exact comparison, due to the different conditions (cf. above) and geometry. Moreover, the exact calculated values are subject to uncertainties, due to uncertainties in the reaction rate coefficients [48,49]. Hence, they should be interpreted merely based on trends. It is clear, however, that the packed bed reactor can improve the conversion of both CO_2 and CH_4 with more than a factor two, at the same residence time.

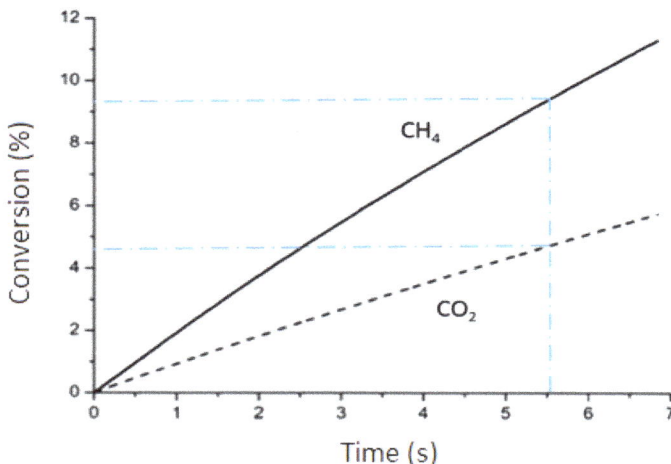

Figure 4. Calculated CH_4 and CO_2 conversion as a function of residence time in a non-packed DBD reactor, adopted from modelling. Adopted with permission from ref. [46]. Copyright 2018 American Chemical Society.

Moreover, the data clearly exhibits that the CH_4 conversion is always higher than the CO_2 conversion, both in the model and in the experiments (both for non-packed and packed reactor). In addition, the model predicts that the CH_4 conversion is typically twice as high as the CO_2 conversion, in good agreement with our results for the non-packed reactor, while the packed bed reactors reveal a ratio of CH_4/CO_2 conversion varying between 1.5 and 2.2, with the exception of the largest SiO_2 beads, where the ratio is only 1.2 (see Table 3), indicating a vast impact of the packing materials on the conversion process. The underlying reasons for these differences in conversion are difficult to link to specific material properties, as the materials diverge in many properties and there is no direct (linear) correlation in the trends in properties that coincide with the trends in conversions (see material characteristics in the Supplementary Materials). Hence, more research will be needed, using materials that are modified, in a controlled way, in specific material properties that are expected to play a key role.

2.3. Comparison Studies α/γ-Al_2O_3

To obtain more insight in the effect of material parameters, we made a comparison between α-Al_2O_3 and γ-Al_2O_3 spheres of 2.0–2.24 mm. The CO_2, CH_4 and total conversion are depicted in Figure 5.

The CO_2 conversion appears a factor 1.7 higher for the α-Al_2O_3 spheres than for the γ-Al_2O_3 spheres (i.e., 22.5% vs. 13.4%), while the CH_4 conversion is only a factor 1.05 higher (i.e., 32.8% vs. 31.2%). The total conversion is a factor 1.24 higher for α-Al_2O_3 (i.e., 27.7% vs. 22.3%). These results show a clear impact of the bead material properties and/or surface area on conversion, possibly due to a higher BET-surface, a difference in crystallinity, acidity, higher porosity and/or total open pore volume of the γ-Al_2O_3, as shown in the Supplementary Materials (Table S1). However, to understand the underlying reasons for this effect, more detailed (operando) surface experiments would be needed, which are outside the scope of this paper. In conclusion, these differences show the importance of indicating as much as possible the material properties of packing materials applied, something that is not systematically done in the majority of the plasma catalysis papers.

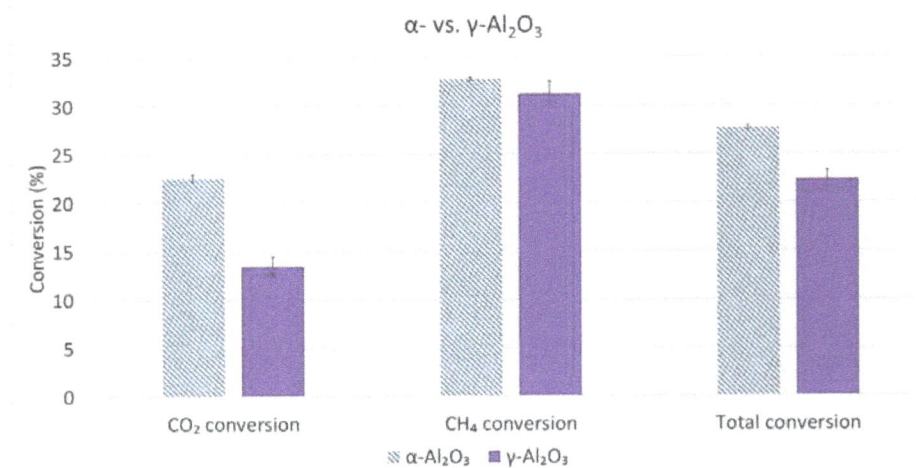

Figure 5. Comparison of the CO_2, CH_4 and total conversion between γ-Al_2O_3 and α-Al_2O_3 (2.0–2.24 mm spheres).

2.4. Carbon, Hydrogen and Oxygen Balances

To determine whether all products have been identified by the GC (Gas Chromatograph), we present the mass balances for carbon, hydrogen and oxygen in Figure 6. Important to note here is that part of the deficit is possibly caused by the gas expansion, as explained above (see materials and methods). As can be seen, the carbon, hydrogen and oxygen balances seldom reach 100%. The largest deficit (between 20% and 30% loss of product) is in the hydrogen balance of the non-packed reactor at 50 mL/min, as well as for the $BaTiO_3$ spheres of 1.6–1.8 mm, the α-Al_2O_3 spheres of 1.2–1.4 mm and the ZrO_2 spheres of 2.0–2.4 mm. In all other cases, less than 20% product remains unaccounted for. Moreover, the oxygen and carbon balances reach much higher values: close to 90% (and even up to 95%) and thus less than 10% loss. It thus suggests that mainly products with more than one hydrogen atom are not taken into account in the converted products. We presume that mostly the formation of H_2O and the sum of less abundant (oxygenated) hydrocarbons, that were not calibrated on the GC, lie at the basis of these incomplete balances. Indeed, the deficit in the hydrogen balance is for the majority of the experiments double of the deficit in the oxygen balance, suggesting the formation of H_2O. An example of a chromatogram, showing the number (and type) of products that have not been calibrated and accounted for in the mass balances, is shown in Supplementary Materials (Figure S15). In addition, also coke deposition could be at the basis of carbon losses. When looking at the Raman measurements (see Supplementary Materials; Figures S16–S23), it is clear that SiO_2 and to a limited extent also α-Al_2O_3 and ZrO_2 suffer from coking at the sphere's surface, unlike the γ-Al_2O_3 and $BaTiO_3$ spheres. To visually show the amount of cokes deposited on the spheres, photos are added in the Supplementary Materials (Figure S24).

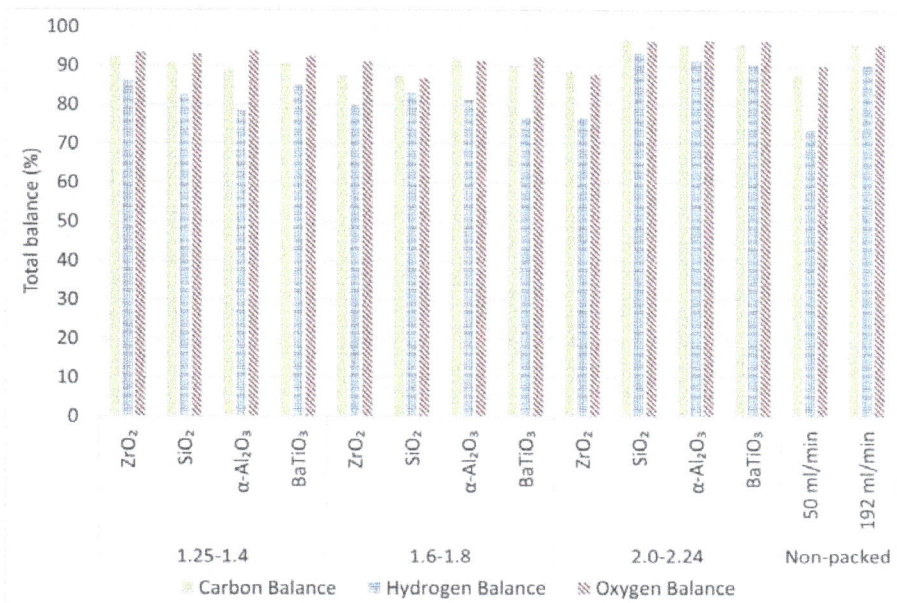

Figure 6. Carbon, Hydrogen and oxygen balance for different sphere sizes and materials, as well as the non-packed reactor.

More detailed carbon, hydrogen and oxygen balances (with the contribution of the different components identified and calibrated by the GC) are shown in the Supplementary Materials (Figures S25–S27 for the carbon balance, Figures S28–S30 for the hydrogen balance and Figures S31–S33 for the oxygen balance). They allow a clear view on all identified products in the treated gas stream, as well as their relative contribution to the total converted products. From these balances, clear differences in product fractions also become apparent when comparing different packing materials. These are discussed in more detail in the following part.

2.5. Product Fractions

As explained in the materials and methods section, the calculation of selectivities and balances induces an uncertainty, caused by the gas expansion. Therefore, we calculated the product fractions in this work (see Equation (4)), as these values only show the relative contribution of each product in the total identified product mixture, which is not subject to the gas expansion. The product fractions are plotted in Figure 7, to provide a general overview and are also listed in Table 4, to better compare the trends, based on quantitative data.

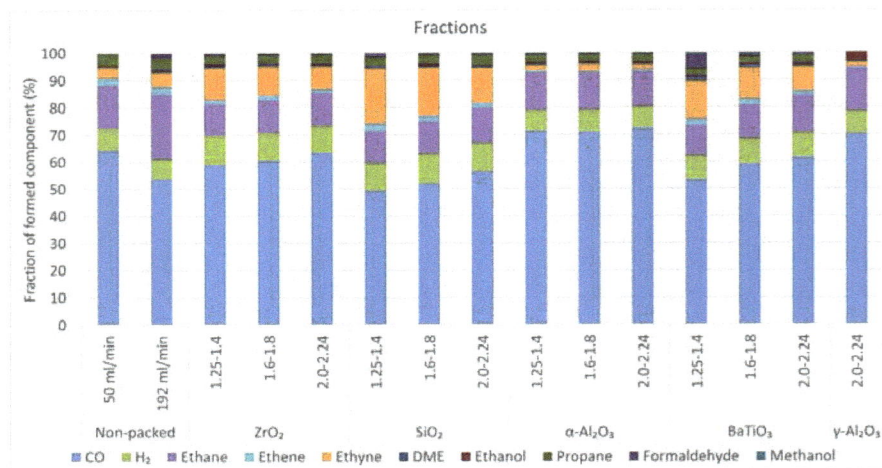

Figure 7. Product fractions for different sphere sizes and materials, as well as for the non-packed reactor.

Before going into more detail on differences for the different packing materials, we can make a general observation for the non-packed reactor. Indeed, it seems that the product fraction is to some extent determined by the flow rate, although the ratio of CH_4 over CO_2 conversion is very similar (see Table 3). Mainly the formation of CO, ethane, ethyne, DME (Dimethylether) and formaldehyde seem to be affected by this. This can be attributed to different formation rates of different products, as explained in the Discussion section, because the different flow rate yields a different residence time.

Table 4 and Figure 7 clearly show that CO is always the largest fraction, for all packing materials and for the non-packed reactors. Moreover, by altering the flow rate (non-packed reactor) or packing materials, the relative amount of CO versus higher hydrocarbons or oxygenates can be altered. Indeed, the CO product fraction can vary from about 53% up to 72%. Therefore, we list in Table 3 also the obtained CO/H_2 ratio for the different sphere sizes and materials. This value ranges from 4.7 to above 9, which is quite striking, because the ratio of CH_4 over CO_2 conversion is always between roughly 1 and 2. It indicates that the majority of C (especially of CO_2) is converted into CO, while the H (originating from CH_4) preferentially takes part in the formation of many products, not only for H_2 but also for higher hydrocarbons.

Furthermore, it is clear from Table 4 and Figure 7 that the type of packing material has a vast impact on the product fractions. Moreover, in case of $BaTiO_3$ and SiO_2, also the sphere size seems to have a clear impact, while this is much less visible for ZrO_2 and α-Al_2O_3. For example, when high fractions of ethyne are envisioned, the smallest size of the SiO_2 spheres seems to be the best choice. The α- or γ-Al_2O_3 packing seems to produce the highest CO/H_2 ratios (see also Table 3), while at the same time producing substantially less dehydrogenated hydrocarbons (ethene and ethyne).

When comparing the different types of Al_2O_3 supports (non-porous α- and porous γ-Al_2O_3), we do not only see differences in conversion (cf. Figure 5 and Table 3), causing a large discrepancy in CH_4/CO_2 conversion ratio (i.e., 1.5 vs. 2.3, respectively) but also interesting changes in the product fractions. Indeed, although the CO fraction is similar, a larger fraction of ethane and ethanol is obtained for the γ-Al_2O_3 packing, while the fractions of ethyne and propane are lower and formaldehyde, DME and methanol do not even reach the detection limits.

Table 4. Product fractions for different sphere sizes and materials (the highest fractions for each component are highlighted).

		CO	H₂	C₂H₆ Ethane	C₂H₄ Ethene	C₂H₂ Ethyne	C₃H₈ Propane	C₂H₆O DME	C₂H₅OH Ethanol	CH₂O Formaldehyde	CH₃OH Methanol	CO/H₂ Ratio
Non-packed (50 mL/min)		64.2	8.1	15.9	2.7	4.1	4.3	0.2	0.22	0.14	0.15	7.9
Non-packed (192 mL/min)		53.5	7.4	23.8	2.8	5.6	4.0	0.7	0.27	1.63	0.28	7.2
ZrO₂	1.25–1.4	58.8	10.8	11.6	1.6	11.9	2.9	0.9	0.33	0.87	0.36	5.5
	1.6–1.8	60.4	10.2	12.2	1.5	10.9	3.1	0.7	0.33	0.42	0.31	5.9
	2.0–2.24	63.4	9.8	12.0	1.2	8.6	3.3	0.5	0.34	0.32	0.32	6.4
SiO₂	1.25–1.4	49.3	10.3	11.7	2.7	20.3	2.9	0.9	0.24	1.44	0.26	4.8
	1.6–1.8	51.9	11.0	11.8	2.0	18.0	3.3	0.8	0.28	0.51	0.26	4.7
	2.0–2.24	56.3	10.6	13.0		13.1	3.7	0.6	0.31	0.41	0.33	5.3
α-Al₂O₃	1.25–1.4	71.3	7.5	14.0	0.5	2.2	2.9	0.5	0.28	0.29	0.43	9.5
	1.6–1.8	70.9	8.0	13.6	0.6	2.7	2.7	0.4	0.25	0.36	0.41	8.8
	2.0–2.24	72.2	8.0	12.9	0.5	2.2	2.9	0.4	0.29	0.29	0.46	9
BaTiO₃	1.25–1.4	53.1	8.9	11.1	2.5	13.8	2.4	1.6	0.40	5.28	0.86	6
	1.6–1.8	59.0	9.3	12.4	2.1	11.4	2.5	1.3	0.37	0.96	0.69	6.3
	2.0–2.24	61.4	8.9	13.7	1.8	9.0	2.8	0.8	0.33	0.59	0.57	6.9
γ-Al₂O₃	2.0–2.24	70.1	8.5	15.4	0.6	2.0	0.4	0.0	3.0	0.0	0.0	8.3

Furthermore, the $BaTiO_3$ packing with smallest bead size is the only material able to produce a substantial fraction of formaldehyde and produces overall relatively more oxygenated products, including higher amounts of DME, compared to the other materials.

When looking more closely to the results, four different trends can be observed when taking into account the four largest component fractions (excluding CO, which is always the largest fraction):

- For the non-packed reactor at 50 mL/min and all α-Al_2O_3 spheres, the order is: ethane > H_2 > propane > ethyne.
- For the non-packed reactor at 192 mL/min, the order is: ethane > H_2 > ethyne > propane > ethene.
- For the smallest ZrO_2 and $BaTiO_3$ spheres and all SiO_2 spheres, the order is: ethyne ≅ ethane > H_2 > propane (formaldehyde in case of $BaTiO_3$).
- For the two largest $BaTiO_3$ spheres and the intermediate ZrO_2 spheres, the order is: ethane > ethyne ≅ H_2 > propane > ethene.

Also the oxygenated fractions, which are much smaller, show clear differences depending on the packing material and size, as detailed in Table 4.

3. Discussion

The results of the non-packed reactor show an interesting way of tuning product fractions. By reducing the residence time (higher flow rate), the ratio of CO_2/CH_4 conversion is similar but the fraction of the products can be altered. Indeed, shorter residence times seem to produce less CO and more oxygenates, hinting towards a kinetic effect that will determine the product fractions. Indeed, model calculations predict that the rates of formation of different products are different [50]: some products rise quickly, while others rise more slowly as a function of time or go over a maximum, because they are converted into another product. Hence, depending on the residence time (and thus flow rate), the product fractions can be altered.

Not only the residence time in the plasma/reactor has an influence on the conversion and product fractions but also the residence time of species in contact with the packing material's surfaces. Indeed, according to the Sabatier principle, the residence time and binding energy between the adsorbing molecule and the surface should be long/strong enough for conversion to take place, while the residence time and binding energy between the products and the surface should be short/weak, so that the product can easily desorb. However, in case of plasma-assisted conversion, also many other underlying mechanisms, both physical and chemical, that take place simultaneously, can influence the reactions (both partial chemical equilibrium and kinetics) and thus conversion and product distribution.

Indeed, based on the results, also packing materials clearly influence the plasma chemistry, as can be deduced from the different CO_2/CH_4 conversion ratios and product fractions. The difference in the CO_2/CH_4 conversion ratio can be caused by many factors, such as differences in discharge type, the number and transferred energy of the streamers, the streamer propagation, electric field enhancement, electron temperature difference, surface adsorption effects and so forth. We present the electrical characteristics for the different sphere sizes and materials in the Supplementary Materials but they do not reveal clear trends that can explain the observed differences in conversion ratios. Probably it is a combination of different effects. Similarly, no clear correlation can be made to the material properties (also presented in the Supplementary Materials). Indeed, all these differences influence the CO_2 and CH_4 conversion and thus the resulting products formed, due to differences in gain and loss reactions. In our previous work for pure CO_2 splitting, we could correlate the impact of bead size and material to differences in number of contact points, size of void spaces and to some extent the dielectric constant of the material but it could not explain all data, so other underlying mechanisms must be present as well [32]. Even though we expect differences induced by changes in the discharge mode and discharge properties, due to the differences in for example, dielectric constants of the packing materials, the data extracted from the electrical characterisation (Supplementary Materials: Table S2) display no straightforward correlation to the observed differences in CO_2/CH_4 conversion.

Nevertheless, not all differences in discharge behaviour can be measured. For example, modelling has revealed important differences in streamer propagation and/or streamer versus surface discharge behaviour, positive restrikes and local discharges, for packed bed reactors, depending on the dielectric constant of the packing material [51]. Moreover, the same modelling study showed that the impact of the discharge mode will be different for different chemical species and thus its impact on CO_2 and CH_4 conversion, as well as on the intermediate species and products, might vary, resulting in the observed differences in CO_2/CH_4 conversion and product distribution. This complex interplay induced by the packing is too complex to postulate the underlying mechanisms for the observed differences in the data [51] and requires much more extended research, focused on materials with systematically altered properties, as well as extensive modelling.

Furthermore, some packing materials, such as Al_2O_3, behave superior to the others, both in case of CO_2 and CH_4 conversion, indicating that the observed results are not only related to the dielectric constant and its effects on the electrical properties of the plasma. Indeed, otherwise, $BaTiO_3$ (which has the highest dielectric constant) would provide the best results, which is clearly not the case. Moreover, if the results would only be correlated to the dielectric constant of the material, α- and γ- Al_2O_3, both having the same dielectric constant, would yield the same conversion. This indicates that other effects, like for example, the surface area and/or the surface acidity, may lie at the base of this difference. Nevertheless, the fact that $BaTiO_3$ performs worse than the other materials can also be correlated to some extent to the electrical properties, because Wang et al. predicted by modelling that materials with higher dielectric constant constrain the discharge to the contact points of the packing materials. They suggested that this can limit surface activation due to a lower surface area in contact with the discharge [51]. On the other hand, materials with a higher dielectric constant result in a higher electric field enhancement, which will also be beneficial for CO_2 and CH_4 conversion [52,53]. Hence, these are opposite effects and this could explain why Al_2O_3 is a superior material, having an "intermediate" dielectric constant of 9, while $BaTiO_3$ (with a dielectric constant of ~4000 [54]) is performing worse. It should be noted that $BaTiO_3$ gave the best results in pure CO_2 splitting, indicating that the effect of electric field enhancement was in that case more important than the effect of the surface discharges. The role of surface discharge behaviour on CH_4 conversion (and vice versa) thus seems important, although this is only a hypothesis.

Other literature reports also support this careful hypothesis, suggesting a difference in behaviour of CH_4 and CO_2 conversion. Indeed, Snoeckx et al. predicted by modelling that CO_2 is not only converted during the microdischarge filaments in a DBD reactor but is also able to react further in the afterglow (both in between filaments as well as post-plasma), whereas CH_4 is mainly converted during the filaments and is being formed again (by recombination of reaction products) in the afterglow [46]. Nevertheless, the effect of different packing materials and sizes on the CH_4/CO_2 conversion ratios might be more complicated, as a result of several other mechanisms as well, so it is not possible to explain all differences in detail. Thus, due to the complex and intertwined nature of the chemistry and physical effects at play, extensive modelling would be needed to confirm or reject this first hypothesis as part of the possible underlying mechanisms.

In addition to the above possible mechanisms, also other interesting hypotheses can be made, based on the surprising result of the difference in performance of $BaTiO_3$ in DRM versus pure CO_2 splitting.

Based on the results of pure CO_2 splitting, it is possible that $BaTiO_3$ strongly promotes the equilibrium of CO_2 splitting towards CO and O. In combination with a high CH_4 conversion (CH_4/CO_2 conversion ratio of 2), which results in a high fraction of H atoms, the O atoms might recombine with H atoms into OH. The latter can further react towards oxygenated components (explaining the higher fractions of oxygenates in the presence of $BaTiO_3$), as well as towards H_2O (and possibly HO_2 and H_2O_2). The trapping of O atoms into OH radicals and H_2O, when small amounts of CH_4 are added to CO_2 streams, has been predicted by modelling [55]. In the latter paper, it was described as a positive effect, because it allowed easier separation of the produced gas mixture but the study was only applied for a few % of CH_4 addition to CO_2. Due to the high performance of $BaTiO_3$ towards CO_2 splitting, as demonstrated in our previous work [32], a much higher concentration of OH radicals might be present here, engaging in other (more negative) reactions, lowering the conversion. Indeed, recent modelling studies of CH_4/O_2 mixtures have indicated a preferential formation of H_2O from OH radicals [50]. These H_2O molecules will promote the back reaction of CO into CO_2, as suggested based on CO_2/H_2O models [56]. This can explain the lower CO_2 conversion in DRM for a $BaTiO_3$ packing, compared to pure CO_2 splitting. We cannot measure H_2O with our GC but the deficits in the oxygen and hydrogen balance (see Figure 6) suggest that indeed a large amount of H_2O might be formed. However, more research is needed to verify the above hypotheses. Note that the high amounts of OH radicals can not only cause back reactions of CO into CO_2 but can also explain the higher oxygenate content in case of the $BaTiO_3$ packing, compared to the other materials. It is thus advised, when aiming for a suitable catalyst for plasma-based DRM, to search for a material that benefits the reaction of OH towards CHO or further towards CH_3O_2 instead of towards H_2O. The different reaction pathways mentioned in this reasoning, are shown in the Supplementary Materials (Figures S34—S36).

Nevertheless, the above reasoning is only a first hypothesis, as other materials exhibiting a lower CO_2 conversion in case of DRM versus pure CO_2 splitting (i.e., ZrO_2 with bead size of 2.0–2.24 mm) do not result in a higher fraction of oxygenated products. This might be due to a difference in kinetics between the back reaction of H_2O with CO_2 versus oxygenate formation. However, much more experimental and modelling work is needed to substantiate this hypothesis.

Finally, the CH_4 conversion is always higher than the CO_2 conversion, due to the lower C–H bond dissociation energy compared to C=O bond dissociation energy, for all packing materials and sphere sizes. However, the CH_4/CO_2 conversion ratio varies from 1.2 to 2.3 (see Table 3), so the difference is more pronounced for some materials than for others. This suggests that for those packing materials with a lower CH_4/CO_2 conversion ratio (e.g., 1.25–1.4 mm α-Al_2O_3 and $BaTiO_3$ and 2.0–2.24 mm SiO_2 and α-Al_2O_3; see Table 3), the situation is more complicated, for example, a back reaction or an impact on the kinetics of CH_4 conversion or CO_2 conversion is taking place.

4. Materials and Methods

We applied the same setup as described in our previous work [32]. It comprises two concentric electrodes: a grounded inner electrode made of stainless steel and the live outer electrode (10 cm) consisting of a stainless steel mesh, wrapped around the dielectric barrier. The dielectric barrier forms the reactor tube that encloses the gap with the inner electrode and is made of Al_2O_3. The gap is confined between the inner electrode (8 mm outer diameter) and the dielectric barrier (inner diameter 17 mm, thickness 2.4 mm), resulting in a gap size of 4.5 mm. In this gap, we inserted the spherical dielectric packing material. The packing spans the full discharge volume with a length of 10 cm (the outer electrode length). To prevent the spherical packing from shifting, the beads were secured with glass wool at both ends of the discharge zone. The high voltage was supplied by a generator, a transformer and a power supply (AFS GmbH, Horgau, Germany). The voltage was measured with a high voltage probe (Tektronix P6015A, Beaverton, OR, USA), while the current was measured with a Rogowski coil (Pearson 4100, London, UK) and the condenser (10 nF) measures the charge. The electrical signals were recorded with an oscilloscope (PicoScope 6402 A, Tyler, TX, USA).

Plotting Q versus U results in Q-U Lissajous figures, giving insight in the electrical characteristics. Analysing the Lissajous data and the oscillograms with Matlab yields six different data. The plasma power is calculated by multiplying the measured current and voltage. The burning voltage (U_{bur}) and peak-to-peak voltage (U_{pp}) are calculated from the Lissajous graphs. Furthermore, the root-mean-square current (I_{RMS}), number of micro discharges per period and displaced charge per micro discharge are extracted from the oscillograms. More information about how these data are obtained from the Lissajous plots and oscillograms can be found in ref. [32,57]. The data are summarised in Table S2 of the Supplementary Materials. All packing materials result in a lower burning voltage, which has already been observed before [58], as well as more micro discharges per period and a larger root-mean-square current.

The reaction conditions and packing materials tested in this work are listed in Table 5.

Table 5. Operating conditions and materials used in this work.

Parameter	Specification
Gap (mm)	4.5
Frequency (kHz)	23.5
Power (Watt)	100
Gas flow rate (mL/min)	50 (or 192, for non-packed, to have the same residence time as in the packed reactor)
Type of material	Non-packed reactor versus SiO_2, α-Al_2O_3, γ-Al_2O_3, ZrO_2 and $BaTiO_3$
Diameter spheres (mm) [a]	1.25–1.4; 1.6–1.8; 2.0–2.24
CO_2/CH_4 ratio	1/1
Temperature	Ambient (no external heating)
Pressure	Atmospheric ($\pm$1.2 atm)

[a] The γ-Al_2O_3 spheres were only tested for a diameter of 2.0–2.24 mm.

Five different spherical packing materials were used in this work, that is, SiO_2 (SiLiBeads, Warmensteinach, Germany), Y-stabilised ZrO_2 (SiLiBeads, Warmensteinach, Germany), $BaTiO_3$ (Catal, Sheffield, UK), γ-Al_2O_3 (BASF) and α-Al_2O_3 (in-house formulated by droplet coagulation at VITO (Vlaamse Instelling voor Technologisch Onderzoek—Flemish institute for technological research), with the α-Al_2O_3 being purchased from Almatis, Rotterdam, The Netherlands [32]). The different physical and chemical characteristics of each material are reported in the Supplementary Materials: Table S1 and Section 1 (Figures S2–S14 and Table S4). The stability of the materials is also discussed in the Supplementary Materials. (Section 2, Figures S16–S24), which focuses on coking resistance.

The gas feed flow rates for both CO_2 and CH_4 are regulated with thermal mass flow controllers (Bronkhorst, Ruurlo, The Netherlands) and the outlet gas is analysed with a custom made online gas chromatograph (Trace GC 1310, Interscience, Bretèche, France). The GC is equipped with a TCD (thermal conductivity detector) and an FID (flame ionization detector) with a methanizer. The separation of the gasses is accomplished with four columns: a Molsieve 5A, 2 RT-Q-bonds and a RTX-f column.

The experiments are carried out as follows: the reactor is always packed with fresh packing. A vibration step is applied during packing to ensure dense packing of the reactor and uniform void spaces. Subsequently, the gas is flushed through the reactor for 10 min, followed by a *blanc* (i.e., without plasma) measurement, consisting of four consecutive GC measurements and electrical measurements, confirming the feed concentration $CO_{2,in}$ and $CH_{4,in}$ (a constant CO_2/CH_4 ratio of 1/1 is applied in this study). Then, the plasma is ignited and stabilised for a duration of 40 min, followed by four consecutive GC and electrical measurements. This *plasma* measurement is repeated three times, each time with fresh packing. This way, the uncertainties introduced by packing the reactor are included in the final result. The error bars on the data-points below are thus based on the 12 measurements executed as explained above.

Based on the peak areas of the GC chromatogram obtained from the *plasma* measurements ($CO_{2,out}$ and $CH_{4,out}$) and the *blanc* measurement, the conversions for CO_2 and CH_4 are calculated (Equations (1) and (2)). The total conversion is calculated using the fractions of both gasses in the inlet gas flow (in our case, both 50%; Equation (3)).

$$X_{CO_2} = \frac{CO_{2,in} - CO_{2,out}}{CO_{2,in}} * 100\% \tag{1}$$

$$X_{CH_4} = \frac{CH_{4,in} - CH_{4,out}}{CH_{4,in}} * 100\% \tag{2}$$

$$X_{Total} = \frac{X_{CO_2} + X_{CH_4}}{2} \tag{3}$$

As explained in our previous work [32], the conversion of gasses into a larger number of molecules leads to an expansion of the volume of the gas, causing a pressure increase. As the GC depressurizes the gas to 1 bar upon sampling (sample loop volume of 100 μL), some converted volume could thus be lost upon depressurizing, relative to the blanc experiment executed at a constant pressure of 1 bar. The formation of higher hydrocarbons, on the other hand, would lead to an increase in density. The extent of conversion, the type of products formed (density) and the product distribution, thus determine the extent of pressure increase and thus the possible loss of converted gas upon sampling. For CO_2 splitting, this can be easily accounted for, as demonstrated in refs. [32,45,57,59,60]. However, for DRM, it is nearly impossible to take this into account, because a plethora of products can be formed, which are not a priori known or can even not all be identified in the GC. Yet, it is still important to know that this process can play a role and expansion of the gas can influence (slightly overestimate) the conversions. More details about the extent of its impact on the obtained results can be found in the work of Pinhão et al. [59]. As it does not only affect the conversions but also the way to calculate the product selectivities, we report the data as relative fractions of products to the total of identified products. Indeed, these product fractions are not affected by the gas expansion. The relative product fractions are defined as follows (shown for H_2 as example):

$$F_{H_2} = \frac{H_2}{CO + H_2 + C_2H_6 + C_2H_4 + C_2H_2 + C_2H_6O + C_2H_5OH + C_3H_8 + CH_2O + CH_3OH} \quad (4)$$

For information of the reader, the yields and selectivities (albeit with the uncertainties due to the gas expansion) are also calculated and shown in the Supplementary Materials (Tables S5 and S6).

Next to the conversion, also the carbon, hydrogen and oxygen balances (CB, HB, OB) were calculated, to give insights in the presence of products not yet identified in the analysis (or not possible to identify in our analysis, for example, H_2O) or losses such as in cokes. These are calculated as follows:

$$CB\ (\%) = \frac{CO_{2,\ out} + CH_{4,\ out} + CO + 2*C_2H_6 + 2*C_2H_4 + 2*C_2H_2 + 2*C_2H_5OH + 2*C_2H_6O + 3*C_3H_8 + CH_2O + CH_3OH}{CO_{2,\ in} + CH_{4,\ in}} \quad (5)$$

$$HB\ (\%) = \frac{2*H_2 + 4*CH_{4,\ out} + 6*C_2H_6 + 4*C_2H_4 + 2*C_2H_2 + 6*C_2H_5OH + 6*C_2H_6O + 8*C_3H_8 + 2*CH_2O + 4*CH_3OH}{4*CH_{4,\ in}} \quad (6)$$

$$OB\ (\%) = \frac{2*CO_{2,\ out} + CO + C_2H_5OH + C_2H_6O + CH_2O + CH_3OH}{2*CO_{2,\ in}} \quad (7)$$

In Formulas (5)–(7), the terms in the nominator are subject to the gas expansion explained before, whereas the terms in the denominator are not. Hence, the mass balance percentage might be slightly under- or overestimated, depending on the product mix.

The SEI (specific energy input) is defined as

$$SEI\ \left(\frac{kJ}{L}\right) = \frac{Plasma\ power\ (kW)}{Total\ gas\ flow\ rate\ \left(\frac{L}{min}\right)} * 60\ \left(\frac{s}{min}\right) \quad (8)$$

The total gas flow rate is the sum of the flow rates of CO_2 and CH_4. For all experiments in the packed bed reactor, this value is 50 mL/min, while in the non-packed reactor, we use a flow rate of 50 mL/min or 192 mL/min. Indeed, the experiments with a non-packed reactor at 50 mL/min provide comparison with the packed bed reactor at equal flow rate, while the experiments at 192 mL/min compare at the same residence time. This way, the reduction in the reactor volume caused by the addition of the packing (estimated as 74% volume, independent of the packing size [61]), is accounted for.

The plasma power in the above formula is the power generated in the plasma reactor, calculated based on the measured voltage and current and not the power that is set on the power supply (typically there is a power loss of ~40%, from 100 Watt to 60 Watt). The analysis of the obtained Lissajous data gives a more correct value of the actual power that is supplied to the plasma (see Supplementary Materials: Table S2).

5. Conclusions

The aim of this research was to study the influence of different packing materials on the conversion and product fractions formed in the dry reforming of CH_4 in a packed bed DBD reactor and to compare this to our previous work on CO_2 splitting.

For this purpose, five different packing materials in three different sizes, that were not explicitly activated with catalytically active elements but could be catalytic in nature, were compared. The following conclusions can be drawn:

The highest CO_2, CH_4 and total conversion obtained in the packed bed reactor was 22.5%, 32.8% and 27.7%, respectively, for α-Al_2O_3 spheres with a diameter of 2.0–2.24 mm. In the non-packed reactor at equal flow rate, the CH_4 and total conversion yielded still higher values of 37.3% and 28.5%, respectively, due to the longer residence time. Analysis of the packing materials before and after plasma confirmed that most of the packing materials have a high resistance to coking, although SiO_2 showed clear D and G bands.

It was clearly evidenced that the type and size of packing materials cannot only influence the overall conversion but also the CH_4/CO_2 conversion ratio and the product fractions, even without being activated with catalytic elements. This emphasizes the importance of studying all essential aspects of a catalyst in case of plasma catalysis, including the non-catalytically activated support material.

Depending on the packing material applied, very high CO/H_2 ratios can be obtained, hinting to mechanisms where the H atoms (originating from CH_4) are mainly involved in the formation of hydrocarbons or oxygenated products, rather than into H_2.

By studying two types of Al_2O_3 (α and γ), with the same dielectric constant, we can conclude that apart from differences in electrical characteristics and discharge behaviour, other materials chemistry or structural (e.g., porosity) related features have a vast impact on product formation, leading to a very different product distribution, in case of α-Al_2O_3 versus γ-Al_2O_3. It has to be noted that γ-Al_2O_3 results in the highest product selectivity (higher than α-Al_2O_3), with no detectable fractions of oxygenated products, except for a 10-fold higher ethanol formation (fraction of 3%), in combination with a high CO content (~70%), the latter being similar to α-Al_2O_3.

Another interesting observation was the discrepancy between the high CO_2 conversion of $BaTiO_3$ for CO_2 splitting, in contrast to the low CO_2 conversion in case of DRM. A possible explanation for this was put forward, based on models that hint towards the recombination of O and H atoms into OH and possibly enhanced back reactions. However, further studies, including both extensive modelling and plasma catalysis with materials with systematically altered properties, are required to confirm the complicated interplay of the different mechanisms.

In general, we can conclude that, even without a catalytic activation, the packing material already has a vast effect on the conversions and product fractions. This indicates the importance of studying all materials aspects in case of plasma catalysis, including the non-activated packing materials. Furthermore, it shows that more research is needed, combining extensive modelling with material research, to unravel the mechanisms at play. Finally, it exemplifies the tremendous future opportunities to create catalysts with true synergy in packing material and active element, that can significantly impact both conversion and selective production of chemicals, allowing to steer DRM to different types of products, ranging from oxygenates to higher hydrocarbons in a one-step process, making plasma-catalytic DRM competitive with thermal DRM in the future.

Supplementary Materials: The following are available online at http://www.mdpi.com/2073-4344/9/1/51/s1. Figure S1: UV-DR spectra of SiO_2 before (blue graph) and after (red graph) plasma exposure (milled spheres). Figure S2: UV-DR spectra for ZrO_2 before (blue graph) and after (red graph) plasma exposure (milled spheres). Figure S3: UV-DR spectra for $BaTiO_3$ before (blue graph) and after (red graph) plasma exposure (milled spheres). Figure S4: Nitrogen Sorption for SiO_2. Figure S5: Nitrogen Sorption for ZrO_2. Figure S6: Nitrogen Sorption for α-Al_2O_3. Figure S7: Nitrogen Sorption for γ-Al_2O_3. Figure S8: Nitrogen Sorption for $BaTiO_3$. Figure S9: Hg-porosimetry for SiO_2. Figure S10: Hg-porosimetry for ZrO_2. Figure S11: Hg-porosimetry α-Al_2O_3. Figure S12: Hg-porosimetry γ-Al_2O_3. Figure S13: Hg-porosimetry $BaTiO_3$. Figure S14: Raman spectrum for SiO_2, before and after plasma exposure. Figure S15: Raman spectrum for ZrO_2, before and after plasma exposure. Figure S16: Raman spectrum for α-Al_2O_3, before and after plasma exposure. For both spheres (before and after plasma), 2 spectra are recorded: one with 90% of the light filtered out and one with 99% of the light filtered out. Figure S17: Zoomed-in (at coking regions) Raman spectrum for α-Al_2O_3, before and after plasma exposure. For both spheres (before and after plasma), 2 spectra are recorded: one with 90% of the light filtered out and one with 99% of the light filtered out. Figure S18: Raman spectrum for γ-Al_2O_3, before and after plasma exposure. Figure S19: Zoomed-in (at coking regions) Raman spectrum for γ-Al_2O_3, before and after plasma exposure. Figure S20: Raman spectrum for $BaTiO_3$, before and after plasma exposure. Figure S21: Zoomed-in (at coking regions) Raman spectrum for $BaTiO_3$, before and after plasma exposure. Figure S22: visual image of the spheres before and after plasma treatment. Figure S23: CO_2, CH_4 and total conversion for different sphere sizes and materials, compared to the results for the non-packed reactor, at the same flow rate (50 mL/min) and at the same residence time (5.52 s; flow rate of 192 mL/min). Figure S24: Part of a gas chromatogram obtained in this work, zoomed in on the baseline. Figure S25: Total carbon balance for different sphere sizes and materials. Figure S26: Detailed carbon balance for different sphere sizes and materials, without CO_2 and CH_4 contribution. Figure S27: Normalized carbon balance for different sphere sizes and materials, without CO_2 and CH_4 contribution. Figure S28: Total hydrogen balance for different sphere sizes and materials. Figure S29: Detailed hydrogen balance for different sphere sizes and materials, without CH_4 contribution. Figure S30: Normalized hydrogen balance for different sphere sizes and materials, without CH_4 contribution. Figure S31: Total oxygen balance for different sphere sizes and materials. Figure S32: Detailed oxygen balance for different sphere sizes and materials, without CO_2 contribution. Figure S33: Normalized oxygen balance for different sphere sizes and materials, without CO_2 contribution. Figure S34: Reaction scheme to illustrate the main pathways for the conversions of CH_4 and O_2 and their interactions. Adopted with permission from ref. [17]. Copyright 2018 American Chemical Society. Figure S35: Reaction scheme to illustrate the main pathways for dry reforming of methane. Adopted with permission from ref. [17]. Copyright 2018 American Chemical Society. Figure S36: Reaction scheme to illustrate the main pathways for the conversions of CO_2 and H_2O and their interactions. Adopted with permission from ref. [18]. Copyright 2018 Wiley-VCH. Table S1: Electrical characterisation for all experiments. Table S2: Physical and chemical characteristics of the packing materials. Table S3: Specifics of the equipment for all characterization techniques. Table S4: SEM-EDX measurements for all spheres before and after plasma, measured at 3 points per sphere. Table S5: Identified products, ranked in decreasing order of their yields, for the different packing materials and the non-packed reactor. The components highlighted are present for more than 1%, the others for more than 100 ppm. Table S6: Product selectivities (%) for the different packing materials and sizes and for the non-packed reactor. The highest selectivities for each component are highlighted.

Author Contributions: Data curation, I.M.; Formal analysis, I.M.; Investigation, I.M. and Y.U.; Supervision, A.B. and V.M.; Writing–original draft, I.M.; Writing–review & editing, A.B. and V.M.

Funding: The authors acknowledge financial support from the Institute for the Promotion of Innovation by Science and Technology in Flanders (IWT Flanders) for I. Michielsen (IWT-141093), from an IOF-SBO project from the University of Antwerp, from the Fund for Scientific Research (FWO; grant number: G.0254.14 N) and from the European Fund for Regional Development through the cross-border collaborative Interreg V program Flanders-the Netherlands (project EnOp).

Acknowledgments: We would also like to thank the Judith Pype and Bart Michielsen from VITO (Vlaamse Instelling voor Technologisch Onderzoek—Flemish institute for technological research) for the shaping of the α-Al_2O_3 beads, the characterisation thereof and the Hg porosimetry and XRD measurements and Jeremy Mertens and François Reniers from ULB for the profilometry measurements.

Conflicts of Interest: The authors declare no conflict of interest

References

1. Snoeckx, R.; Bogaerts, A. Plasma technology—A novel solution for CO_2 conversion? *Chem. Soc. Rev.* **2017**, *46*, 5805–5863. [CrossRef] [PubMed]
2. Song, C. Global challenges and strategies for control, conversion and utilization of CO_2 for sustainable development involving energy, catalysis, adsorption and chemical processing. *Catal. Today* **2006**, *115*, 2–32. [CrossRef]
3. Chung, W.C.; Chang, M.B. Review of catalysis and plasma performance on dry reforming of CH4 and possible synergistic effects. *Renew. Sustain. Energy Rev.* **2016**, *62*, 13–31. [CrossRef]
4. Usman, M.; Wan Daud, W.M.A.; Abbas, H.F. Dry reforming of methane: Influence of process parameters—A review. *Renew. Sustain. Energy Rev.* **2015**, *45*, 710–744. [CrossRef]
5. Jarvis, S.M.; Samsatli, S. Technologies and infrastructures underpinning future CO_2 value chains: A comprehensive review and comparative analysis. *Renew. Sustain. Energy Rev.* **2018**, *85*, 46–68. [CrossRef]
6. Chung, W.; Pan, K.; Lee, H.; Chang, M. Dry Reforming of Methane with Dielectric Barrier Discharge and Ferroelectric Packed-Bed Reactors. *Energy Fuels* **2014**, *28*, 7621–7631. [CrossRef]

7. Arkatova, L.A. The deposition of coke during carbon dioxide reforming of methane over intermetallides. *Catal. Today* **2010**, *157*, 170–176. [CrossRef]
8. Pakhare, D.; Spivey, J. A review of dry (CO$_2$) reforming of methane over noble metal catalysts. *Chem. Soc. Rev.* **2014**, *43*, 7813–7837. [CrossRef] [PubMed]
9. Samukawa, S.; Hori, M.; Rauf, S.; Tachibana, K.; Bruggeman, P.; Kroesen, G.; Whitehead, J.C.; Murphy, A.B.; Gutsol, A.F.; Starikovskaia, S.; et al. The 2012 Plasma Roadmap. *J. Phys. D Appl. Phys.* **2012**, *45*, 253001. [CrossRef]
10. Lavoie, J.-M. Review on dry reforming of methane, a potentially more environmentally-friendly approach to the increasing natural gas exploitation. *Front. Chem.* **2014**, *2*, 81. [CrossRef] [PubMed]
11. Kogelschatz, U. Dielectric-barrier discharges: Their history, discharge physics, and industrial applications. *Plasma Chem. Plasma Process.* **2003**, *23*, 1–46. [CrossRef]
12. Wang, L.; Yi, Y.; Wu, C.; Guo, H.; Tu, X. One-Step Reforming of CO$_2$ and CH$_4$ into High-Value Liquid Chemicals and Fuels at Room Temperature by Plasma-Driven Catalysis. *Angew. Chem. Int. Ed.* **2017**, *56*, 13679–13683. [CrossRef] [PubMed]
13. Neyts, E.C.; Bogaerts, A. Understanding plasma catalysis through modelling and simulation—A review. *J. Phys. D Appl. Phys.* **2014**, *47*, 224010. [CrossRef]
14. Aerts, R.; Somers, W.; Bogaerts, A. Carbon Dioxide Splitting in a Dielectric Barrier Discharge Plasma: A Combined Experimental and Computational Study. *ChemSusChem* **2015**, *8*, 702–716. [CrossRef] [PubMed]
15. Paulussen, S.; Verheyde, B.; Tu, X.; De Bie, C.; Martens, T.; Petrovic, D.; Bogaerts, A.; Sels, B. Conversion of carbon dioxide to value-added chemicals in atmospheric pressure dielectric barrier discharges. *Plasma Sources Sci. Technol.* **2010**, *19*, 034015. [CrossRef]
16. Ozkan, A.; Bogaerts, A.; Reniers, F. Routes to increase the conversion and the energy efficiency in the splitting of CO$_2$ by a dielectric barrier discharge. *J. Phys. D Appl. Phys.* **2017**, *50*, 084004. [CrossRef]
17. Zhang, A.-J.; Zhu, A.-M.; Guo, J.; Xu, Y.; Shi, C. Conversion of greenhouse gases into syngas via combined effects of discharge activation and catalysis. *Chem. Eng. J.* **2010**, *156*, 601–606. [CrossRef]
18. Zheng, X.; Tan, S.; Dong, L.; Li, S.; Chen, H. Plasma-assisted catalytic dry reforming of methane: Highly catalytic performance of nickel ferrite nanoparticles embedded in silica. *J. Power Sources* **2015**, *274*, 286–294. [CrossRef]
19. Zheng, X.; Tan, S.; Dong, L.; Li, S.; Chen, H. LaNiO$_3$@SiO$_2$ core–shell nano-particles for the dry reforming of CH$_4$ in the dielectric barrier discharge plasma. *Int. J. Hydrog. Energy* **2014**, *39*, 11360–11367. [CrossRef]
20. Karuppiah, J.; Manoj Kumar Reddy, P.; Linga Reddy, E.; Subrahmanyam, C. Catalytic non-thermal plasma reactor for decomposition of dilute chlorobenzene. *Plasma Process. Polym.* **2013**, *10*, 1074–1080. [CrossRef]
21. Krawczyk, K.; Młotek, M.; Ulejczyk, B.; Schmidt-Szałowski, K. Methane conversion with carbon dioxide in plasma-catalytic system. *Fuel* **2014**, *117*, 608–617. [CrossRef]
22. Wang, Q.; Cheng, Y.; Jin, Y. Dry reforming of methane in an atmospheric pressure plasma fluidized bed with Ni/γ-Al$_2$O$_3$ catalyst. *Catal. Today* **2009**, *148*, 275–282. [CrossRef]
23. Zeng, Y.; Zhu, X.; Mei, D.; Ashford, B.; Tu, X. Plasma-catalytic dry reforming of methane over-Al$_2$O$_3$ supported metal catalysts. *Catal. Today* **2015**, *256*, 80–87. [CrossRef]
24. Zhang, K.; Mukhriza, T.; Liu, X.; Greco, P.P.; Chiremba, E. A study on CO$_2$ and CH$_4$ conversion to synthesis gas and higher hydrocarbons by the combination of catalysts and dielectric-barrier discharges. *Appl. Catal. A Gen.* **2015**, *502*, 138–149. [CrossRef]
25. Tu, X.; Gallon, H.J.; Twigg, M.V.; Gorry, P.A; Whitehead, J.C. Dry reforming of methane over a Ni/Al$_2$O$_3$ catalyst in a coaxial dielectric barrier discharge reactor. *J. Phys. D Appl. Phys.* **2011**, *44*, 274007. [CrossRef]
26. Sentek, J.; Krawczyk, K.; Młotek, M.; Kalczewska, M.; Kroker, T.; Kolb, T.; Schenk, A.; Gericke, K.-H.; Schmidt-Szałowski, K. Plasma-catalytic methane conversion with carbon dioxide in dielectric barrier discharges. *Appl. Catal. B Environ.* **2010**, *94*, 19–26. [CrossRef]
27. Pham, M.H.; Goujard, V.; Tatibouët, J.M.; Batiot-Dupeyrat, C. Activation of methane and carbon dioxide in a dielectric-barrier discharge-plasma reactor to produce hydrocarbons-Influence of La$_2$O$_3$/γ-Al$_2$O$_3$ catalyst. *Catal. Today* **2011**, *171*, 67–71. [CrossRef]
28. Gallon, H.J.; Tu, X.; Whitehead, J.C. Effects of Reactor Packing Materials on H$_2$ Production by CO$_2$ Reforming of CH$_4$ in a Dielectric Barrier Discharge. *Plasma Process. Polym.* **2012**, *9*, 90–97. [CrossRef]
29. Song, H.K.; Choi, J.-W.; Yue, S.H.; Lee, H.; Na, B.-K.; Songu, H.K. Synthesis gas production via dielectric barrier discharge over Ni/γ-Al$_2$O$_3$ catalyst. *Catal. Today* **2004**, *89*, 27–33. [CrossRef]

30. Tu, X.; Whitehead, J.C. Plasma-catalytic dry reforming of methane in an atmospheric dielectric barrier discharge: Understanding the synergistic effect at low temperature. *Appl. Catal. B Environ.* **2012**, *125*, 439–448. [CrossRef]

31. Wang, Q.; Yan, B.; Jin, Y.; Cheng, Y. Dry Reforming of Methane in a Dielectric Barrier Discharge Reactor with Ni/Al$_2$O$_3$ Catalyst: Interaction of Catalyst and Plasma. *Energy Fuels* **2009**, *23*, 4196–4201. [CrossRef]

32. Michielsen, I.; Uytdenhouwen, Y.; Pype, J.; Michielsen, B.; Mertens, J.; Reniers, F.; Meynen, V.; Bogaerts, A. CO$_2$ dissociation in a packed bed DBD reactor: First steps towards a better understanding of plasma catalysis. *Chem. Eng. J.* **2017**, *326*, 477–488. [CrossRef]

33. Vandenbroucke, A.M.; Morent, R.; De Geyter, N.; Leys, C. Non-thermal plasmas for non-catalytic and catalytic VOC abatement. *J. Hazard. Mater.* **2011**, *195*, 30–54. [CrossRef] [PubMed]

34. Futamura, S.; Zhang, A.; Einaga, H.; Kabashima, H. Involvement of catalyst materials in nonthermal plasma chemical processing of hazardous air pollutants. *Catal. Today* **2002**, *72*, 259–265. [CrossRef]

35. Kim, H.-H. Nonthermal Plasma Processing for Air-Pollution Control: A Historical Review, Current Issues, and Future Prospects. *Plasma Process. Polym.* **2004**, *1*, 91–110. [CrossRef]

36. Subrahmanyam, C.; Magureanu, M.; Renken, A.; Kiwi-Minsker, L. Catalytic abatement of volatile organic compounds assisted by non-thermal plasma. *Appl. Catal. B Environ.* **2006**, *65*, 150–156. [CrossRef]

37. Kim, H.H.; Ogata, A. Interaction of Nonthermal Plasma with Catalyst for the Air Pollution Control. *Int. J. Plasma Environ. Sci. Technol.* **2012**, *6*, 43–48.

38. Francke, K.-P.; Miessner, H.; Rudolph, R. Plasmacatalytic processes for environmental problems. *Catal. Today* **2000**, *59*, 411–416. [CrossRef]

39. Pasquiers, S. Removal of pollutants by plasma catalytic processes. *Eur. Phys. J. Appl. Phys* **2004**, *28*, 319–324. [CrossRef]

40. Guaitella, O.; Thevenet, F.; Puzenat, E.; Guillard, C.; Rousseau, A. C$_2$H$_2$ oxidation by plasma/TiO$_2$ combination: Influence of the porosity, and photocatalytic mechanisms under plasma exposure. *Appl. Catal. B Environ.* **2008**, *80*, 296–305. [CrossRef]

41. Van Durme, J.; Dewulf, J.; Leys, C.; Van Langenhove, H. Combining non-thermal plasma with heterogeneous catalysis in waste gas treatment: A review. *Appl. Catal. B Environ.* **2008**, *78*, 324–333. [CrossRef]

42. Aerts, R.; Somers, W.; Bogaerts, A. A detailed description of the CO$_2$ splitting by dielectric barrier discharges. *ChemSusChem* **2015**, *8*, 702–716. [CrossRef] [PubMed]

43. Yu, Q.; Kong, M.; Liu, T.; Fei, J.; Zheng, X. Characteristics of the Decomposition of CO$_2$ in a Dielectric Packed-Bed Plasma Reactor. *Plasma Chem. Plasma Process.* **2012**, *32*, 153–163. [CrossRef]

44. Mei, D.; Zhu, X.; He, Y.-L.Y.Y.; Yan, J.D.; Tu, X. Plasma-assisted conversion of CO$_2$ in a dielectric barrier discharge reactor: Understanding the effect of packing materials. *Plasma Sources Sci. Technol.* **2015**, *24*, 15011. [CrossRef]

45. Ramakers, M.; Michielsen, I.; Aerts, R.; Meynen, V.; Bogaerts, A. Effect of argon or helium on the CO$_2$ conversion in a dielectric barrier discharge. *Plasma Process. Polym.* **2015**, *12*, 755–763. [CrossRef]

46. Snoeckx, R.; Aerts, R.; Tu, X.; Bogaerts, A. Plasma-Based Dry Reforming: A Computational Study Ranging From Nanoseconds to Seconds Timescale. *J. Phys. Chem.* **2013**, *117*, 4957–4970. [CrossRef]

47. Atkins, P.; Jones, L. *Chemical Principles: The Quest for Insight*, 4th ed.; Craig Bleyer: New York, NY, USA, 2008.

48. Wang, W.; Berthelot, A.; Zhang, Q.; Bogaerts, A. Modelling of plasma-based dry reforming: How do uncertainties in the input data affect the calculation results? *J. Phys. D Appl. Phys.* **2018**, *51*, 204003. [CrossRef]

49. Berthelot, A.; Bogaerts, A. Modeling of CO$_2$ plasma: Effect of uncertainties in the plasma chemistry. *Plasma Sources Sci. Technol.* **2017**, *26*, 115002. [CrossRef]

50. De Bie, C.; Van Dijk, J.; Bogaerts, A. The Dominant Pathways for the Conversion of Methane into Oxygenates and Syngas in an Atmospheric Pressure Dielectric Barrier Discharge. *J. Phys. Chem. C* **2015**, *119*, 22331–22350. [CrossRef]

51. Wang, W.; Kim, H.H.; Van Laer, K.; Bogaerts, A. Streamer propagation in a packed bed plasma reactor for plasma catalysis applications. *Chem. Eng. J.* **2018**, *334*, 2467–2479. [CrossRef]

52. Van Laer, K.; Bogaerts, A. Influence of Gap Size and Dielectric Constant of the Packing Material on the Plasma Behaviour in a Packed Bed DBD Reactor: A Fluid Modelling Study. *Plasma Process. Polym.* **2017**, *14*, e1600129. [CrossRef]

53. Van Laer, K.; Bogaerts, A. How bead size and dielectric constant affect the plasma behaviour in a packed bed plasma reactor: A modelling study. *Plasma Sources Sci. Technol.* **2017**, *26*, 085007. [CrossRef]

54. Butterworth, T.D. The Effects of Particle Size on CO_2 reduction in Packed Bed Dielectric Barrier Discharge Plasma Reactors. Ph.D. Thesis, University of Sheffield, Sheffield, UK, 2015.

55. Aerts, R.; Snoeckx, R.; Bogaerts, A. In-Situ Chemical Trapping of Oxygen in the Splitting of Carbon Dioxide by Plasma. *Plasma Process. Polym.* **2014**, *11*, 985–992. [CrossRef]

56. Snoeckx, R.; Ozkan, A.; Reniers, F.; Bogaerts, A. The Quest for Value-Added Products from Carbon Dioxide and Water in a Dielectric Barrier Discharge: A Chemical Kinetics Study. *ChemSusChem* **2017**, *10*, 409–424. [CrossRef] [PubMed]

57. Uytdenhouwen, Y.; Van Alphen, S.; Michielsen, I.; Meynen, V.; Cool, P.; Bogaerts, A. A packed-bed DBD micro plasma reactor for CO_2 dissociation: Does size matter? *Chem. Eng. J.* **2018**, *348*, 557–568. [CrossRef]

58. Whitehead, J.C. Plasma–catalysis: The known knowns, the known unknowns and the unknown unknowns. *J. Phys. D Appl. Phys.* **2016**, *49*, 243001. [CrossRef]

59. Pinhão, N.; Moura, A.; Branco, J.B.; Neves, J. Influence of gas expansion on process parameters in non-thermal plasma plug-flow reactors: A study applied to dry reforming of methane. *Int. J. Hydrog. Energy* **2016**, *41*, 9245–9255. [CrossRef]

60. Snoeckx, R.; Heijkers, S.; Van Wesenbeeck, K.; Lenaerts, S.; Bogaerts, A. CO_2 conversion in a dielectric barrier discharge plasma: N_2 in the mix as a helping hand or problematic impurity? *Energy Environ. Sci.* **2016**, *9*, 30–39. [CrossRef]

61. Dullien, F.A. *Porous Media-Fluid Transport and Pore Structure*; Academic Press: Cambridge, MA, USA, 1991; ISBN 9780122236518.

catalysts

MDPI

Article

Plasma-Catalytic Mineralization of Toluene Adsorbed on CeO$_2$

Zixian Jia [1,*] **, Xianjie Wang** [1]**, Emeric Foucher** [1]**, Frederic Thevenet** [2] **and Antoine Rousseau** [1,*]

1   LPP, Ecole Polytechnique, UPMC, CNRS, Université Paris-Sud 11, 91128 Palaiseau CEDEX, France;
     xianjie.wang@lpp.polytechnique.fr (X.W.); Emeric.FOUCHER@gmail.com (E.F.)
2   IMT Lille Douai, University Lille, SAGE, 59000 Lille, France; frederic.thevenet@imt-lille-douai.fr
*   Correspondence: zixian.jia@lspm.cnrs.fr (Z.J.); antoine.rousseau@lpp.polytechnique.fr (A.R.);
     Tel.:+33-1-6933-5963 (Z.J.)

Received: 6 July 2018; Accepted: 24 July 2018; Published: 27 July 2018

Abstract: In the context of coupling nonthermal plasmas with catalytic materials, CeO$_2$ is used as adsorbent for toluene and combined with plasma for toluene oxidation. Two configurations are addressed for the regeneration of toluene saturated CeO$_2$: (i) in plasma-catalysis (IPC); and (ii) post plasma-catalysis (PPC). As an advanced oxidation technique, the performances of toluene mineralization by the plasma-catalytic systems are evaluated and compared through the formation of CO$_2$. First, the adsorption of 100 ppm of toluene onto CeO$_2$ is characterized in detail. Total, reversible and irreversible adsorbed fractions are quantified. Specific attention is paid to the influence of relative humidity (RH): (i) on the adsorption of toluene on CeO$_2$; and (ii) on the formation of ozone in IPC and PPC reactors. Then, the mineralization yield and the mineralization efficiency of adsorbed toluene are defined and investigated as a function of the specific input energy (SIE). Under these conditions, IPC and PPC reactors are compared. Interestingly, the highest mineralization yield and efficiency are achieved using the in-situ configuration operated with the lowest SIE, that is, lean conditions of ozone. Based on these results, the specific impact of RH on the IPC treatment of toluene adsorbed on CeO$_2$ is addressed. Taking into account the impact of RH on toluene adsorption and ozone production, it is evidenced that the mineralization of toluene adsorbed on CeO$_2$ is directly controlled by the amount of ozone produced by the discharge and decomposed on the surface of the coupling material. Results highlight the key role of ozone in the mineralization process and the possible detrimental effect of moisture.

Keywords: toluene; CeO$_2$; mineralization; in plasma-catalysis; post plasma-catalysis; relative humidity

1. Introduction

Volatile organic compounds (VOCs), from both natural sources and human activities such as transport, organic solvents and solvent-containing products, production processes and combustion processes [1], have environmental and health impacts [2,3]. Toluene is widespread in the environment owing to its use in a wide variety of household and commercial products [4]. In indoor environments, toluene levels are higher than outdoor; this confinement effect is clearly enhanced by specific sources such as tobacco smoke [5]. Therefore, the abatement of VOCs has motivated research toward an efficient and economical approach. Nonthermal plasma (NTP) technology, as an alternative to conventional VOC abatement techniques, received increasing interest during recent decades [6–15]. However, the application of NTP for VOC abatement has three main drawback: first, the incomplete oxidation of primary pollutants with unwanted side-product emissions; second, the low mineralization rate of organic pollutants; and third, the low energy efficiency [16]. In order to overcome these weaknesses, an alternative relies on the combination of plasma with catalysts. Plasma-catalytic systems can be divided into two categories depending on the location of the coupling material with respect to the dielectric barrier discharge (DBD) reactor. If the catalyst is directly placed inside the discharge zone, it is referred to as in plasma catalysis (IPC) [17,18]. If the catalyst is placed downstream the DBD reactor, it is referred to as post plasma catalysis (PPC) [18,19].

In the PPC configuration, the main role of the plasma is to generate reactive chemical species, mainly ozone, to convert the pollutants residing on the surface of the coupling catalyst. The role of the catalyst is to enhance the process selectivity and efficiency as well as to remove undesired by-products released by the plasma, such as NOx or O_3. In the IPC configuration, coupling materials are directly inserted in the discharge zone. In 2003, Ogata et al. [20] confirmed: (i) the positive impact of porous and high specific surface materials; and (ii) the possible activation of catalytic surfaces under plasma exposure. The activation of various catalytic surfaces, among them metal loaded catalysts, was also confirmed by Hammer et al. [21], Kirkpatrick et al. [22] and Ayrault et al. [23]. Various papers pointing out the synergetic effects between plasma and catalysts in the IPC configuration were published and tried to raise and validate hypotheses to explain this positive interaction. In their review on the removal of VOCs by single-stage and two-stage plasma catalytic systems, Chen et al. [18] rigorously attempted to distinguish the influence of the catalyst on the plasma processes from the plasma influence on the catalytic processes in the IPC configuration.

Besides the catalyst position, air relative humidity plays a key role into the VOC oxidation process. Even in the absence of a catalyst, humidity strongly affects the plasma characteristics and among them the ozone production. Some studies indicated [24–26] that a moderate relative humidity present in the gas mixture has a favorable effect on the toluene decomposition in the plasma without catalysts. However, in plasma-catalytic system, a negative effect both in in-situ and post-situ plasma configurations was reported for different catalysts [26–28]. Indeed, the adsorption of water molecules on the catalyst surface could hinder the sorptive and reactive sites [25] leading to a reduced catalyst activity. It has also been reported that the various behaviors of the VOCs on TiO_2 surface are directly related to their adsorption modes and parameters and to the VOC–water interactions in adsorbed phase [29]. More generally, it is required to investigate the performances of any air treatment process under humid conditions: first to address realistic conditions, second to be able to assess the effective role of moisture on the process performance which can be twofold. Although water can be dissociated to produce HO• radicals beneficial for VOC oxidation, high humidity levels are detrimental to O_3 by decreasing O concentration on the catalyst surface [30].

In a previous study [31], adsorption and oxidation of toluene on CeO_2 was reported under plasma exposure using two different configurations and only surface monitoring diagnostics: IPC was studied using Sorbent-TRACK device [32]; PPC was studied using Diffuse Reflectance Infrared Fourier Transform Spectroscopy—DRIFTS [33]. The formations and the temporal evolutions of organic reaction intermediates onto CeO_2 surface were discussed. Unlike the former study was centered on surface processes, this paper focuses on the performances of CeO_2 coupled to NTP in the oxidation of toluene through gas phase characterization. As a VOC oxidation process, the main performance parameter of plasma-catalysis is the formation of CO_2, that is, the mineralization. In that regard, specific attention has been paid to the monitoring of CO_2 formation. The key process parameters such as specific input energy of the plasma, level of relative humidity in the air flow and configurations of the reactor (IPC and PPC) are evaluated in order to determine how they influence the mineralization performances of the plasma-catalytic system. Obtained results aim (i) at determining the optimal conditions for toluene mineralization and (ii) at understanding the limiting steps and the performances of such an air treatment process.

2. Results and Discussion

2.1. Characterization of the Adsorption of 100 ppm of Toluene on CeO$_2$

The adsorption capacity of CeO$_2$ regarding 100 ppm of toluene is determined using the experimental method reported in previous study [34] and described in the first three steps of the experimental protocol reported in Section 3.5. Figure 1 displays the quantification of the (i) total, (ii) reversible and (iii) irreversible fractions of toluene adsorbed on CeO$_2$ surface and expressed in µmol/m^2 for different RH levels (0–80%). Every experiment is repeated at least three times and the corresponding standard errors are calculated by dividing standard deviations by the square root of the number of experiments. Under 0% RH, a total of 3.4 µmol/m^2 of toluene are adsorbed on CeO$_2$ right after toluene breakthrough. The flushing step under dry air leads to the desorption of 2.0 µmol/m^2 of toluene which accounts for the reversibly adsorbed fraction. The amount of irreversibly adsorbed toluene on CeO$_2$ is given by the subtraction of these two values, that is, 1.4 ± 0.2 µmol/m^2. F. Batault. [35] reported the irreversible fraction of toluene as 0.77 ± 0.29 µmol/m^2 onto P25-Degussa TiO$_2$ for the same concentration range of toluene. Bouzaza et al. [36] reported that the total adsorbed amount of toluene on TiO$_2$ is 3.9 µmol/m^2. Similarly, Takeuchi et al. [37] reported values ranging from 2 to 3 µmol/m^2 for toluene adsorption on zeolites. In spite of the contrasted chemical natures of these materials, values obtained for the adsorption of ca. 100 ppm of toluene range within the same order of magnitude onto metal oxides.

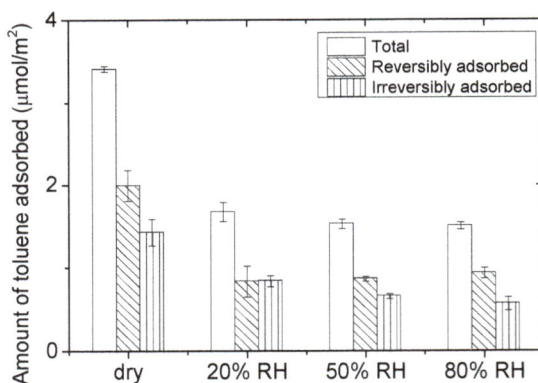

Figure 1. Total, reversible and irreversible adsorbed amounts of toluene on CeO$_2$ for relative humidity ranging from dry to 80% RH (flow: 0.5 L/min, 70 mg CeO$_2$, 100 ppm of toluene, $P = 101.3$ kPa, and $T = 298$ K).

In the presence of moisture, the total amount of toluene adsorbed acutely drops from 3.4 µmol/m^2 under dry condition to 1.7 µmol/m^2 under 20% RH. Between 20, 50 and 80% RH, the total amount of toluene adsorbed remains unchanged while the amount of irreversibly adsorbed toluene slightly decreases with increasing RH. Goss et al. [38] have shown that mineral surfaces, that is, metal oxide surfaces, exhibit monolayer coverage by water molecules as the relative humidity reaches 20–30%. Below 20%, as the surface coverage of the material by water molecule is lower than 1, the VOCs can either directly adsorb on the metal oxide surface or interact with the incomplete water layer. Nevertheless, even with a RH lower than 20%, the adsorption of water molecules is favored because of (i) the high affinity of water for metal oxide polar surfaces [38] and (ii) the high partial pressure of water, ca. 6300 ppm for 20% RH under room temperature, compared to that of the considered VOC (100 ppm). This explains the significant decrease in the tota amount of toluene adsorbed from dry conditions to 20% RH. As RH increases over 20–30%, VOCs necessarily interact with water layers present on the surface of the metal oxide. Adsorption at the interface between the adsorbent and the water layer is thermodynamically unfavorable for non-polar compounds such as toluene [39,40].

2.2. Comparison of PPC and IPC Performances under Dry Air Conditions

2.2.1. Performance Criteria

The performances of the PPC and IPC configurations are compared regarding their respective abilities to regenerate CeO_2 surface saturated with toluene. The main criterion relies in the mineralization yield, that is, the formation yield of CO_2 proceeding from toluene oxidation. The mineralization yield, denoted by ρ (t) in the following, represents the evolution of the quantity of CO_2 produced per Joule of injected energy (nmol CO_2/J) as a function of time t Equation (1).

$$\rho(t) = \frac{q_{CO_2}(t)}{P_{inj}} \tag{1}$$

In Equation (1), $q_{CO_2}(t)$ represents the quantity of CO_2 produced as a function of time t and expressed in nmol. P_{inj} represents the discharge injected power (J/s). The integration of ρ (t) on the time interval $[0, t]$ enables the calculation of the mineralization efficiency η as reported in Equation (2).

$$\eta = \frac{\int_0^t \rho}{t} \tag{2}$$

2.2.2. Formation of Ozone Using PPC and IPC Configurations

First, the ability of each reactor configuration to generate ozone has been addressed. Figure 2 reports ozone concentration as a function of the specific input energy for both PPC and IPC reactors, under dry and humid conditions, with and without ceria. On the one hand, no difference in ozone production by the PPC and IPC configuration reactor is noticeable in the absence of ceria. Both reactors provide the same increasing ozone generation with respect to specific input energy (SIE) up to 55 J/L. This behavior evidences the ability of both experimental devices to generate the same flow of ozone for the same specific input energy. On the other hand, when ceria is coupled to IPC and PPC reactors, no ozone is detected at the reactor outlets regardless of IPC or PPC configuration. Under both configurations, CeO_2 provides a complete conversion of the ozone molecules produced whether the material is placed inside or downstream the discharge. Moreover, CeO_2 has higher oxygen storage/transport capacity combined with the ability to shift easily between reduced and oxidized states (i.e., Ce^{3+}–Ce^{4+}) which results in an oxygen vacancies. The ozone decomposition to reactive oxygen species on CeO_2 has been proposed by Mao et al. [41] in the following reaction:

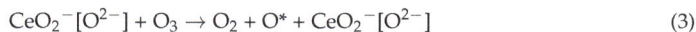

$$CeO_2{}^-[O^{2-}] + O_3 \rightarrow O_2 + O^* + CeO_2{}^-[O^{2-}] \tag{3}$$

The mechanism of toluene decomposition could be initiated by the reaction of toluene with O* and form CO_2 and H_2O.

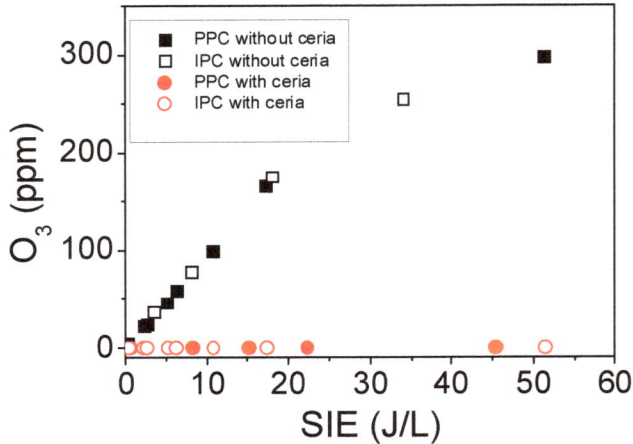

Figure 2. Evolution of ozone concentration as a function of the specific input energy (SIE) in post plasma-catalysis (PPC) and in plasma-catalysis (IPC) configurations, with and without ceria, under dry condition (dry air flow: 0.5 L/min, $P = 101.3$ kPa, and $T = 298$ K).

2.2.3. Influence of the SIE on CO_2 Formation Yield: ρ (t)

Once the adsorption equilibrium of toluene is reached on CeO_2 and the reversible fraction has been removed by flushing, plasma is ignited either in the IPC or PPC configuration. The ignition time of the discharge corresponds to $t = 0$ on Figure 3a–d. The temporal evolutions of CO_2 concentration (ppm) monitored at the reactor outputs for the PPC and IPC configurations respectively are reported in Figure 3a,c. It has to be noted that CO_2 is considered the single mineral by-product of toluene mineralization, since CO is formed in an extent that remains lower than 1% of CO_2 concentration in all experiments.

At low input energy, that is, 1.9 J/L for IPC (Figure 3a) and 2.1 J/L for PPC (Figure 3c), CO_2 production is constant along the whole plasma phase. As the input energy increases, a peak of CO_2 is first observed, but CO_2 production gradually decreases with of treatment time. The same phenomenon is noticeable whether CeO_2 is located inside (IPC) or downstream the discharge (PPC). The observed decrease in CO_2 production with the treatment time could be contributed by (i) the depletion of toluene or (ii) the poisoning of the CeO_2 surface. The by-products (benzyl alcohol, phenol, benzoate-like species) resulting for incomplete oxidation of toluene and reported on ceria surface in our previous study [31] using in-situ infrared spectroscopies, may induce a surface poisoning effect. Similar observations have been reported in the literature [42,43] related to toluene oxidation and support the second hypothesis.

Interestingly, it is observed on Figure 3b,d that the efficiency of toluene conversion into CO_2, characterized by ρ (t), is higher by a factor of ca. 3 for a lower SIE for both IPC (Figure 3b) and PPC (Figure 3d) configurations. At low SIE, it can be suggested that the number of toluene molecules involved in the oxidation process per time unit is lower, but the advancement of the oxidation is enhanced. As a result, in spite of the fact that the removal of toluene from the sorbent surface is decreased by a lower SIE, the formation of adsorbed side-products is limited. Thus, on the time scale of the reported experiments the depletion of toluene is not reached and no poisoning of the surface is induced. As a consequence, CO_2 production appears as a constant process and CO_2 yield is optimal. On the contrary it can be suggested that an increase in SIE chiefly initiates the oxidation of a larger number of adsorbed toluene molecules, but it appears to be detrimental to the advancement of the oxidation process and the mineralization.

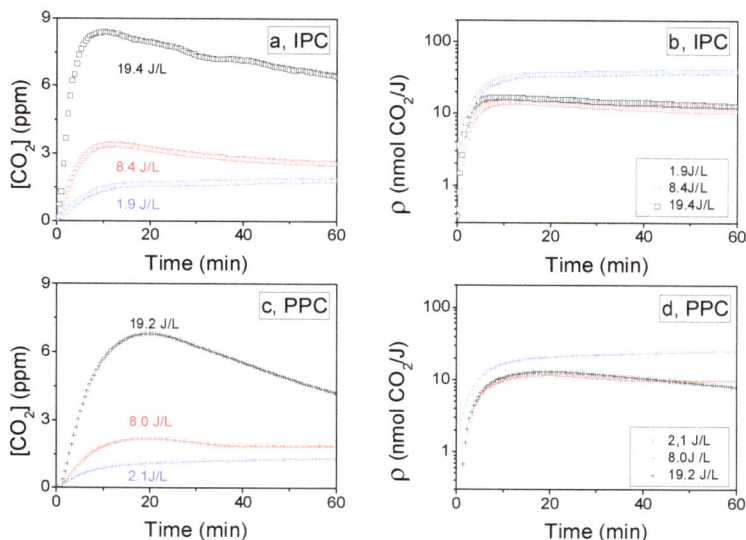

Figure 3. Evolution of CO_2 concentration as a function of time at the output of IPC reactor (**a**) and PPC reactor (**c**) for different specific input energies (SIE); evolution of CO_2 formation yield as a function of time for IPC reactor (**b**) and PPC reactor (**d**) for different specific input energies (SIE) (dry air flow: 0.5 L/min, 60 min plasma treatment, 70 mg ceria, 1.4 ± 0.2 µmol/m^2 toluene initially adsorbed, $P = 101.3$ kPa, and $T = 298$ K).

2.2.4. Influence of the SIE on the Mineralization Efficiency: η

The mineralization efficiency (η) of adsorbed toluene has been calculated over 60 min plasma treatments for both IPC and PPC configurations using different SIE as reported in Figure 4. Two mains observations can be retrieved from Figure 4: (i) the mineralization efficiency of adsorbed toluene into CO_2 is significantly promoted by a decrease in SIE below 2 J/L; under this condition, η is enhanced by 65% in IPC compared to PPC; and (ii) above 8 J/L, no difference is observed between IPC and PPC.

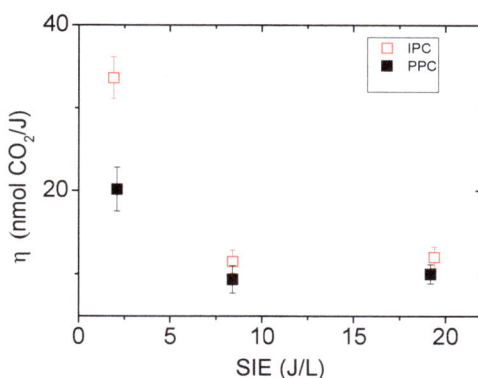

Figure 4. Evolution of the mineralization efficiency η as a function of the specific input energy (SIE) for the IPC and PPC reactors (dry air flow: 0.5 L/min, 60 min plasma treatment, 70 mg ceria, 1.4 ± 0.2 µmol/m^2 toluene initially adsorbed, $P = 101.3$ kPa, and $T = 298$ K).

As observed in Figure 2, all ozone molecules produced in IPC or PPC configuration are decomposed on the surface of CeO_2. Each of them leads to the formation of one O_2 molecule and one active oxygen atoms. Assuming that these oxygen atoms are involved in the mineralization of adsorbed toluene, the Reaction is considered.

$$C_6H_5CH_3 + 18\,O \rightarrow 7\,CO_2 + 4\,H_2O \tag{4}$$

Based on Reaction the theoretical stoichiometric ratio of ozone to toluene is 18:1. Both IPC and PPC reactors have the same capacities to generate ozone. The amounts of ozone produced during 60 min plasma using IPC or PPC are reported in Table 1 for different SIE. They vary from 24 µmol of ozone, for a SIE of 2 J/L, to 204 µmol, for a SIE of 19 J/L. Considering that the minimum amount of toluene adsorbed onto CeO_2 surface present in the IPC or PPC reactor is 7.1 µmol, the ratios of ozone to toluene are calculated for different SIE and reported in Table 1.

Table 1. Quantities of ozone produced during a 60min-plasma for different SIE under IPC or PPC, and corresponding ozone/toluene ratios (dry air: 0.5 L/min, 70 mg ceria, a minimum of 7.1 µmol of toluene initially adsorbed, $P = 101.3$ kPa, and $T = 298$ K).

SIE (J/L)	Ozone Produced During 60 min (µmol)	Ratio of Ozone to Toluene
2	24	3:1
8	120	17:1
19	204	29:1

Based on Table 1, the specific input energy of 8 J/L appears as the minimum SIE to reach the stoichiometric conditions of toluene oxidation within 60 min. Indeed, with a SIE of 2J/L, the ratio of ozone to toluene is only 3:1 which signifies an import lack of ozone. With a specific input energy of 19 J/L the amount of ozone available within 60 min highly exceeds the stoichiometry of Reaction (II). Irrespectively of the IPC or PPC configuration, the SIE directly controls the amount of ozone provided to the catalytic surface. As the amount of ozone exceeds the stoichiometry required for toluene mineralization (SIE > 8 J/L) no difference is noticed between IPC and PPC regarding their mineralization efficiency (Figure 4). The higher mineralization efficiency of the IPC configuration is observed as ozone is provided in sub-stoichiometric conditions. Remarkably, Figures 3 and 4 and Table 1 evidence that the in-situ configuration (IPC) offers the highest mineralization yield and efficiency under lean conditions of ozone. These findings are supported by the works of Harling et al. [44], using Ag/TiO_2 and Ag/Al_2O_3, and Van Durme et al. [45], using TiO_2, who similarly report an increase in the energy efficiency of toluene degradation for lower SIE as the coupling material is used in-situ (IPC). The main hypothesis proposed to explain the higher efficiency of IPC configuration relies on the presence of additional short lived species.

2.3. Influence Relative Humidity on the Performances of IPC

IPC is evidenced as the optimal configuration because it provides the highest mineralization yield for the lowest SIE, thus this section aims at investigating more in details this configuration and especially the most impacting process parameter: the relative humidity (RH).

2.3.1. Influence of Relative Humidity on O_3 Formation

The evolution of ozone concentration as a function of the specific input energy in IPC reactor is reported in Figure 5. The data series of Figure 5 differ through (i) the relative humidity levels used and (ii) the absence or presence of CeO_2 in the reactor. In the absence of ceria, the concentration of ozone produced by the discharge decreases as the relative humidity increases; while in the presence of ceria, no ozone is detected at reactor outlet.

Depending on the relative humidity condition, the major oxidants produced on the surface of the coupling material are different. On the one hand, under dry conditions, and using CeO_2/γ-Al_2O_3 catalyst, Wu et al. [46] reported that atomic oxygen (O*) is massively generated from the surface decomposition of O_3 on the catalyst. On the other hand, under wet conditions and using cerium-based mixed oxides in the ozonation of oxalic acid, Orge et al. [47] concluded that the oxidizing species are predominantly HO radicals formed by the interaction between ozone and Ce (III) centers on the water-covered catalyst surface. Consequently, moving from dry to wet conditions turns the key oxidizing species from atomic oxygen to hydroxyl radicals.

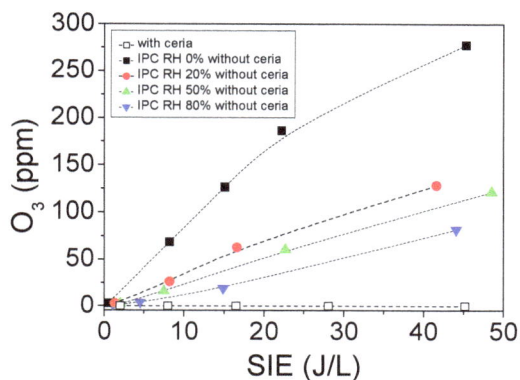

Figure 5. Evolution of ozone produced as a function of the specific input energy (SIE) in the IPC configuration, with and without ceria, and with different relative humidity levels (flow: 0.5 L/min, injected power from 0 to 50 J/L, P = 101.3 kPa, and T = 298 K).

2.3.2. Influence of Relative Humidity on the Mineralization of Adsorbed Toluene

Investigating the influence of relative humidity on the mineralization of adsorbed toluene requires first to recall the impact of RH on toluene adsorption. The adsorption of toluene on CeO_2 has been quantified varying RH from 0 to 80%. Based on Figure 1, the irreversibly adsorbed fractions of toluene are reported in Table 2 for different relative humidity levels. As discussed in Section 2.1, the increase of RH induces a significant lessening of the irreversible fraction of toluene adsorbed on the surface of CeO_2.

Table 2. Quantities of irreversibly adsorbed toluene on CeO_2 expressed in $\mu mol/m^2$ for different relative humidity levels (%) (flow: 0.5 L/min, 70 mg ceria, P = 101.3 kPa, [Tol.] = 100 ppm and T = 298 K).

RH (%)	Amount of Irreversibly Adsorbed Toluene ($\mu mol/m^2$)
0	1.40
20	0.80
50	0.66
80	0.56

Considering that the concentration of ozone produced by the discharge varies with RH (Figure 5) and that the amount of toluene adsorbed on CeO_2 depends on RH as well (Table 2), the parameter ε (t) is proposed to enable accurate comparisons of toluene mineralization from one RH level to another. This parameter is defined by Equation (3) as the ratio of CO_2 concentration (ppm), monitored at the outlet of the IPC reactor, to the amount of toluene adsorbed ($\mu mol/m^2$) and the concentration of ozone (ppm) at the corresponding RH. As a consequence, in the following, ε (t) is referred to as the normalized mineralization.

$$\varepsilon(t) = \frac{[CO_2]}{[\text{adsorbed Toluene}] \times [O_3]} \tag{5}$$

Figure 6a,c report the temporal evolution of CO_2 concentration at the IPC reactor outlet during 60 min plasma treatments of irreversibly adsorbed toluene on CeO_2. Experiments have been performed with RH levels of 0, 20, 50 and 80%. Figure 6a,c differ by their specific input energies, respectively 2 J/L and 19 J/L. Under both SIE conditions, the CO_2 concentration dramatically falls down as RH increase. The most significant drop being observed as RH is increased from 0 to 20%. This behavior could be related to the modification of the major oxidizing species or the decrease in toluene adsorption. In order to propose more accurate discussions, Figure 6b,d report the temporal evolutions of the normalized mineralization ε (t) taking into account the initial amount of toluene adsorbed and the quantity of ozone produced. Interestingly, under both SIE conditions, after a short transient regime, values of ε (t) tend to converge towards similar values, irrespectively of (i) the RH level, (ii) the initial amount of toluene adsorbed and (iii) the concentration of O_3 produced by the discharge. Under 2J/L, ε tends to 0.07 $m^2/\mu mol$ while it tends to 0.04 $m^2/\mu mol$ under 19 J/L. The convergence is noticeably faster under 2 J/L compared to 19 J/L. This difference could be related to higher concentrations of O_3 produced using 19 J/L. However, it has to be noted that the normalized mineralization (ε) is always higher using the lowest SIE, irrespectively of the RH level, confirming our former observations on the more effective use of low SIE by IPC.

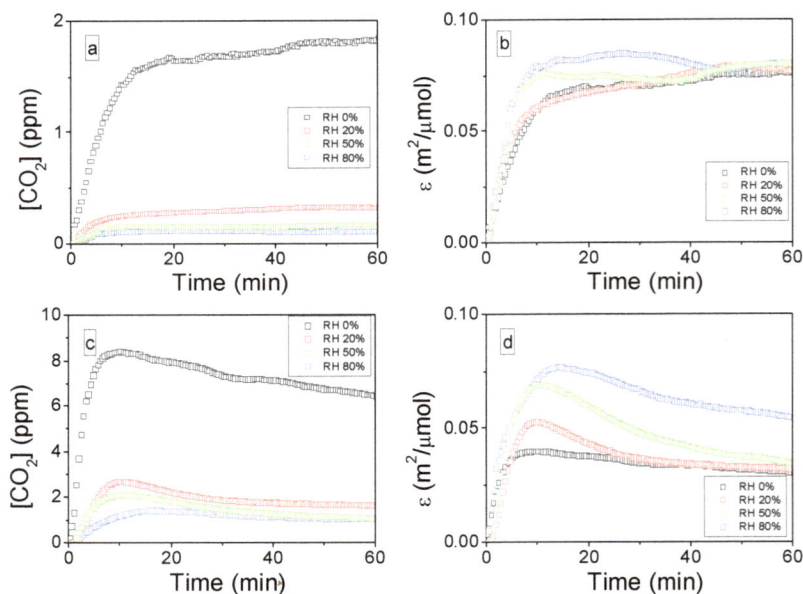

Figure 6. Temporal evolutions of CO_2 concentration at the IPC reactor outlet during à 60-min plasma exposure of toluene adsorbed on CeO_2 under different RH levels with SIE of 2 J/L (**a**) and 19 J/L (**c**); and temporal evolutions of the normalized mineralization of toluene ε(t) under different RH levels with SIE of 2 J/L (**d**) and 19 J/L (**b**) (flow: 0.5 L/min, 70 mg ceria, $P = 101.3$ kPa, and $T = 298$ K).

These observations reveal that the mineralization is directly proportional to the amount of toluene adsorbed on the surface of CeO_2. It means that on the investigated range of surface concentration of toluene, the concentrations of the oxidizing species produced by the discharge are not the limiting factor of the process. Besides, Figure 6b,d evidence that the mineralization is also directly proportional to the concentration of ozone produced by the discharge. In the presence of moisture, the main oxidizing species turns from oxygen atoms to hydroxyl radicals. However, oxygen atoms produced from O_3 decomposition appear as the most effective oxidizing agents to lead to toluene mineralization. The convergence of ε parameter irrespectively of the RH levels indicates that the role of water molecules consists of preventing toluene adsorption and ozone formation, while the generation of hydroxyl radical does not impact the process.

3. Materials and Methods

3.1. Gas Flow Preparation

Certified gas cylinders are supplied by Air Liquid. In all experiments, the inlet concentration of toluene is 100 ppm. This concentration insures that the adsorption equilibrium of toluene onto CeO_2 catalyst is achieved within a time span of typically 90 min. The regulation of gas flows is insured using Brooks mass flow controllers. The mass flow controller attributed to the main dilution line can go up to 2000 mL/min ± 1%, that is., it is less accurate below an imposed flow of 20 mL/min; while the one used for VOC cylinder regulation has a maximum of 10 mL/min and can be used accurately down to 0.1 mL/min. As shown in Figure 7, experiments are performed with a constant total gas flow of 500 mL/min. A 494.3 mL/min air flow is sent through gas line (1) connected with the water bubbler. Relative humidity (RH) can be adjusted from 0% to 80% using water bubbler containing de-ionized water by adjusting the bubbler temperature from 274 K to 294 K with a Huber® cryostat. For dry conditions, the cryostat containing water is by-passed. A 5.7 mL/min air flow is sent through gas line (2) connected to the toluene bubbler. The toluene concentration at the toluene bubbler outlet is 8960 ppm. Once both lines are mixed together the concentration of toluene is 100 ± 2 ppm at the reactor inlet.

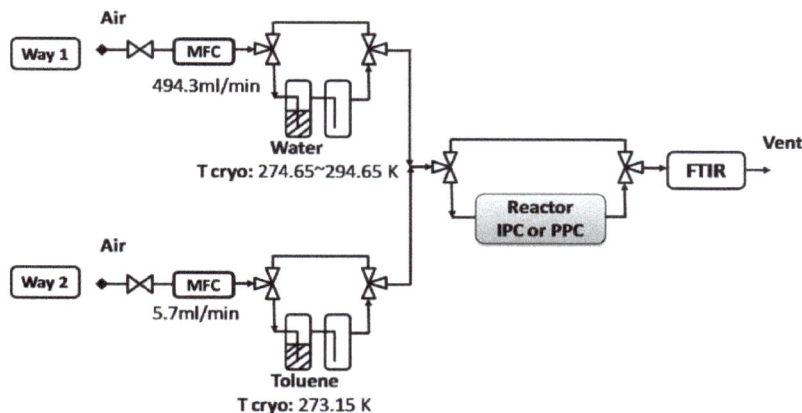

Figure 7. General scheme of the experimental set-up.

3.2. Materials

70 mg of CeO_2 powder (CAS 1306-38-3, Sigma-Aldrich, Lyon, France are used as coupling material. The specific surface of the material is determined as 84 ± 5 m^2/g. The CeO_2 powder is first pressed, and then split up into smaller pieces. These pieces have been first passed through a 12 mesh screen and then trapped by a 16 mesh screen that allows producing CeO_2 agglomerates with particle sizes ranging from 1.2 mm to 1.6 mm. Such particle size in experimental reactor could minimize the gas

by-passes and pressure-drops. Complementary specific surface measurements evidenced that the abovementioned agglomeration of CeO_2 particles do not impact the specific surface of the material.

3.3. FTIR Monitoring

The detection and the quantification of the gas phase species are performed using a high-resolution Nicolet 6700 Fourier Transform Infrared spectrometer (FTIR) equipped with a 10 m long optical-path White cell and a cooled mercury cadmium telluride (MCT) detector. The cell is heated at 50 °C to ensure thermal regulation and to prevent adsorption onto the walls. Two spectra per minute are collected with 16 scans per spectrum by Omnic software and a spectral resolution of 0.5 cm^{-1}. Figure 8 shows the gas phase spectra acquired during calibration of the main detected species. CO quantification is carried out using the average of two rotational absorption peaks (2159.2–2157.2 cm^{-1}, 2166.9–2164.2 cm^{-1}) associated to the C–O stretch vibration. The regions 2388–2348 cm^{-1} and 1021.6–1000.4 cm^{-1} are respectively selected for CO_2 and O_3 quantification. The presence of toluene is detected in the region (732.0–724.0 cm^{-1}), corresponding to molecule-specific aromatic C–H out of plane vibrations [48] where the water signal is minimal. The detection limits of the analytical instrument have been determined as two times the signal/noise ratio in the regions of interest; they are reported in Table 3.

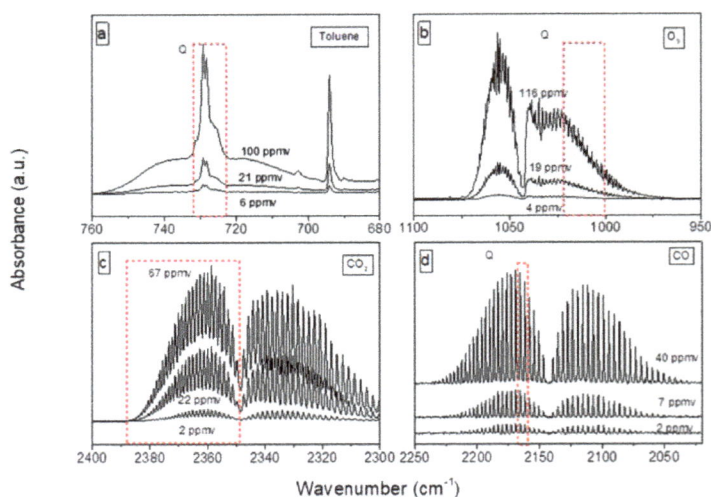

Figure 8. Fourier transfer infrared (FTIR) gas phase spectra acquired during toluene (**a**), O_3 (**b**), CO_2 (**c**) and CO (**d**) calibrations. The framed regions (Q) indicated the portions of the spectra are used to quantify the corresponding species.

Table 3. FTIR detection limits of the main detected species.

Compound	Detection Limit (ppbv)
Toluene	130
CO_2	20
CO	10
O_3	15
H_2O	2800

3.4. Post vs in Plasma–Catalytic Reactors

The two following sections describe the plasma-catalytic reactors under both configurations: post-situ and in-situ. For the post-situ configuration, the system consists of two distinct reactors whereas for in-situ configuration, the catalyst is placed inside the dielectric barrier discharge (DBD) reactor. Injected power is calculated using the Lissajous figure method, also known as "Manley method"; calculations are described in reference [31]. Specific input energy (SIE), expressed in J/L, is defined as the ratio of the injected power to the gas flow rate.

3.4.1. Post Plasma Reactor–PPC

In the post-situ configuration, the material is placed in a U-shape Pyrex tube (Figure 9a) in the form of CeO_2 agglomerates as described above. The DBD reactor is placed upstream the U-shape reactor. It consists of a Pyrex glass (dielectric) tube of 1.7 mm inner diameter, 3.3 mm outside diameter and 235 cm length. The inner electrode consists of a tungsten (W) wire of 0.2 mm diameter placed in the middle of the tube. The outer electrode is a copper sheet of 16 mm wrapped around the dielectric Pyrex tube. The U-shaped reactor containing the catalyst is a Pyrex tube with an external diameter of 8 mm and wall thickness of 2.2 mm. The Pyrex tube is conceived with glass tips in the middle to stabilize the material, yet avoiding any pressure drop in the gas flow.

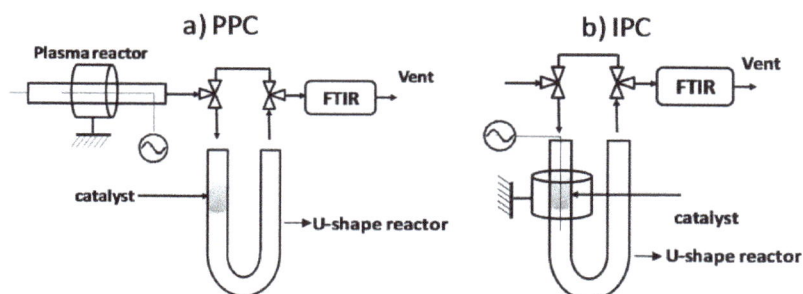

Figure 9. Scheme of the plasma-catalytic reactors in the PPC (**a**) and IPC (**b**) configurations.

3.4.2. In Plasma Reactor—IPC

In the in-situ configuration, the catalyst is directly placed inside the DBD, more precisely, in the discharge zone. The catalyst is placed in the same U-shaped reactor used for the post-plasma configuration. The inner electrode consists of a tungsten (W) wire of 0.2 mm thickness placed in main axis of the Pyrex tube. The counter electrode is a hollow brass cylinder, 16 mm high, screwed along the catalyst bed area of the U-shape reactor. The height of the catalyst bed is ca. 8 mm which is inferior to the counter electrode height. In order to compare the results obtained in IPC and PPC configuration, the charge used for both reactors is the same.

3.5. Typical Experimental Protocole

The experimental procedure is composed of five steps irrespectively of PPC or IPC configurations:

- Pretreatment of CeO_2 sample under dry air at 400 °C to remove water and other adsorbed species and ensure repeatability of experiments.
- Adsorption of 100 ppm toluene on CeO_2 until breakthrough: the air flow with different RH levels (0–80%) containing toluene breaks through the CeO_2 bed and is gradually adsorbed on the catalyst surface until equilibration of the sorption sites.
- Flushing of CeO_2 under synthetic dry air flow to remove the reversibly adsorbed fraction of toluene. Flushing the sorbent bed under air desorbs the molecules with the weakest heats of adsorption, that is, physisorbed species, leaving only the irreversibly adsorbed toluene molecules on CeO_2 surface.

- Surface exposure under air flow with different RH levels (0–80%) by switching on the nonthermal plasma for 60 minutes for IPC or PPC. Once the plasma is turned off, the system is purged with synthetic air flow until ozone concentration returns to zero.
- Temperature programmed oxidation (TPO) under synthetic air flow with different RH levels (0–80%) at 400 °C is performed in order to remove the remaining adsorbed species and regenerate the CeO_2 surface.

4. Conclusions

The oxidation of toluene adsorbed on CeO_2 has been studied using two configurations: (i) in plasma-catalysis (IPC) and (ii) post plasma-catalysis (PPC), respectively. The mineralization yield ρ (t) and the mineralization efficiency η were used to compare the performance of IPC and PPC configurations with different injected powers and relative humidity levels. Under dry air condition, CO_2 production appears as a constant process and CO_2 yield ρ (t) was optimal with a lower SIE (2J/L) for both IPC and PPC configuration. Moreover, η was enhanced by 65% in IPC compared to PPC with a lower SIE but no difference was observed for a SIE above 8J/L.

Both ozone concentrations without ceria and amount of irreversibly adsorbed toluene on ceria decreased with the increasing of the relative humidity levels. The parameter ε (t) was proposed to enable accurate comparisons of toluene mineralization considering these two variable values. Interestingly, under both SIE conditions, after a short transient regime, values of ε (t) tended to converge towards similar values, irrespectively of: (i) the RH level; (ii) the initial amount of toluene adsorbed; and (iii) the concentration of O_3 produced by the discharge. Results highlight the key role of ozone in the mineralization process and the possible detrimental effect of moisture.

Author Contributions: Z.J., A.R. and F.T. conceived and designed the experiments; Z.J., X.W. and E.F. performed the experiments; A.R., F.T. and Z.J. contributed to the data interpretation, the discussions and the writing of the paper.

Acknowledgments: Authors acknowledge the China Scholarship Council (CSC) for the PhD grant attributed to Xianjie Wang, Labex Plas@Par and Ecole Polytechnique. This work has been done within the French DGA project of 2012 60 013 00470 7501, and the LABEX PLAS@PAR project by the Agence Nationale de la Recherche under thereference ANR-11-IDEX-0004-02. IMT Lille Douai participates in the Labex CaPPA project funded by the ANR through the PIA under contract ANR-11-LABX-0005-01, and in the CLIMIBIO project, funded by the "Hauts-de-France" Regional Council and the European Regional Development Fund (ERDF).

Conflicts of Interest: The authors declare no conflict of interest.

References

1. Theloke, J.; Friedrich, R. Compilation of a database on the composition of anthropogenic VOC emissions for atmospheric modeling in Europe. *Atmos. Environ.* **2007**, *41*, 4148–4160. [CrossRef]
2. Coates, J.D.; Chakraborty, R.; Lack, J.G.; O'Connor, S.M.; Cole, K.A.; Bender, K.S.; Achenbach, L.A. Anaerobic benzene oxidation coupled to nitrate reduction in pure culture by two strains of Dechloromonas. *Nature* **2001**, *411*, 1039–1043. [CrossRef] [PubMed]
3. Benignus, V.A. Health effects of toluene: A review. *Neurotoxicology* **1981**, *2*, 567–588. [PubMed]
4. Fishbein, L. An overview of environmental and toxicological aspects of aromatic hydrocarbons II. Toluene. *Sci. Total Environ.* **1985**, *42*, 267–288. [CrossRef]
5. Lebret, E.; Van de Wiel, H.J.; Bos, H.P.; Noij, D.; Boleij, J.S.M. Volatile organic compounds in Dutch homes. *Environ. Int.* **1986**, *12*, 323–332. [CrossRef]
6. Feng, X.; Liu, H.; He, C.; Shen, Z.; Wang, T. Synergistic effects and mechanism of a non-thermal plasma catalysis system in volatile organic compound removal: A review. *Catal. Sci. Technol.* **2018**, *8*, 936–954. [CrossRef]
7. Zhu, X.; Zhang, S.; Yang, Y.; Zheng, C.; Zhou, J.; Gao, X.; Tu, X. Enhanced performance for plasma-catalytic oxidation of ethyl acetate over $La_{1-x}Ce_xCoO_{3+\delta}$ catalysts. *Appl. Catal. B Environ.* **2017**, *213*, 97–105. [CrossRef]

8. Wang, W.; Kim, H.-H.; Van Laer, K.; Bogaerts, A. Streamer propagation in a packed bed plasma reactor for plasma catalysis applications. *Chem. Eng. J.* **2018**, *334*, 2467–2479. [CrossRef]

9. Veerapandian, S.; Leys, C.; De Geyter, N.; Morent, R. Abatement of VOCs Using Packed Bed Non-Thermal Plasma Reactors: A Review. *Catalysts* **2017**, *7*, 113. [CrossRef]

10. Ye, Z.; Giraudon, J.-M.; De Geyter, N.; Morent, R.; Lamonier, J.-F. The Design of MnOx Based Catalyst in Post-Plasma Catalysis Configuration for Toluene Abatement. *Catalysts* **2018**, *8*. [CrossRef]

11. Xu, X.; Wang, P.; Xu, W.; Wu, J.; Chen, L.; Fu, M.; Ye, D. Plasma-catalysis of metal loaded SBA-15 for toluene removal: Comparison of continuously introduced and adsorption-discharge plasma system. *Chem. Eng. J.* **2016**, *283*, 276–284. [CrossRef]

12. Thevenet, F.; Sivachandiran, L.; Guaitella, O.; Barakat, C.; Rousseau, A. Plasma-catalyst coupling for volatile organic compound removal and indoor air treatment: A review. *J. Phys. D Appl. Phys.* **2014**, *47*. [CrossRef]

13. Jia, Z.; Ben Amar, M.; Yang, D.; Brinza, O.; Kanaev, A.; Duten, X.; Vega-González, A. Plasma catalysis application of gold nanoparticles for acetaldehyde decomposition. *Chem. Eng. J.* **2018**, *347*, 913–922. [CrossRef]

14. Jia, Z.; Vega-Gonzalez, A.; Amar, M.B.; Hassouni, K.; Tieng, S.; Touchard, S.; Kanaev, A.; Duten, X. Acetaldehyde removal using a diphasic process coupling a silver-based nano-structured catalyst and a plasma at atmospheric pressure. *Catal. Today* **2013**, *208*, 82–89. [CrossRef]

15. Jia, Z.; Barakat, C.; Dong, B.; Rousseau, A. VOCs Destruction by Plasma Catalyst Coupling Using AL-KO PURE Air Purifier on Industrial Scale. *J. Mater. Sci. Chem. Eng.* **2015**, *3*, 19–26. [CrossRef]

16. Vandenbroucke, A.M.; Morent, R.; De Geyter, N.; Leys, C. Non-thermal plasmas for non-catalytic and catalytic VOC abatement. *J. Hazard. Mater.* **2011**, *195*, 30–54. [CrossRef] [PubMed]

17. Christensen, P.A.; Mashhadani, Z.T.A.W.; Md Ali, A.H.B.; Carroll, M.A.; Martin, P.A. The Production of Methane, Acetone, "Cold" CO and Oxygenated Species from IsoPropyl Alcohol in a Non-Thermal Plasma: An In-Situ FTIR Study. *J. Phys. Chem. A* **2018**, *122*, 4273–4284. [CrossRef] [PubMed]

18. Chen, H.L.; Lee, H.M.; Chen, S.H.; Chang, M.B.; Yu, S.J.; Li, S.N. Removal of Volatile Organic Compounds by Single-Stage and Two-Stage Plasma Catalysis Systems: A Review of the Performance Enhancement Mechanisms, Current Status, and Suitable Applications. *Environ. Sci. Technol.* **2009**, *43*, 2216–2227. [CrossRef] [PubMed]

19. Norsic, C.; Tatibouët, J.-M.; Batiot-Dupeyrat, C.; Fourré, E. Methanol oxidation in dry and humid air by dielectric barrier discharge plasma combined with MnO_2-CuO based catalysts. *Chem. Eng. J.* **2018**, *347*, 944–952. [CrossRef]

20. Ogata, A.; Einaga, H.; Kabashima, H.; Futamura, S.; Kushiyama, S.; Kim, H.-H. Effective Combination of Nonthermal Plasma and Catalysts for Decomposition of Benzene in Air. *Appl. Catal. B Environ.* **2003**, *46*, 87–95. [CrossRef]

21. Hammer, T.; Kappes, T.; Baldauf, M. Plasma catalytic hybrid processes: Gas discharge initiation and plasma activation of catalytic processes. *Catal. Today* **2004**, *89*, 5–14. [CrossRef]

22. Kirkpatrick, M.J.; Finney, W.C.; Locke, B.R. Plasma-Catalyst Interactions in the Treatment of Volatile Organic Compounds and NOx with Pulsed Corona Discharge and Reticulated Vitreous Carbon Pt/Rh-Coated Electrodes. *Catal. Today* **2004**, *89*, 117–126. [CrossRef]

23. Ayrault, C.; Barrault, J.; Blin-Simiand, N.; Jorand, F.; Pasquiers, S.; Rousseau, A.; Tatibouët, J.M. Oxidation of 2-heptanone in air by a DBD-type plasma generated within a honeycomb monolith supported Pt-based catalyst. *Catal. Today* **2004**, *89*, 75–81. [CrossRef]

24. Du, C.M.; Yan, J.H.; Cheron, B. Decomposition of toluene in a gliding arc discharge plasma reactor. *Plasma Sources Sci. Technol.* **2007**, *16*, 791–797. [CrossRef]

25. Guo, Y.; Ye, D.; Tian, Y.; Chen, K. Humidity Effect on Toluene Decomposition in a Wire-plate Dielectric Barrier Discharge Reactor. *Plasma Chem. Plasma Process.* **2006**, *26*, 237–249. [CrossRef]

26. Van Durme, J.; Dewulf, J.; Demeestere, K.; Leys, C.; Van Langenhove, H. Post-plasma catalytic technology for the removal of toluene from indoor air: Effect of humidity. *Appl. Catal. B Environ.* **2009**, *87*, 78–83. [CrossRef]

27. Thevenet, F.; Guaitella, O.; Puzenat, E.; Guillard, C.; Rousseau, A. Influence of water vapour on plasma/photocatalytic oxidation efficiency of acetylene. *Appl. Catal. B Environ.* **2008**, *84*, 813–820. [CrossRef]

28. Chang, T.; Shen, Z.; Huang, Y.; Lu, J.; Ren, D.; Sun, J.; Cao, J.; Liu, H. Post-plasma-catalytic removal of toluene using MnO_2–Co_3O_4 catalysts and their synergistic mechanism. *Chem. Eng. J.* **2018**, *348*, 15–25. [CrossRef]

29. Batault, F.; Thevenet, F.; Hequet, V.; Rillard, C.; Le Coq, L.; Locoge, N. Acetaldehyde and acetic acid adsorption on TiO₂ under dry and humid conditions. *Chem. Eng. J.* **2015**, *264*, 197–210. [CrossRef]

30. Huang, H.; Ye, D. Combination of photocatalysis downstream the non-thermal plasma reactor for oxidation of gas-phase toluene. *J. Hazard. Mater.* **2009**, *171*, 535–541. [CrossRef] [PubMed]

31. Jia, Z.; Wang, X.; Thevenet, F.; Rousseau, A. Dynamic probing of plasma-catalytic surface processes: Oxidation of toluene on CeO₂. *Plasma Process. Polym.* **2017**, *14*. [CrossRef]

32. Jia, Z.; Rousseau, A. Sorbent track: Quantitative monitoring of adsorbed VOCs under in-situ plasma exposure. *Sci. Rep.* **2016**, *6*. [CrossRef] [PubMed]

33. Stere, C.E.; Anderson, J.A.; Chansai, S.; Delgado, J.J.; Goguet, A.; Graham, W.G.; Hardacre, C.; Taylor, S.F.R.; Tu, X.; Wang, Z.; et al. Non-Thermal Plasma Activation of Gold-Based Catalysts for Low-Temperature Water–Gas Shift Catalysis. *Angew. Chem. Int. Ed. Engl.* **2017**, *56*, 5579–5583. [CrossRef] [PubMed]

34. Barakat, C.; Gravejat, P.; Guaitella, O.; Thevenet, F.; Rousseau, A. Oxidation of isopropanol and acetone adsorbed on TiO₂ under plasma generated ozone flow: Gas phase and adsorbed species monitoring. *Appl. Catal. B Environ.* **2014**, *147*, 302–313. [CrossRef]

35. Batault, F. Effect of Adsorption and Operating Parameters on Volatile Organic Compounds Mixture Photocatalytic Treatment in Indoor Air Conditions. Ph.D. Thesis, University of Lille, Lille, France, 2014.

36. Bouzaza, A.; Laplanche, A. Photocatalytic degradation of toluene in the gas phase: Comparative study of some TiO₂ supports. *J. Photochem. Photobiol. A Chem.* **2002**, *150*, 207–212. [CrossRef]

37. Takeuchi, M.; Hidaka, M.; Anpo, M. Efficient removal of toluene and benzene in gas phase by the TiO₂/Y-zeolite hybrid photocatalyst. *J. Hazard. Mater.* **2012**, *237–238*, 133–139. [CrossRef] [PubMed]

38. Goss, K.-U. The Air/Surface Adsorption Equilibrium of Organic Compounds Under Ambient Conditions. *Crit. Rev. Environ. Sci. Technol.* **2004**, *34*, 339–389. [CrossRef]

39. Fowkes, F.M. Attractive Forces at Interfaces. *Ind. Eng. Chem. Res.* **1964**, *56*, 40–52. [CrossRef]

40. Goss, K.-U.; Eisenreich, S.J. Adsorption of VOCs from the Gas Phase to Different Minerals and a Mineral Mixture. *Environ. Sci. Technol.* **1996**, *30*, 2135–2142. [CrossRef]

41. Mao, L.; Chen, Z.; Wu, X.; Tang, X.; Yao, S.; Zhang, X.; Jiang, B.; Han, J.; Wu, Z.; Lu, H.; et al. Plasma-catalyst hybrid reactor with CeO₂/Γ-Al₂O₃ for benzene decomposition with synergetic effect and nano particle by-product reduction. *J. Hazard. Mater.* **2018**, *347*, 150–159. [CrossRef] [PubMed]

42. Blount, M.C.; Falconer, J.L. Steady-state surface species during toluene photocatalysis. *Appl. Catal. B Environ.* **2002**, *39*, 39–50. [CrossRef]

43. Augugliaro, V.; Coluccia, S.; Loddo, V.; Marchese, L.; Martra, G.; Palmisano, L.; Schiavello, M. Photocatalytic oxidation of gaseous toluene on anatase TiO₂ catalyst: Mechanistic aspects and FT-IR investigation. *Appl. Catal. B Environ.* **1999**, *20*, 15–27. [CrossRef]

44. Harling, A.M.; Demidyuk, V.; Fischer, S.J.; Whitehead, J.C. Plasma-catalysis destruction of aromatics for environmental clean-up: Effect of temperature and configuration. *Appl. Catal. B Environ.* **2008**, *82*, 180–189. [CrossRef]

45. Van Durme, J.; Dewulf, J.; Sysmans, W.; Leys, C.; Van Langenhove, H. Efficient toluene abatement in indoor air by a plasma catalytic hybrid system. *Appl. Catal. B Environ.* **2007**, *74*, 161–169. [CrossRef]

46. Wu, J.; Huang, Y.; Xia, Q.; Li, Z. Decomposition of Toluene in a Plasma Catalysis System with NiO, MnO₂, CeO₂, Fe₂O₃, and CuO Catalysts. *Plasma Chem. Plasma Process.* **2013**, *33*, 1073–1082. [CrossRef]

47. Orge, C.A.; Órfão, J.J.M.; Pereira, M.F.R.; Duarte de Farias, A.M.; Fraga, M.A. Ceria and cerium-based mixed oxides as ozonation catalysts. *Chem. Eng. J.* **2012**, *200–202*, 499–505. [CrossRef]

48. Diehl, J.W.; Finkbeiner, J.W.; DiSanzo, F.P. Determination of benzene, toluene, ethylbenzene, and xylenes in gasolines by gas chromatography/deuterium isotope dilution Fourier transform infrared spectroscopy. *Anal. Chem.* **1993**, *65*, 2493–2496. [CrossRef]

catalysts

MDPI

Article

Destruction of Toluene, Naphthalene and Phenanthrene as Model Tar Compounds in a Modified Rotating Gliding Arc Discharge Reactor

Xiangzhi Kong [1], Hao Zhang [1,*] ⬥, Xiaodong Li [1,*], Ruiyang Xu [1], Ishrat Mubeen [2], Li Li [1] and Jianhua Yan [1]

[1] State Key Laboratory of Clean Energy Utilization, Zhejiang University, Hangzhou 310027, China; lanklarde@126.com (X.K.); kaybyren@163.com (R.X.); 21727034@zju.edu.cn (L.L); yanjh@zju.edu.cn (J.Y.)
[2] Institute of Energy and Power Engineering, College of Mechanical Engineering, Zhejiang University of Technology, Hangzhou 310013, China; ishrat.farman@gmail.com
*   Correspondence: zhang_hao@zju.edu.cn (H.Z.); lixd@zju.edu.cn (X.L.); Tel.: +86-571-87952438 (H.Z.); +86-571-87952037 (X.L.)

Received: 30 November 2018; Accepted: 21 December 2018; Published: 28 December 2018

Abstract: Tar removal is one of the greatest technical challenges of commercial gasification technologies. To find an efficient way to destroy tar with plasma, a rotating gliding arc (RGA) discharge reactor equipped with a fan-shaped swirling generator was used for model tar destruction in this study. The solution of toluene, naphthalene and phenanthrene is used as a tar surrogate and is destroyed in humid nitrogen. The influence of tar, CO_2 and moisture concentrations, and the discharge current on the destruction efficiency is emphasized. In addition, the combination of Ni/γ-Al$_2$O$_3$ catalyst with plasma was tested for plasma catalytic tar destruction. The toluene, naphthalene and phenanthrene destruction efficiency reached up to 95.2%, 88.9%, and 83.9% respectively, with a content of 12 g/Nm^3 tar, 12% moisture, 15% CO_2 and a flow rate of 6 NL/min, whereas 9.3 g/kW·h energy efficiency was achieved. The increase of discharge current is advantageous in terms of decreasing black carbon production. The participation of Ni/γ-Al$_2$O$_3$ catalyst shows considerable improvement in destruction efficiency, especially at a relatively high flow rate (over 9 NL/min). The major liquid by-products are phenylethyne, indene, acenaphthylene and fluoranthene. The first two are majorly converted from toluene, acenaphthylene is produced by the co-reaction of toluene and naphthalene in the plasma, and fluoranthene is converted by phenanthrene.

Keywords: rotating gliding arc plasma; tar destruction; toluene; naphthalene; phenanthrene; catalyst

1. Introduction

Owing to the rapid consumption of fossil fuels and increasing emphasis on environmental issues, the use of biomass and municipal solid waste (MSW) as alternative fuel has received considerable attention. Gasification is considered one of the most flexible fuel conversion processes [1–3], which has the advantages of low secondary pollution compared with the traditional incineration method. The products of gasification can not only be used as fuel in gas turbines and gas engines, but also offer some basic building blocks for producing valuable chemicals, liquid fuels and hydrogen [4–6].

However, besides the useful fuels, tar exists in the syngas as an undesirable by-product, which can cause fouling, clogging, and corrosion problems in downstream equipment after cooling and condensing [1,6–9]. The secondary pollutants produced by the aromatic hydrocarbons in the tar component may also pose serious harm to the environment and human health. Tar removal, conversion and destruction is considered to be one of the greatest technical challenges, which must be overcome for the successful development of commercial gasification technologies [9–13]. Aromatic compounds are the most important contents in the tar, which are not only in large amounts but are chemically stable in the gasifier environment [14]. The traditional methods for the removal or destruction of tar can be categorized into (i) mechanical separation (by Venturi scrubbers, water scrubbers, electrostatic precipitators (ESPs), rotational particle separators, or cyclones), which is unattractive since it causes secondary pollution and wastage of chemical energy contained in tar while shows low efficiency in tar removal [1,15,16]; (ii) thermal cracking, which normally requires 1250 °C and takes a few seconds to get good results, and has the drawbacks of high cost and production of heavier products and agglomerated soot particles [17,18]; (iii) and catalytic cracking, which enables obtaining higher yields of valuable H_2 and CO but has a significant limitation on the catalyst lifetime [2].

Non-thermal plasma methods, on the other hand, can tackle the drawbacks mentioned above. Various non-thermal plasma sources have been applied for the tar destruction, such as dielectric barrier discharges (DBDs) [19], corona discharges [20,21], microwave plasma [2,22–24], arc plasma [25,26], and gliding arc discharges (GAD) [11,17,18,27–30]. In non-thermal plasmas, the electrons have a high average temperature of 1-10 eV, which are highly excited to activate reactant molecules to form reactive species, such as radicals, excited species, ions, and photons, by electron impact excitation, ionization, and dissociation, etc., while the bulk gas temperature remains at low level (e.g., 800 °C) [11]. In addition, the flexibility (high specific productivity, instant on/off, high energy density, power scalability) of non-thermal plasma systems provides a high adaptability to different application cases [31].

Recently, gliding arc discharge plasma has received much attention as an application to destroy or reform heavy hydrocarbons, and is considered to be a promising route to destroy tars from gasification. Pemen et al. [32] removed tars produced by biomass gasification with a gliding arc discharge. Nam et al. [27] used a three-blade gliding arc reactor to destruct benzene and naphthalene and obtained a destruction efficiency of 95% and 79% respectively. Gliding arc is a unique plasma, featuring the merits of both thermal plasmas (high energy density) and non-thermal plasmas (high energy efficiency) [33]. Most of the gliding arc power consumed (75%–80%) is dissipated in the nonequilibrium zone meanwhile the specific energy input (SEI) can be up to 1 kW·h/m^2 [34], which means a good reaction capability can coexist with a high energy efficiency.

In this study, a modified atmospheric pressure rotating gliding arc (RGA) discharge reactor is developed for the destruction of tars. A fan-shaped swirling generator is innovatively equipped upstream the inner electrode, which provides a stable swirling flow field that is conducive to the stability of the arc. Unlike the limited planar plasma zone generated by traditional blade-type GAD, the arc generated by the RGA reactor can rotate rapidly and steadily without extinction (with a rotational frequency of up to 100 rotations per second [35]), creating a large three-dimensional plasma region. Previous work by our team has studied the destruction of toluene both in nitrogen and syngas atmospheres, where the destruction efficiency reached over 95% [8,36]. Most of the work done by our team or others chose only one kind of substance as a tar surrogate injected into the reaction at one time (benzene, toluene, or naphthalene). However, the composition of actual tar is very complex, containing substances of monocyclic, diphenyl, and even higher rings simultaneously. There is a lack of data for the study of multi-component tar, especially having a substance with more than two benzene rings.

For this reason, a toluene solution of naphthalene and phenanthrene as simulated tar is used to investigate the reaction of tar compounds in the RGA reactor. Toluene, naphthalene and phenanthrene are among the dominant components of tar with one, two or three benzene rings [25,26,37] so that the result in this study can match more closely to the conditions prevailing in practical application. The effects of concentrations of tar, moisture, carbon dioxide, and discharge current on destruction efficiency, energy efficiency, and main products are studied. Moreover, in order to further enhance the reaction performance, plasma catalysis, i.e., the combination of plasma with catalyst, was investigated for tar destruction, which, to the best of our knowledge, is the first attempt in gliding arc plasma used for tar destruction. Ni/γ-Al$_2$O$_3$ catalyst was used in this study because of its proven performance for tar destruction and its merits of high mechanical strength, easy preparation, and low cost [38].

In the whole experiment, the total flow rate was kept at 6 NL/min, the total tar concentration was kept at 12 g/Nm^3, and the moisture content was kept at 12% unless otherwise mentioned. Naphthalene and phenanthrene were dissolved in toluene with a mass ratio of 1:1:20.

2. Results and Discussions

2.1. Effect of Tar Concentration

The effect of tar concentration on the destruction efficiency is presented in Figure 1. A destruction efficiency of 94.3% for toluene, 87% for naphthalene, and 83.5% for phenanthrene is obtained under a total flow rate of 12 NL/min, moisture concentration of 12%, and tar concentration of 12 g/Nm^3, yielding an energy efficiency of 10.2 g/kW·h. The destruction efficiency decreases remarkably when tar concentration is over 16 g/Nm^3, in contrast to our previous study of toluene and naphthalene (without phenanthrene) destruction, where the destruction efficiency has a high value at a tar concentration of 20 g/Nm^3. This is because, as the experiment shows, phenanthrene addition tends to cause frequent breaking of the gliding arc, especially when both tar and moisture concentrations are high. Although the arc can be regenerated, the breaking decreases the destruction efficiency.

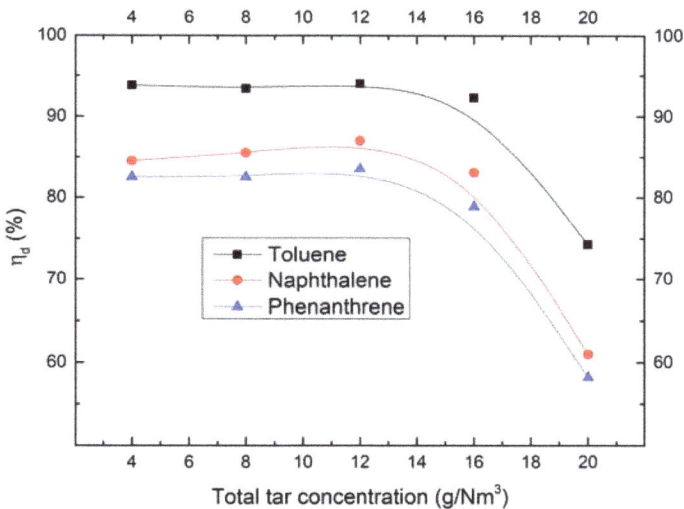

Figure 1. Effect of tar concentration on the destruction efficiency.

The order of destruction efficiency of the three components is consistently η(toluene) > η(Naphthalene) > η(Phenanthrene). The result that η(toluene) > η(Naphthalene) is in line with the results of Zhu et al. [14] and Nunnally et al. by gliding arc [15]. In the simulation work by Fourcault et al., similar results were also obtained in a plasma torch [16]. The order of activation energy of the three component is phenanthrene > naphthalene > toluene, explaining why a substance with more benzene rings is hard to destroy.

Figure 2 shows that the energy efficiency increases as the total tar concentration increases until reaching 16 g/Nm^3, after which the efficiency started decreasing. The instability of the reactor caused by the addition of phenanthrene is probably the factor that limits the maximum energy efficiency to 13 g/kW·h. The selectivity of C_2H_2 and CO rises monotonously with increasing tar concentration, while the selectivity of CO_2 drops (Figure 3), though the absolute production of them as well as of H_2 increases significantly as the tar concentration increases (Figure 4). The reason for this can be explained by water-gas shift reaction (Equation (1)) and reforming reaction of C_2H_2 (Equation (2)). The increase of tar concentration leads to the increase of CO and C_2H_2. The increase of CO and C_2H_2 promote the positive reaction of water-gas shifting and reforming reaction respectively, which increases the production of all the gas products. However, the concentration of H_2O is constant, which limits the conversion rate of CO and C_2H_2 in the case of high tar concentration, leading to a high selectivity of CO and C_2H_2 while lowering the selectivity of CO_2.

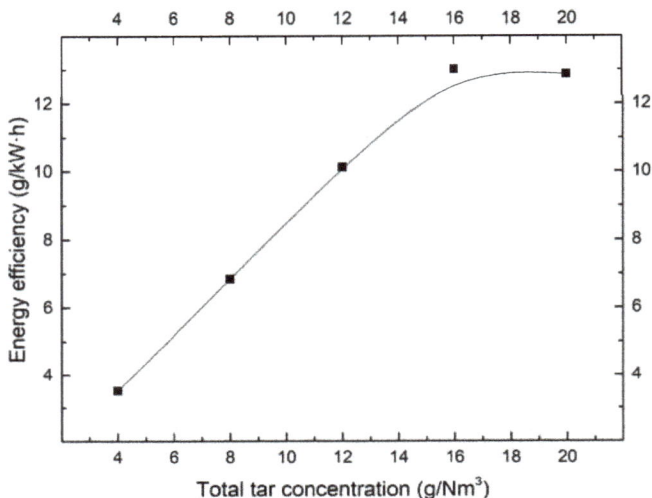

$$CO + H_2O \rightarrow CO_2 + H_2 \tag{1}$$

$$C_2H_2 + 2H_2O \rightarrow 2CO + 3H_2 \tag{2}$$

Figure 2. Effect of f tar concentration on the energy efficiency.

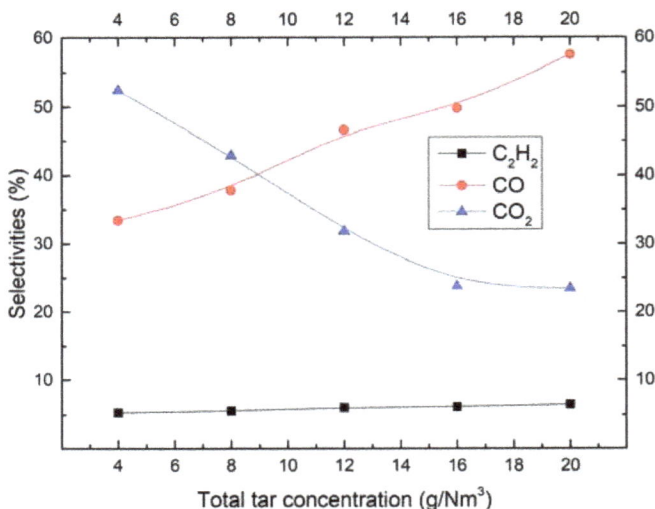

Figure 3. Effect of tar concentration on the selectivity of CO, CO_2 and C_2H_2.

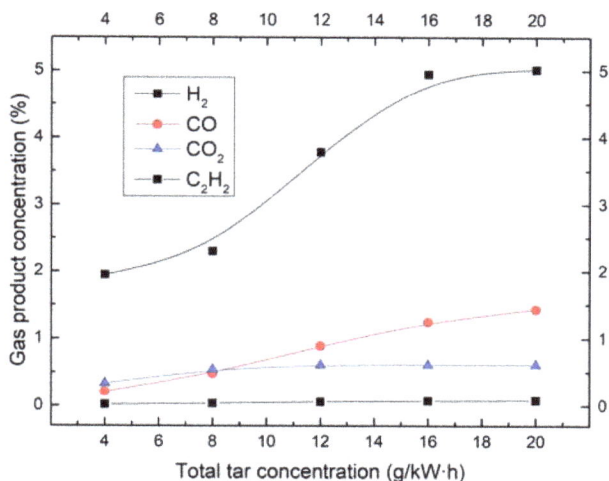

Figure 4. Effect of tar concentration on the composition of the off-gas.

2.2. Effect of Moisture Content

Figure 5 indicates that as the moisture concentration increases, the destruction efficiency of toluene, naphthalene, and phenanthrene increases first to a maximum at a moisture concentration of 8% or 12% and then starts decreasing. Previous studies by Zhu et al. [36] and Chun et al. [39] showed a similar phenomenon for the decomposition of tar surrogate using gliding arc. The proper amount of moisture facilitates the reaction because of the generation of OH radicals [40], while too much water has an adverse effect on tar destruction due to its electronegative characteristics, which limits electron density and quenches reactive species [30]. When the moisture concentration is up to 20%, the destruction efficiency declines suddenly because high concentration of moisture makes the reactor unsteady due to the existence of a high concentration of phenanthrene. The energy efficiency decreases linearly as the moisture concentration increases, as shown in Figure 6.

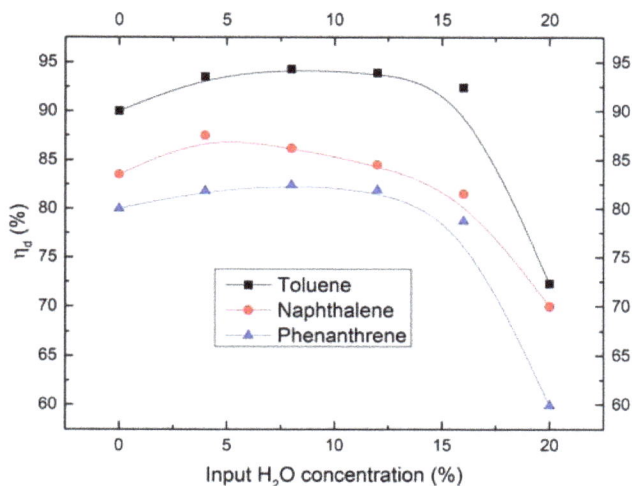

Figure 5. Effect of moisture concentration on destruction efficiency.

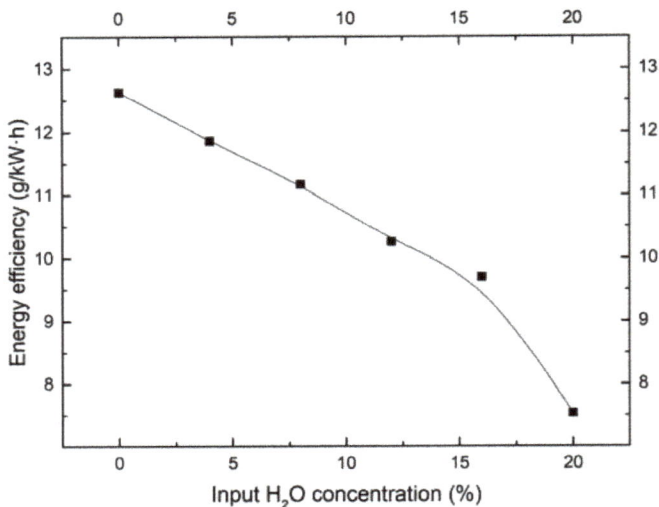

Figure 6. Effect of moisture concentration on energy efficiency.

As Figure 7 presents, the selectivity of CO and CO_2 increases with the increase of the moisture concentration, while the selectivity of C_2H_2 decreases because water brings oxygen to the reaction. The selectivity of CO, CO_2, and C_2H_2 can be over 80% in total if enough water molecules are provided, and reach a maximum of 95% at a moisture concentration of 20%. However, the selectivity of C_2H_2 is 13% and selectivity of CO and CO_2 is 0 under the dry conditions, which means most of the carbon element turns into black carbon in water deficient conditions. In addition, the increase of moisture concentration raises the hydrogen production remarkably, as presented in Figure 8. It is interesting to note that proper concentration can turn the production of solid carbon into gas, especially flammable gas like H_2 and CO, which are highly desirable.

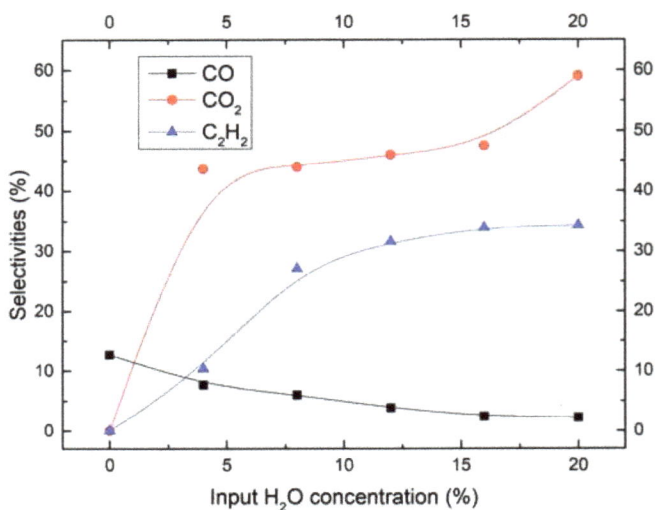

Figure 7. Effect of moisture concentration on the selectivity of CO, CO_2, C_2H_2.

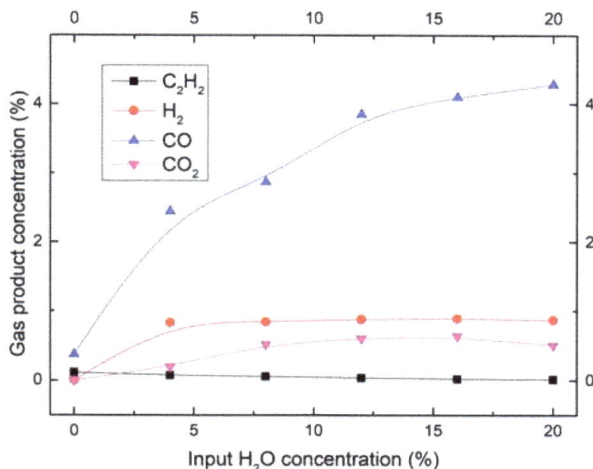

Figure 8. Effect of moisture concentration on the composition of the off-gas.

2.3. Effect of CO_2 Concentration

Figure 9 shows that with increasing CO_2 concentration, the destruction efficiency of the toluene, naphthalene, and phenanthrene increases to a maximum at an input of 15% CO_2 concentration, and then slightly reduced. Uncommon voltage waveform was observed under the condition with proper concentration of CO_2, which can explain why the highest destruction efficiency is obtained in the atmosphere of 15% CO_2. The voltage waveform is shown in Figure 10 under the condition with or without CO_2. Compared with the typical condition without CO_2, the waveform becomes regular in the condition of 10%–20% CO_2. The voltage as well as the input energy have increased about 15% (from 1.80 kV to 2.07 kV of voltage), while the discharge current is the same. When the CO_2 increases to about 25%, the waveform of voltage again becomes similar to the waveform without CO_2 and the average voltage reduces simultaneously but remains higher than that in the condition without the participation of CO_2. The role of oxidation of CO_2 in destruction efficiency is slightly lower compared to H_2O that can provide sufficient O atoms.

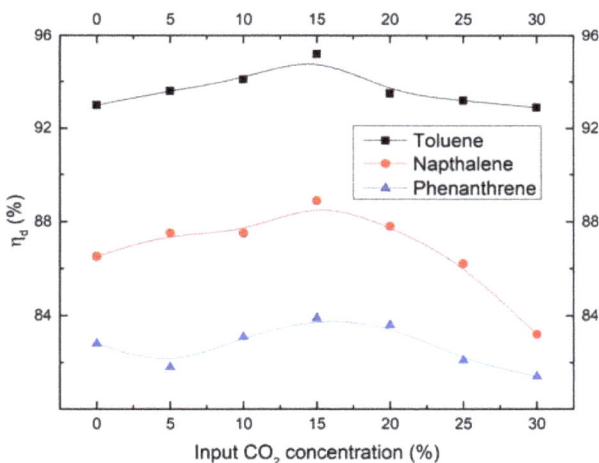

Figure 9. Effect of CO_2 concentration on destruction efficiency.

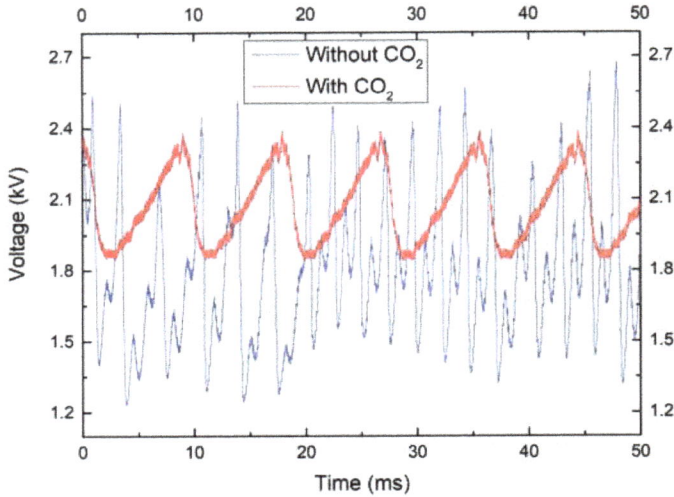

Figure 10. Voltage waveform of the discharge with 15% CO_2 and without CO_2.

When the concentration of CO_2 increases, the amount of CO in the product increases and the amount of H_2 and C_2H_2 decreases because of the reverse reaction of water-gas shift reaction (Equation (1)) and the dry reforming reaction of C_2H_2 (Equation (3)). Although the components of gas production change a lot because of CO_2, the heating value of the fuel after the reaction changes little because the production of CO increases while the production of H_2 decreases. As shown in Figure 11, the negative effect of CO_2 on the destruction is that it can reduce the energy efficiency of the reaction (Figure 12).

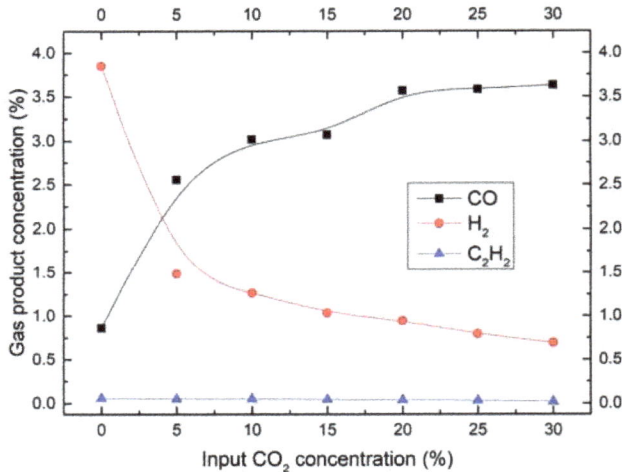

$$C_2H_2 + 2CO_2 \rightarrow 4CO + 2H_2 \tag{3}$$

Figure 11. Effect of CO_2 concentration on the composition of the off-gas.

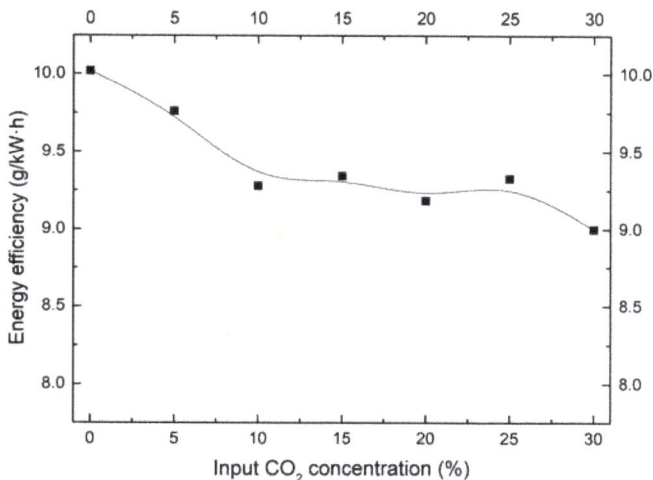

Figure 12. Effect of CO_2 concentration on energy efficiency.

2.4. Effect of Discharge Current

As expected, with an increasing discharge current, destruction efficiency increases (Figure 13). However, compared to this slight improvement in destruction efficiency (about 3%), decreased energy efficiency it causes is much higher (more than 10%, as shown in Figure 14). The production of H_2 and CO_2 increases significantly with the increase of discharge current, which indicates that extra input of energy is probably consumed for splitting of water as discharge current rises (Figure 15). The total selectivity of carbon-contained gases (CO, CO_2 and C_2H_2) increases from 80% to 88% as the discharge current increases from 200 mA to 250 mA (Figure 16), which means increasing discharge current reduces about half of the production of solid carbon. This result is desirable to achieve maximum conversion of carbon to gas components.

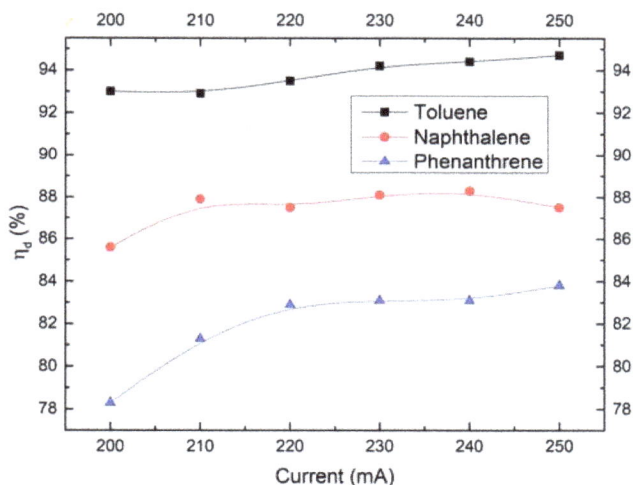

Figure 13. Effect of discharge current on destruction efficiency.

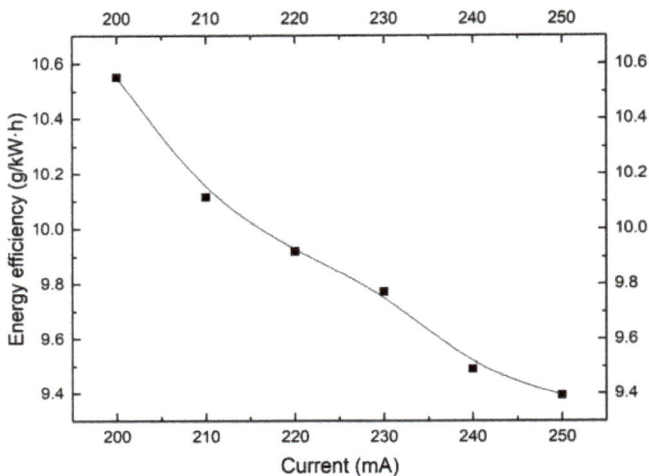

Figure 14. Effect of discharge current on the energy efficiency.

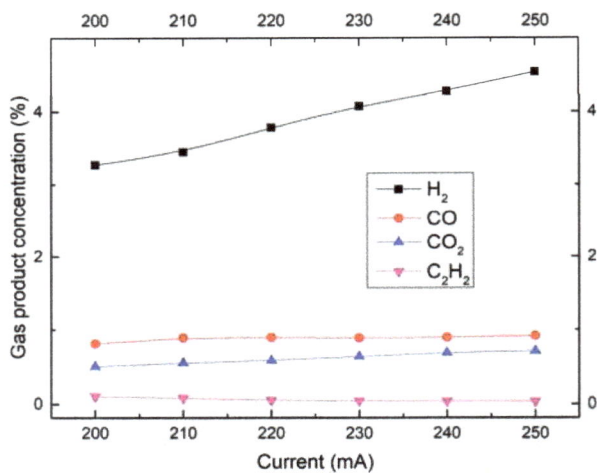

Figure 15. Effect of discharge current on the composition of the off-gas.

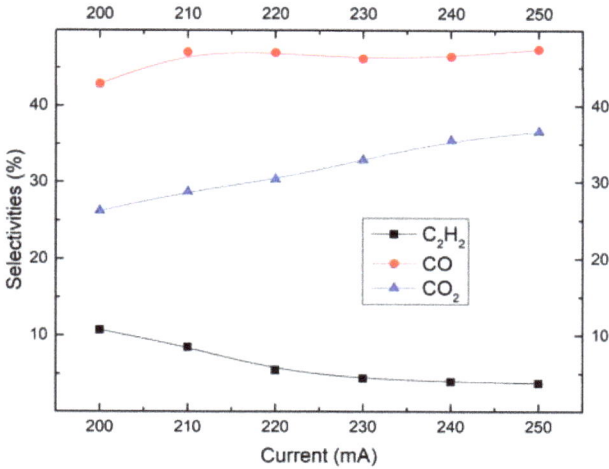

Figure 16. Effect of discharge current on the selectivity of C_2H_2, CO and CO_2.

SEI goes up as the discharge current goes up, but less pronouncedly, because the voltage of the reactor decreases with increase of discharge current. The discharge current increases 25% (from 200 mA to 250 mA) while the SEI increases about 13% (from about 1.05 kW·h/m^3 to about 1.20 kW·h/m^3, as shown in Figure 17).

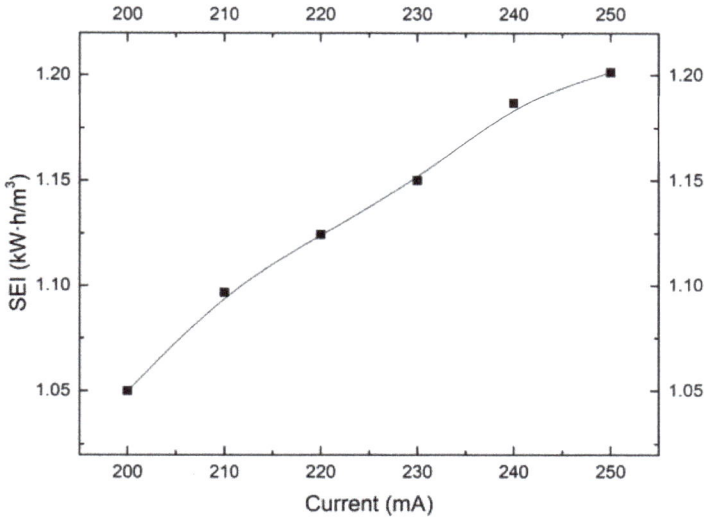

Figure 17. Effect of discharge current on specific energy input (SEI).

2.5. Effect of the Coupling of Catalyst with Plasma

In this section, the influence of catalyst in the reaction is investigated under different flow rates. A tray containing catalyst is placed downstream the plasma area, i.e., 18 mm under the anode. In this distance the rotating gliding arc can just touch the catalyst surface. The diameter of the spherical catalyst particles is 2 mm. Tar and moisture concentration were fixed as 12 g/Nm^3 and 12% respectively.

To understand whether the performance variation in the experiment is caused by the catalysis or flow field, comparative experiments have been performed with glass beads on the tray.

Figure 18 illustrates the influence of catalyst on destruction efficiency under different flow rates. The catalyst has little effect when the flow rate is relatively low (6 NL/min), as shown in Figure 18a. This is possibly because the destruction efficiency of plasma alone is so high that the amount of remaining tar is too low to involve in catalytic reaction. However, the catalyst reflects to be effective and improves the destruction efficiency remarkably under the flow rate of 9 NL/min and 12 NL/min. For instance, for a flow rate of 12 NL/min, the destruction efficiency of toluene, naphthalene, and phenanthrene increases to 90%, 86%, and 76% respectively, with the presence of the catalyst in the tray, from 85%, 80%, and 70% respectively with only glass beads in the tray (Figure 18c). The destruction efficiency in the condition of glass beads in the reactor is similar to that without a tray in the reactor according to Figure 18a–c, which indicates that the flow field changed by the tray contributes little to the change of destruction efficiency. In addition, the catalyst increases the energy efficiency when the flow rate is high, especially at a flow rate of 12 L/min, from 19.1 g/kW·h to 21.3 g/kW·h (Figure 19). As shown in Figure 20, the catalyst increases the production of gas products as well as destruction efficiency, indicating that the tar is definitely destroyed instead of being absorbed on the catalyst. The result illustrates that the Ni/γ-Al_2O_3 catalyst can be highly effective when single plasma cannot reduce the tar concentration to an ideal standard. As a result, plasma coupled with catalyst is worth using when the flow rate is high. It should be noted that a slight carbon deposition was observed on the catalyst after an operation of 1.5 h, which should be considered in the application of plasma catalysis.

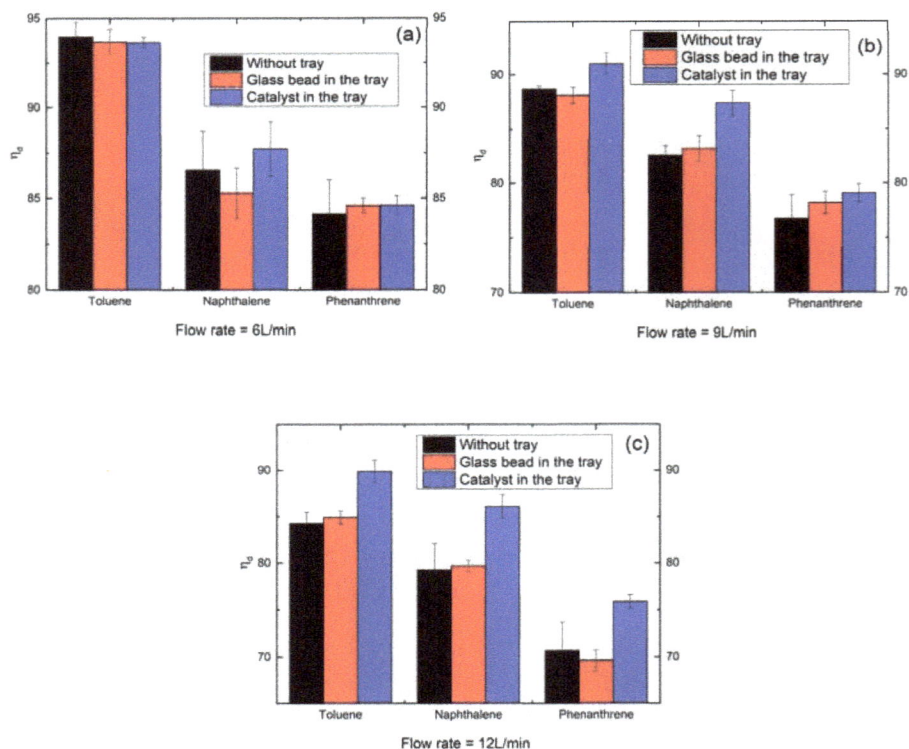

Figure 18. Effect of catalyst on destruction efficiency at flow rates of (**a**) 6 NL/min, (**b**) 9 NL/min and (**c**) 12 NL/min.

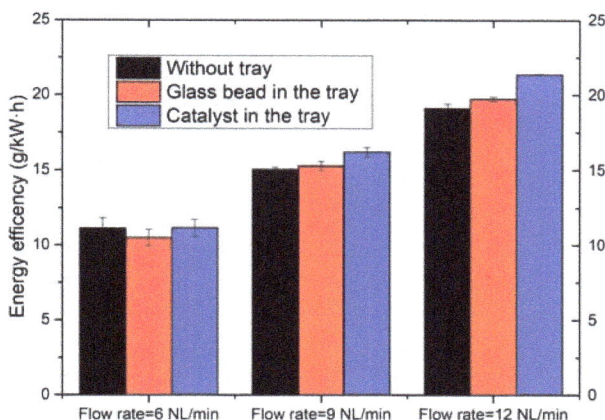

Figure 19. Effect of catalyst on energy efficiency at flow rates of 6 NL/min, 9 NL/min and 12 NL/min.

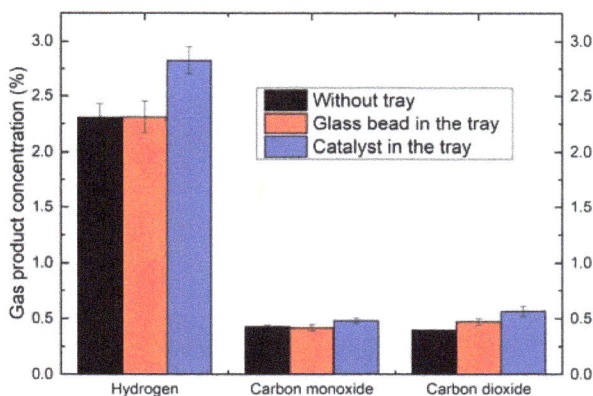

Figure 20. Effect of catalyst on the composition of the off-gas at a flow rate of 12 NL/min.

2.6. Liquid By-Products Analysis

To investigate the reaction mechanisms in the destruction of toluene, naphthalene, and phenanthrene together, liquid by-products collected in the absorption bottle were analyzed using GC/MS. Results in two working conditions are presented in this section. The first one is the typical working condition (a total flow rate of 6 NL/min, tar concentration of 12 g/Nm³, and moisture concentration of 12%), and the second one is similar to the first but there is 15% CO_2 in the atmosphere, occupying the content of nitrogen. The composition of the two liquid products have some differences, as shown in Table 1. The corresponding product structures are schematically shown in Table 2.

The major by-products are phenylethyne, indene, acenaphthylene, and fluoranthene, according to the full scan GC/MS result presented in Table 1, irrespective of the presence or absence of CO_2 in the atmosphere. Among the four major products, phenylethyne and indene were also found in by-products in the previous study of single toluene destruction by gliding arc [8], which indicates that these are probably majorly converted from toluene. The possible reaction pathway is shown in Equations (4)–(10) [22,24]. The yield of acenaphthylene in the previous study of single toluene or single naphthalene destruction by gliding arc was far lower [8,41], compared with present work with very high efficiency of naphthalene toluene solution destruction, which means naphthalene and toluene are both involved in the reaction of acenaphthylene production.

Table 1. Liquid by-products detected by GC/MS (in typical atmosphere with or without CO_2).

No	Retention Time (min)	Species	Area Percentage (%, the Area Percent of Input Naphthalene is 1000%) under Typical Atmosphere	Area Percentage (%, the Area Percent of Input Naphthalene is 1000%) when 15% CO_2 Exists
1	6.590	Ethylbenzene	0.9	0.9
2	6.865	Phenylethyne	17.2	15.3
3	7.109	Styrene	0.7	1
4	8.222	Benzaldehyde	<0.1	0.7
5	8.477	Phenol	<0.1	0.4
6	8.572	Benzonitrile	0.7	0.8
7	9.383	Indene	11.3	16
8	9.685	Phenol, 2-methyl-	<0.1	0.3
9	11.791	Cyclobuta[de]naphthalene	0.9	1.1
10	12.085	Naphthalene, 2-methyl-	1.1	1.3
11	12.240	Naphthalene, 1-methyl-	0.7	1.3
12	12.538	Benzene, 2,4-pentadiynyl-	0.2	0.7
13	12.775	Biphenyl	1.3	1.2
14	13.354	Acenaphthylene	39.8	16.2
15	13.813	Bibenzyl	0.4	1.7
16	14.269	Fluorene	27.9	23.4
17	16.040	4H-Cyclopenta[def]phenanthrene	1.0	0.9
18	16.622	Fluoranthene	2.9	1.1
19	16.847	Pyrene	5.9	1.1

The feasible way is shown in Equations (7), (11), and (12). In the case of fluoranthene, it is not detected by the GC/MS when phenanthrene was not involved in previous study, indicating that it is converted from phenanthrene. The molecular structure of this substance also corresponds to this speculation. The conversion of phenanthrene to fluoranthene is a vital reaction, as the area percentage of fluoranthene is up to 15% compared with unreacted phenanthrene. The feasible way is shown in Equations (13) and (14).

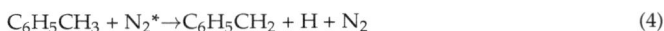

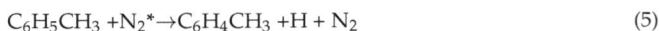

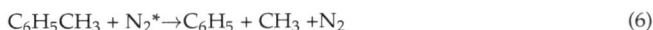

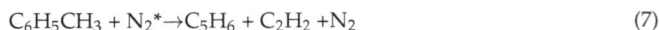

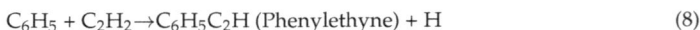

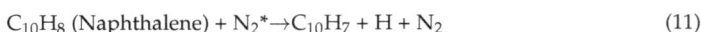

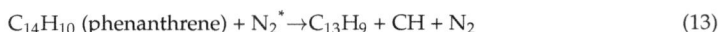

$$C_6H_5CH_3 + N_2^* \rightarrow C_6H_5CH_2 + H + N_2 \tag{4}$$

$$C_6H_5CH_3 + N_2^* \rightarrow C_6H_4CH_3 + H + N_2 \tag{5}$$

$$C_6H_5CH_3 + N_2^* \rightarrow C_6H_5 + CH_3 + N_2 \tag{6}$$

$$C_6H_5CH_3 + N_2^* \rightarrow C_5H_6 + C_2H_2 + N_2 \tag{7}$$

$$C_6H_5 + C_2H_2 \rightarrow C_6H_5C_2H \text{ (Phenylethyne)} + H \tag{8}$$

$$C_6H_5CH_2 + C_2H_2 \rightarrow C_6H_4C_3H_4 \text{ (Indene)} \tag{9}$$

$$C_6H_4CH_3 + C_2H_2 \rightarrow C_6H_4C_3H_4 \text{ (Indene)} \tag{10}$$

$$C_{10}H_8 \text{ (Naphthalene)} + N_2^* \rightarrow C_{10}H_7 + H + N_2 \tag{11}$$

$$C_{10}H_7 + C_2H_2 \rightarrow C_{12}H_{10} \text{ (Acenaphthylene)} + H \tag{12}$$

$$C_{14}H_{10} \text{ (phenanthrene)} + N_2^* \rightarrow C_{13}H_9 + CH + N_2 \tag{13}$$

$$C_{13}H_9 + H \rightarrow C_{13}H_{10} \text{ (Fluorene)} \tag{14}$$

As shown in Table 2, the main difference in products under two working conditions is the content of oxygen-containing by-products (e.g., benzaldehyde, phenol, 2-methyl-) when CO_2 is involved in the reaction compared with no participation of CO_2. This is because CO_2 provides an oxygen element, which is easy to react.

It should be noted that there must be anthracene in the by-product, but the detection of it is interfered as the phenanthrene is not completely pure (95% purity) and most of the impurity in it is anthracene according to the GC/MS result of blank test. Despite this, the impurity in phenanthrene is not high enough to influence other results discussed in this paper. Moreover, the quantity of by-products heavier than phenanthrene molecules do not reach a level worth mentioning.

Previous study indicates that the destruction of naphthalene in this reactor can produce phenylethyne. Therefore, it is expected that the partial cracking of phenanthrene can also lead to the formation of bicyclic or even monocyclic products.

Table 2. Relevant liquid by-product structures.

Monocyclic Products

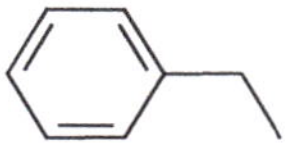

Ethylbenzene

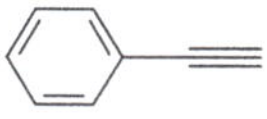

Phenylethyne

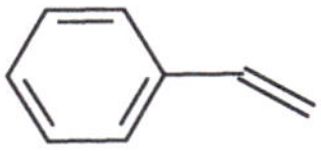

Styrene

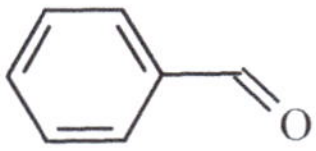

Benzaldehyde

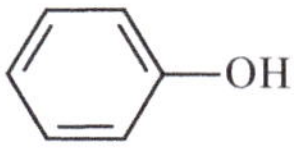

Phenol

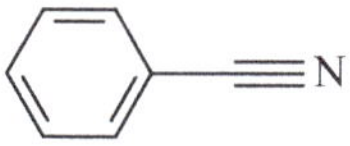

Benzonitrile

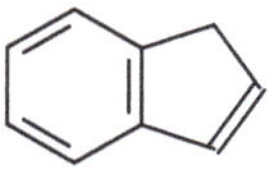

Indene

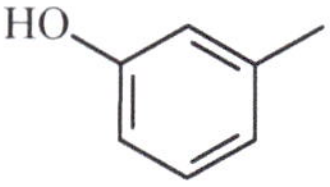

Phenol, 2-methyl-

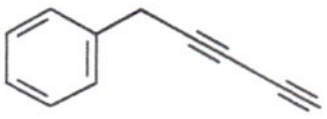

Benzene, 2,4-pentadiynyl-

Bicyclic Products

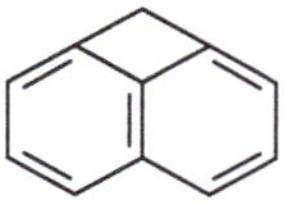

Cyclobuta[de]naphthalene

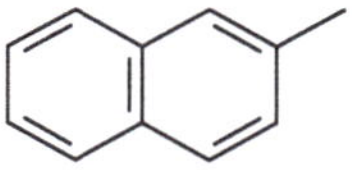

Naphthalene, 2-methyl-

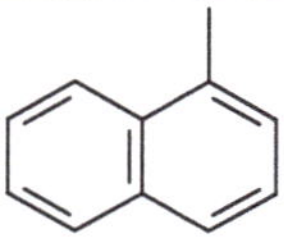

Naphthalene, 1-methyl-

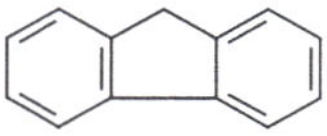

Fluorene

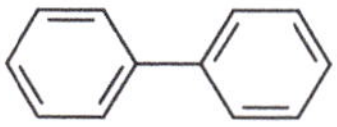

Biphenyl

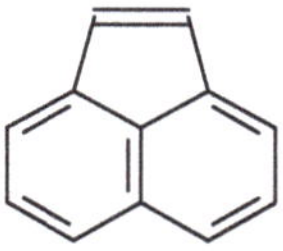

Acenaphthylene

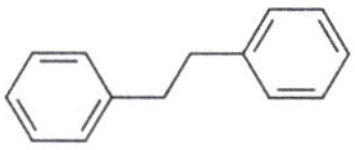

Bibenzyl

Tricyclic and Quadruple Products

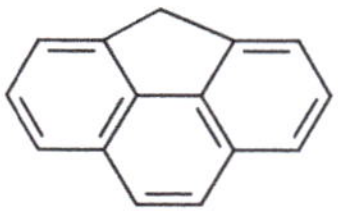

4H-Cyclopenta[def]phenanthrene

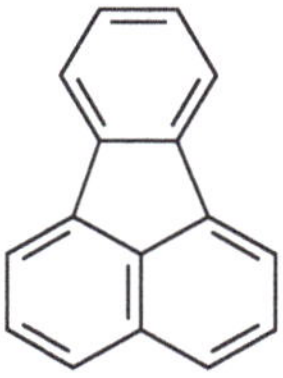

Fluoranthene

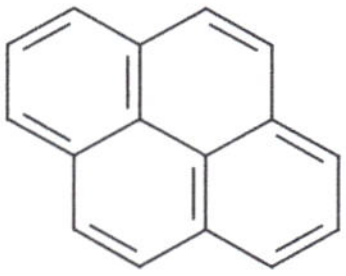

Pyrene

3. Experimental

3.1. Experimental Setup

Figure 21 schematically shows the rotating gliding arc (RGA) experimental setup for model tar destruction. The whole system consists of 4 parts: The Feeding System, the Reactor System, the Circuit System, and the Measurement System.

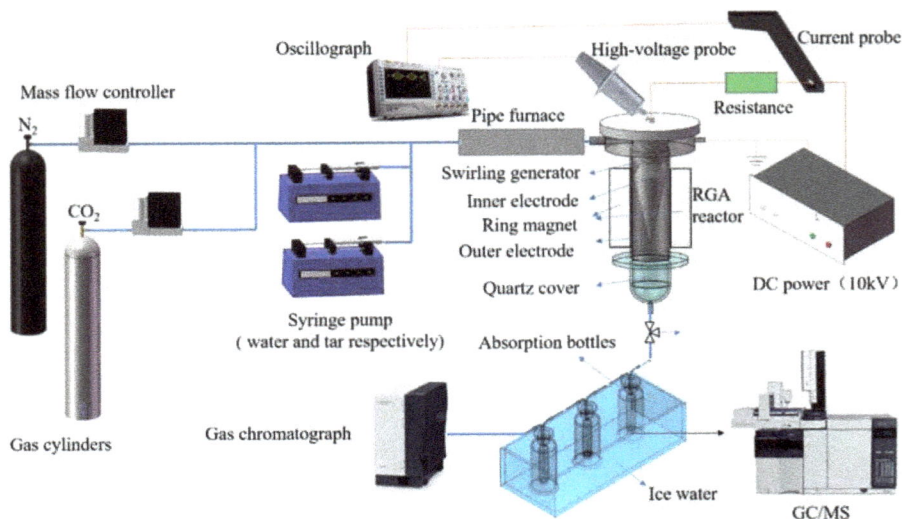

Figure 21. Schematics of the rotating gliding arc (RGA) assisted model tar destruction system.

Nitrogen was used as the carrier gas, and the flow rate was controlled by a mass flow controller (Sevenstar CS 200). Naphthalene and phenanthrene were dissolved in toluene with a mass ratio of 1:1:20 before mixing with H_2O in the carrier gas. Liquid tar and H_2O were injected into the gas line by two high-resolution syringe pumps (Harvard, 11 plus). Heating coils were put around the inlet of the gas line to ensure that the tar and water can be vaporized as soon as they enter into the pipe. All the reactants were mixed and preheated in a pipe furnace (400 °C) before flowing into the reactor. The gas flow after the reaction passed through three absorption bottles successively, the first two containing n-hexane and the third kept empty, to collect the entrained droplets and condensable liquid. The three absorption bottles were placed in an ice water bath.

The reactor consists of a thermostable ceramics base, a conical inner electrode (anode) and a grounded cylindrical outer electrode (cathode) as the wall of the reactor, both made of stainless steel (18/8 Cr–Ni). The upper part of the inner electrode is surrounded by a fan-shaped swirling generator to provide a stable swirling flow field. The outer electrode is surrounded by a ring magnet and with a quartz cover connected with it. The minimum distance between the inner and outer electrodes is 2 mm for the ignition of the discharge. Once the catalyst is used, 10 g of it is contained in a round quartz tray with a diameter slightly smaller than the cathode. The tray is placed downstream (18 mm in vertical distance) the inner electrode. The gaseous reactants flow through the swirling generator to form a vortex flow before pass through the gliding arc. After being ignited initially at the narrowest gap, the arc moves down to the lower point of the conical electrode. Finally, it is anchored at the tip of the inner electrode and rotates rapidly and steadily due to the combined effect of vortex flow and Lorentz force, forming a stable cylindrical plasma zone.

A DC power supply (Teslaman TLP2040, 10 kV) is used in the circuit system. The voltage and discharge current of the reactor are measured by an oscillograph (Tektronix DPO4030B) equipped with a high-voltage probe (Tektronix P6015A) and discharge current probe (Tektronix TCP393) to calculate the electrical power deposit of the reactor.

The effluent gas of the reactor passes through three absorption bottles filled with n-hexane to collect the unreacted toluene, naphthalene, and phenanthrene. Each time the working condition kept changing, before the remaining tar was collected, 10 min passed before the reactor stabilized. A three-way valve is used to control the exhaust into absorption bottles or to the atmosphere. After 10 min collection, the liquid in all three absorption bottles is collected and diluted with n-hexane to a final volume of 100 mL. The gas products are collected in a gas bag after passing through the absorption bottles. The collected gas products are quantitatively analyzed by gas chromatography (Agilent Micro GC 490, Agilent Technologies Inc., Santa Clara, CA, USA) and the liquid products dissolved in the absorption solution are detected by the gas chromatography with mass spectrometry (GC/MS, Agilent 6890N GC/5977B MSD, Agilent Technologies Inc., Santa Clara, CA, USA). The column temperature of GC (Agilent Micro GC 490) is 80 °C. For the GC/MS detection, the initial temperature of the oven is 40 °C, holding time for 4 min; the temperature increases to 140 °C at a rate of 15 °C/min, increases to 210 °C at a rate of 20 °C/min, increases to 280 °C at a rate of 30 °C/min, then holds for 10 min. The injector temperature is set to 300 °C. Helium is used as the carrier gas, with a flow rate of 1.0 mL/min. For the mass spectrometer, the electron impact ionization is maintained at 70 eV, and the ion source temperature is 230 °C. The unreacted toluene, naphthalene, and phenanthrene are quantitatively detected based on the obtained calibration curve, and the other liquid by-products are qualitatively measured.

3.2. Catalyst Preparation

The Ni/γ-Al$_2$O$_3$ catalyst was prepared using the impregnation method reported by Qian et al. [42], in which the γ-Al$_2$O$_3$ particles were added into an aqueous solution of nickel nitrate under stirring. The mixture was kept at room temperature overnight before being evaporated out at 80 °C. The solid was dried at 110 °C for 10 h before final calcination at 700 °C for 3 h. The catalyst was formulated with Ni loadings of 16 wt.%. The γ-Al$_2$O$_3$ particles are spherical with a diameter of 2 mm.

3.3. Reaction Performance Parameters

The destruction efficiency η_d (toluene, naphthalene or phenanthrene), energy efficiency η_e, hydrogen selectivity S(H$_2$), CO, CO$_2$ and C$_2$H$_2$ selectivity S(C$_i$H$_j$O$_k$, C$_i$H$_j$O$_k$ = C$_2$H$_2$, CO, CO$_2$), energy consumption per liter hydrogen production W(H$_2$) is defined as in Equations (15)–(19).

$$\eta_d = \frac{c_{in}(X) - c_{out}(X)}{c_{in}(X)} \times 100\% \tag{15}$$

$$\eta_e = \frac{(\rho_{in}(tar) - \rho_{out}(X)) \times Q/60}{P/3600000} \times 100\% \tag{16}$$

$$S(H_2) = \frac{C_{out}(H_2)}{4 \times (C_{in}(tar) - C_{out}(tar))} \times 100\% \tag{17}$$

$$S(C_iH_jO_k) = \frac{i \times C_{out}(C_iH_jO_k)}{(14 \times (C_{in}(C_{14}H_{10}) - C_{out}(C_{14}H_{10})) + 10 \times (C_{in}(C_{10}H_8) - C_{out}(C_{10}H_8)) + 7 \times (C_{in}(C_7H_8) - C_{out}(C_7H_8))) \times 100\%} \tag{18}$$

$$W(H_2) = \frac{P/1000}{C_{out}(H_2) \times Q/60} \times 100\% \tag{19}$$

where, $c_{in}(X)$ and $C_{out}(X)$ (mol/L) are the molar concentration of substance X in the inflow gas and effluent gas respectively. X is toluene or naphthalene.

$\rho_{in}(tar)$ and $\rho_{out}(tar)$ (g/L) are the mass concentration of simulated tar (toluene and naphthalene) in the inflow gas and effluent gas respectively. Q (L/min) is the total flow rate. P (W) is the power input on the reactor.

$C_{out}(H_2)$, $C_{in}(tar)$ and $C_{out}(tar)$ (mol/L) are the molar concentration of input H$_2$, input simulated tar and output simulated tar respectively.

$C_{out}(C_iH_jO_k)$, $C_{out}(C_{10}H_8)$, $C_{out}(C_7H_8)$, $C_{in}(C_{10}H_8)$ and $C_{in}(C_7H_8)$ (mol/L) are the molar concentration of output C$_i$H$_j$O$_k$, output naphthalene, output toluene, input naphthalene, and input toluene, respectively.

4. Conclusions

In this study, a modified rotating gliding arc discharge reactor equipped with a fan-type swirling generator is developed for the destruction of model tar compounds (toluene, naphthalene, and phenanthrene). The influence of tar concentration, discharge current, different atmospheres (CO_2, H_2O), and the addition of catalyst on the destruction efficiency is investigated in this study. The destruction efficiency of all three tar components decreases as tar concentration increases, not only because of the limitation of the destruction capacity, but also the instability caused by the addition of phenanthrene. Mostly, 8%–12% of moisture benefits the reaction, while a further increase of moisture decreases the destruction efficiencies.

Although the participation of CO_2 increases the energy consumption of the reaction, a moderate amount of it (about 15%) improves the destruction efficiency by stabilizing the reactor, which is desirable. The toluene, naphthalene, and phenanthrene destruction efficiency reached up to 95.2%, 88.9%, and 83.9% respectively, with a tar content of 12 g/Nm^3, 12% moisture content, and 15% CO_2 content, with a flow rate of 6 NL/min. Increasing the discharge current causes the increase of destruction efficiency slightly (by 3%) but lowers the energy efficiency (by more than 10%). It is notable that the increase of the discharge current can increase the selectivity of carbon-contained gas products from 80% to 88%, as the discharge current increases from 200 mA to 250 mA. The increased production of flammable gaseous products may not make up the power consumption in the reaction but decreasing black carbon production by almost 50% is extremely desirable.

The addition of Ni/γ-Al_2O_3 catalyst shows little effect as the flow rate is low, but great improvement in destruction efficiency is achieved when flow rate is high (9 or 12 NL/min), indicating that catalyst coupling with plasma is worth considering in industrial application to deal with large amounts of tars.

Author Contributions: Conceptualization, H.Z., X.L. and J.Y.; methodology, H.Z. and X.K.; investigation, X.K., R.X., L.L. and H.Z.; writing—original draft preparation, X.K.; writing—review and editing, H.Z. and I.M.; supervision, X.L. and J.Y.; funding acquisition, X.L. and J.Y.

Acknowledgments: This work was supported by the National Natural Science Foundation of China [No. 51576174, No. 51706204, No. 51621005] and the Fundamental Research Funds for the Central Universities (No. 2018FZA4010).

Conflicts of Interest: The authors declare no conflict of interest.

References

1. Sun, J.; Wang, Q.; Wang, W.; Song, Z.; Zhao, X.; Mao, Y.; Ma, C. Novel treatment of a biomass tar model compound via microwave-metal discharges. *Fuel* **2017**, *207*, 121–125. [CrossRef]
2. Wnukowski, M.; Jamróz, P. Microwave plasma treatment of simulated biomass syngas: Interactions between the permanent syngas compounds and their influence on the model tar compound conversion. *Fuel Process. Technol.* **2018**, *173*, 229–242. [CrossRef]
3. Luo, H.; Bao, L.; Wang, H.; Kong, L.; Sun, Y. Microwave-assisted in-situ elimination of primary tars over biochar: Low temperature behaviors and mechanistic insights. *Bioresource Technol.* **2018**, *267*, 333–340. [CrossRef] [PubMed]
4. Arena, U. Process and technological aspects of municipal solid waste gasification. A review. *Waste Manag.* **2012**, *32*, 625–639. [CrossRef] [PubMed]
5. Baratieri, M.; Baggio, P.; Bosio, B.; Grigiante, M.; Longo, G.A. The use of biomass syngas in IC engines and CCGT plants: A comparative analysis. *Appl. Therm. Eng.* **2009**, *29*, 3309–3318. [CrossRef]
6. Han, J.; Kim, H. The reduction and control technology of tar during biomass gasification/pyrolysis: An overview. *Renew. Sustain. Energy Rev.* **2008**, *12*, 397–416. [CrossRef]
7. Guan, G.; Kaewpanha, M.; Hao, X.; Wang, Z.; Cheng, Y.; Kasai, Y.; Abudula, A. Promoting effect of potassium addition to calcined scallop shell supported catalysts for the decomposition of tar derived from different biomass resources. *Fuel* **2013**, *109*, 241–247. [CrossRef]
8. Zhu, F.; Li, X.; Zhang, H.; Wu, A.; Yan, J.; Ni, M.; Zhang, H.; Buekens, A. Destruction of toluene by rotating gliding arc discharge. *Fuel* **2016**, *176*, 78–85. [CrossRef]

9. Medeiros, H.S.; Pilatau, A.; Nozhenko, O.S.; Da Silva Sobrinho, A.S.; Petraconi Filho, G. Microwave Air Plasma Applied to Naphthalene Thermal Conversion. *Energy Fuels* **2016**. [CrossRef]

10. Devi, L.; Ptasinski, K.J.; Janssen, F.J.J.G. Pretreated olivine as tar removal catalyst for biomass gasifiers: Investigation using naphthalene as model biomass tar. *Fuel Process. Technol.* **2005**, *86*, 707–730. [CrossRef]

11. Liu, S.; Mei, D.; Wang, L.; Tu, X. Steam reforming of toluene as biomass tar model compound in a gliding arc discharge reactor. *Chem. Eng. J.* **2017**, *307*, 793–802. [CrossRef]

12. Świerczyński, D.; Libs, S.; Courson, C.; Kiennemann, A. Steam reforming of tar from a biomass gasification process over Ni/olivine catalyst using toluene as a model compound. *Appl. Catal. B Environ.* **2009**, *74*, 211–222. [CrossRef]

13. Tursun, Y.; Xu, S.; Wang, G.; Wang, C.; Xiao, Y. Tar formation during co-gasification of biomass and coal under different gasification condition. *J. Anal. Appl. Pyrolysis* **2015**, *111*, 191–199. [CrossRef]

14. Jess, A. Mechanisms and kinetics of thermal reactions of aromatic hydrocarbons from pyrolysis of solid fuels. *Fuel* **1996**, *75*, 1441–1448. [CrossRef]

15. Saleem, F.; Zhang, K.; Harvey, A. Role of CO_2 in the Conversion of Toluene as a Tar Surrogate in a Nonthermal Plasma Dielectric Barrier Discharge Reactor. *Energy Fuels* **2018**, *32*, 5164–5170. [CrossRef]

16. Richardson, Y.; Blin, J.; Julbe, A. A short overview on purification and conditioning of syngas produced by biomass gasification: Catalytic strategies, process intensification and new concepts. *Prog. Energy Combust. Sci.* **2012**, *38*, 765–781. [CrossRef]

17. Nunnally, T.; Tsangaris, A.; Rabinovich, A.; Nirenberg, G.; Chernets, I.; Fridman, A. Gliding arc plasma oxidative steam reforming of a simulated syngas containing naphthalene and toluene. *Int. J. Hydrogen Energy* **2014**, *39*, 11976–11989. [CrossRef]

18. Chun, Y.N.; Kim, S.C.; Yoshikawa, K. Destruction of anthracene using a gliding arc plasma reformer. *Korean J. Chem. Eng.* **2011**, *28*, 1713–1720. [CrossRef]

19. Ashok, J.; Kawi, S. Steam reforming of toluene as a biomass tar model compound over CeO_2 promoted Ni/CaO–Al_2O_3 catalytic systems. *Int. J. Hydrogen Energy* **2013**, *38*, 13938–13949. [CrossRef]

20. Nair, S.A.; Pemen, A.J.M.; Yan, K.; Gompel, F.M.V.; Leuken, H.E.M.V.; Heesch, E.J.M.V.; Ptasinski, K.J.; Drinkenburg, A.A.H. Tar removal from biomass-derived fuel gas by pulsed corona discharges. *Fuel Process. Technol.* **2003**, *84*, 161–173. [CrossRef]

21. Pemen, A.J.M.; Nair, S.A.; Yan, K.; Heesch, E.J.M.V.; Ptasinski, K.J.; Drinkenburg, A.A.H. Pulsed Corona Discharges for Tar Removal from Biomass Derived Fuel Gas. *Plasmas Polym.* **2003**, *8*, 209–224. [CrossRef]

22. Sun, J.; Wang, Q.; Wang, W.; Wang, K. Exploiting the Photocatalytic Effect of Microwave–Metal Discharges for the Destruction of a Tar Model Compound. *Energy Fuels* **2018**, *32*, 241–245. [CrossRef]

23. Jamróz, P.; Kordylewski, W.; Wnukowski, M. Microwave plasma application in decomposition and steam reforming of model tar compounds. *Fuel Process. Technol.* **2018**. [CrossRef]

24. Wnukowski, M. Decomposition of Tars in Microwave Plasma—Preliminary Results. *J. Ecol. Eng.* **2014**, *15*, 23–28.

25. Materazzi, M.; Lettieri, P.; Mazzei, L.; Taylor, R.; Chapman, C. Reforming of tars and organic sulphur compounds in a plasma-assisted process for waste gasification. *Fuel Process. Technol.* **2015**, *137*, 259–268. [CrossRef]

26. Materazzi, M.; Lettieri, P.; Mazzei, L.; Taylor, R.; Chapman, C. Tar evolution in a two-stage fluid bed–plasma gasification process for waste valorization. *Fuel Process. Technol.* **2014**, *128*, 146–157. [CrossRef]

27. Lim, M.; Chun, Y. Light Tar Decomposition of Product Pyrolysis Gas from Sewage Sludge in a Gliding Arc Plasma Reformer. *Environ. Eng. Res.* **2012**, *17*, 89–94. [CrossRef]

28. Yang, Y.C.; Chun, Y.N. Naphthalene destruction performance from tar model compound using a gliding arc plasma reformer. *Korean J. Chem. Eng.* **2011**, *28*, 539–543. [CrossRef]

29. Tippayawong, N.; Inthasan, P. Investigation of Light Tar Cracking in a Gliding Arc Plasma System. *Int. J. Chem. React. Eng.* **2010**. [CrossRef]

30. Du, C.M.; Yan, J.H.; Cheron, B. Decomposition of toluene in a gliding arc discharge plasma reactor. *Plasma Sources Sci. Technol.* **2007**, *16*, 791–797. [CrossRef]

31. Snoeckx, R.; Bogaerts, A. Plasma technology—A novel solution for CO_2 conversion. *Chem. Soc. Rev.* **2017**, *46*, 585–586. [CrossRef]

32. Pemen, A.J.M.; van Paasen, S.V.B.; Yan, K.; Nair, S.A.; van Heesch, E.J.M.; Ptasinski, K.J.; Neeft, J.P.A. Conditioning of biomass derived fuel gas using plasma techniques. In Proceedings of the 12th European Conference on Biomass for Energy, Industry and Climate Protection, Amsterdam, The Netherlands, 17–21 June 2002.

33. Nunnally, T.P.; Gutsol, A.; Fridman, A. Dissociation of H_2S in non-equilibrium gliding arc 'tornado' discharge. *Int. J. Hydrogen Energy* **2009**, *34*, 7618–7625. [CrossRef]

34. Fridman, A.; Nester, S.; Kennedy, L.A.; Saveliev, A.; Mutaf-Yardimci, O. Gliding arc gas discharge. *Prog. Energy Combust. Sci.* **1999**, *25*, 211–231. [CrossRef]

35. Zhang, H.; Zhu, F.; Li, X.; Cen, K.; Du, C.; Tu, X. Rotating Gliding Arc Assisted Water Splitting in Atmospheric Nitrogen. *Plasma Chem. Plasma Process.* **2016**, *36*, 813–834. [CrossRef]

36. Zhu, F.; Zhang, H.; Yang, J.; Yan, J.; Ni, M.; Li, X. Plasma-assisted Toluene Destruction in Simulated Producer Gas. *Chem. Lett.* **2017**, *46*, 1341–1343. [CrossRef]

37. Materazzi, M.; Lettieri, P.; Taylor, R.; Chapman, C. Performance analysis of RDF gasification in a two-stage fluidized bed–plasma process. *Waste Manag.* **2016**, *47*, 256–266. [CrossRef]

38. Ma, W.; Han, L.; Zhang, L.; Lu, W. Thermal-reforming of toluene over core-shell Ni/γ-Al_2O_3 catalysts. In Proceedings of the IEEE International Conference on Materials for Renewable Energy & Environment, Chengdu, China, 19–21 August 2014.

39. Chun, Y.N.; Kim, S.C.; Yoshikawa, K. Decomposition of benzene as a surrogate tar in a gliding arc plasma. *Environ. Prog. Sustain.* **2013**, *32*, 837–845. [CrossRef]

40. Gao, J.; Zhu, J.; Ehn, A.; Aldén, M.; Li, Z. In-Situ Non-intrusive Diagnostics of Toluene Removal by a Gliding Arc Discharge Using Planar Laser-Induced Fluorescence. *Plasma Chem. Plasma Process.* **2017**, *37*, 433–450. [CrossRef]

41. Yan, X.; Li, X.D.; Zhu, F.S.; Kong, X.Z.; Yan, J.H. Decomposition of naphthalene as tar model compound from the gasification of municipal solid waste by rotating gliding arc plasma. *Chem. Ind. Eng. Prog.* **2018**, *37*, 1174–1180.

42. Qian, N.; Zhang, L.; Ma, W.; Zhao, X.; Han, L.; Lu, W. Core–Shell Al_2O_3-Supported Ni for High-Performance Catalytic Reforming of Toluene as a Model Compound of Tar. *Arab. J. Sci. Eng.* **2014**, *39*, 6671–6678. [CrossRef]

catalysts

MDPI

Article

High-Efficiency Catalytic Conversion of NO$_x$ by the Synergy of Nanocatalyst and Plasma: Effect of Mn-Based Bimetallic Active Species

Yan Gao [1,2,3,*] , Wenchao Jiang [1], Tao Luan [4,*], Hui Li [1,2,3], Wenke Zhang [1,2,3], Wenchen Feng [4] and Haolin Jiang [4]

[1]   Department of Thermal Engineering, Shandong Jianzhu University, Jinan 250101, China;
      jiang_wc@126.com (W.J.); lihui_sdjzu@sina.com (H.L.); zhangwk10@126.com (W.Z.)
[2]   Key Laboratory of Renewable Energy Building Utilization Technology of Ministry of Education,
      Shandong Jianzhu University, Jinan 250101, China
[3]   Key Laboratory of Renewable Energy Building Application Technology of Shandong Province,
      Shandong Jianzhu University, Jinan 250101, China
[4]   Engineering Laboratory of Power Plant Thermal System Energy Saving of Shandong Province,
      Shandong University, Jinan 250061, China; wcfeng18@126.com (W.F.); haolin_jiang@hotmail.com (H.J.)
*   Correspondence: gaoyan.sdu@hotmail.com (Y.G.); prof.luantao@gmail.com (T.L.);
     Tel.: +86-138-6415-4887 (Y.G.); +86-175-1531-9316 (T.L.)

Received: 30 November 2018; Accepted: 16 January 2019; Published: 18 January 2019

Abstract: Three typical Mn-based bimetallic nanocatalysts of Mn−Fe/TiO$_2$, Mn−Co/TiO$_2$, Mn−Ce/TiO$_2$ were synthesized via the hydrothermal method to reveal the synergistic effects of dielectric barrier discharge (DBD) plasma and bimetallic nanocatalysts on NO$_x$ catalytic conversion. The plasma-catalyst hybrid catalysis was investigated compared with the catalytic effects of plasma alone and nanocatalyst alone. During the catalytic process of catalyst alone, the catalytic activities of all tested catalysts were lower than 20% at ambient temperature. While in the plasma-catalyst hybrid catalytic process, NO$_x$ conversion significantly improved with discharge energy enlarging. The maximum NO$_x$ conversion of about 99.5% achieved over Mn−Ce/TiO$_2$ under discharge energy of 15 W·h/m^3 at ambient temperature. The reaction temperature had an inhibiting effect on plasma-catalyst hybrid catalysis. Among these three Mn-based bimetallic nanocatalysts, Mn−Ce/TiO$_2$ displayed the optimal catalytic property with higher catalytic activity and superior selectivity in the plasma-catalyst hybrid catalytic process. Furthermore, the physicochemical properties of these three typical Mn-based bimetallic nanocatalysts were analyzed by N$_2$ adsorption, Transmission Electron Microscope (TEM), X-ray diffraction (XRD), H$_2$-temperature-programmed reduction (TPR), NH$_3$-temperature-programmed desorption (TPD), and X-ray photoelectron spectroscopy (XPS). The multiple characterizations demonstrated that the plasma-catalyst hybrid catalytic performance was highly dependent on the phase compositions. Mn−Ce/TiO$_2$ nanocatalyst presented the optimal structure characteristic among all tested samples, with the largest surface area, the minished particle sizes, the reduced crystallinity, and the increased active components distributions. In the meantime, the ratios of Mn^{4+}/(Mn^{2+} + Mn^{3+} + Mn^{4+}) in the Mn−Ce/TiO$_2$ sample was the highest, which was beneficial to plasma-catalyst hybrid catalysis. Generally, it was verified that the plasma-catalyst hybrid catalytic process with the Mn-based bimetallic nanocatalysts was an effective approach for high-efficiency catalytic conversion of NO$_x$, especially at ambient temperature.

Keywords: NO$_x$ conversion; DBD plasma; Manganese; bimetal; nanocatalyst

1. Introduction

Nitrogen oxides (NO_x) are regard as the main air pollutant contributing to acid rain, photochemical smog, greenhouse effects, and ozone depletion [1]. Selective catalytic reduction (SCR) of NO_x by NH_3 or urea is proposed to be the highly effective and completely developed method to eliminate NO_x pollution [2]. In coal fired power plants, the commercial catalyst of V_2O_5-$WO_3(MoO_3)$/TiO_2 is used for its excellent catalytic performance in the typical standard SCR reaction [3]:

$$4NO + 4NH_3 + O_2 \rightarrow 4N_2 + 6H_2O \tag{1}$$

While the $V_2O_5-WO_3(MoO_3)$/TiO_2 catalysts demand a strict temperature window of 300–400 °C, which limit the arrangement flexibility of this kind of catalyst. The vanadium-based catalysts can not reach satisfactory efficiency of eliminating NO_x when the reaction temperature is lower than 250 °C. In recent years, the fast selective catalytic reduction (fast SCR) attracted the attention of many research groups due to its lower reaction temperature and higher reaction efficiency [4]:

$$NO + NO_2 + 2NH_3 \rightarrow 2N_2 + 3H_2O \tag{2}$$

The catalysts appropriate to low temperature SCR are strongly desired, which could be located at downstream electrostatic precipitator and desulfurizer suitably [5]. However, the fast SCR still needs reaction temperature within 150–300 °C to achieve high efficiency of NO_x elimination [4,6]. Furthermore, the mole ratio of $NO:NO_2$ maintained at 1:1 is difficult in the real flue gas. Hence, it is necessary to develop an effective approach to eliminate NO_x with light concentration of NO_2 at low temperature region, which could be beneficial to the $deNO_x$ device arrangement, as well as the SO_2 resistance.

Plasma-catalyst hybrid catalysis has been proved as an efficient technology to unite the high reactivity of plasma and the high selectivity of catalyst [7–9]. During the plasma-catalyst hybrid process, the plasma modifies not only the chemical properties and morphologies of the catalysts, but also changes the reaction pathway of an original catalytic process [10]. Plasma is confirmed to form an abundance of active species, such as O and O_3 radicals, which could oxidize NO into NO_2, further promoting catalysis via the fast SCR approach, especially at low temperature [4]. For the plasma-catalyst hybrid catalysis, the catalysts of $V_2O_5-WO_3$/TiO_2 [11], Ag/r-Al_2O_3 [12], Cu-ZSM-5 [13], and $Mn-Ce$/ZSM5-MWCNTs [4] have presented acceptable NO_x conversion efficiency under relatively low specific input energy. While the NO_x conversion maximum could still be further promoted at lower reaction temperature and smaller energy consume. Among the various transition metal elements applied in the catalysts for NO_x reduction, manganese displays superior activity especially at the low temperature, which can be attributed to the multifarious types of labile oxygen and high mobility of valence states [1]. Meanwhile, it has been found that iron, cobalt, and cerium species can combine with manganese to produce bimetallic catalysts, which contain abundant oxygen vacancies on the catalyst surface, forming strong interaction bands at atomic scale, such as Mn-O-Fe [14], Mn-O-Co [15], and Mn-O-Ce [16]. Moreover, the active metal species of FeO_x, CoO_x, and CeO_x are also regarded as the three typical promoters for NO_x conversion, which serve as core catalyst components of active metal oxides, supplying surface oxygen to accelerate NO_x elimination [14,15,17]. However, the effects of Mn-based bimetallic catalysts on the plasma-catalyst hybrid catalysis, especially the $Mn-Fe$/TiO_2, $Mn-Co$/TiO_2, and $Mn-Ce$/TiO_2 nanocatalysts have not been explored clearly.

In this study, we systematically synthesized three typical Mn-based bimetallic nanocatalysts of $Mn-Fe$/TiO_2, $Mn-Co$/TiO_2, and $Mn-Ce$/TiO_2. The synergistic effects of non-thermal plasma and Mn-based bimetallic nanocatalysts on NO_x catalytic conversion were investigated compared with the catalytic effects of plasma alone and nanocatalysts alone. Meanwhile, the influence factors of reaction temperature and discharge energy were taken into consideration during studying the synergetic mechanisms focusing on NO_x conversion of plasma and bimetallic nanocatalysts hybrid system. Furthermore, the physicochemical properties of these three typical Mn-based bimetallic nanocatalysts were analyzed by Brunauer-Emmett-Teller (BET), transmission electron microscopy (TEM), X-ray diffraction (XRD), H_2-temperature-programmed reduction (TPR), NH_3-temperature-programmed desorption (TPD) and X-ray photoelectron spectroscopy (XPS), in order to expose the relationship between structures and activities. The purpose of this work was mean to explore the synergistic reinforcement mechanism of plasma-catalysis hybrid catalytic process over Mn-based bimetallic nanocatalysts for NO_x elimination with high catalytic efficiency and satisfied catalytic selectivity, especially at atmospheric temperature.

2. Results and Discussion

2.1. NO_x Conversion of Catalyst Alone Catalytic Process

The NO_x catalytic conversion and the catalytic selectivity of three typical Mn-based bimetallic nanocatalysts of Mn−Fe/TiO_2, Mn−Co/TiO_2, and Mn−Ce/TiO_2 were exhibited in Figure 1, and the catalytic ability of Mn/TiO_2 catalyst was also depicted as a contrast. For all the tested Mn-based bimetallic nanocatalysts, the NO_x conversion increased significantly with the temperature rising from 25 °C to 250 °C and presented excellent performance (>90%, above 150 °C). Compared with the Mn/TiO_2 catalyst, the catalytic activities of Mn-based bimetallic nanocatalysts were remarkably improved at the whole temperature range, potentially due to the strong interaction of Mn−O−X bond (X refered to Fe, Co or Ce), the improvement of Brønsted acid sites and Lewis acid sites, and the enhancement of Eley-Rideal (E-R) mechanism reaction [18], which could be further verified by the following physicochemical properties. As shown in Figure 1a, Mn−Ce/TiO_2 nanocatalyst achieved higher catalytic activity than the other samples in the temperature range of 25~200 °C. The Mn−Fe/TiO_2 nanocatalyst showed the minimum NO_x conversion among these three Mn-based bimetallic nanocatalysts, while still much larger than that of Mn /TiO_2 sample. However, the catalytic selectivity of Mn−Fe/TiO_2 nanocatalyst was lower than that of Mn−Co/TiO_2 and Mn−Ce/TiO_2 within 175~250 °C, as exhibited in Figure 1b. Furthermore, it could be easy to find there was no obvious difference of NO_x conversion or catalytic selectivity over these three Mn-based bimetallic nanocatalysts at ambient temperature, which was proposed to be due to the low catalytic activities for all the tested catalysts.

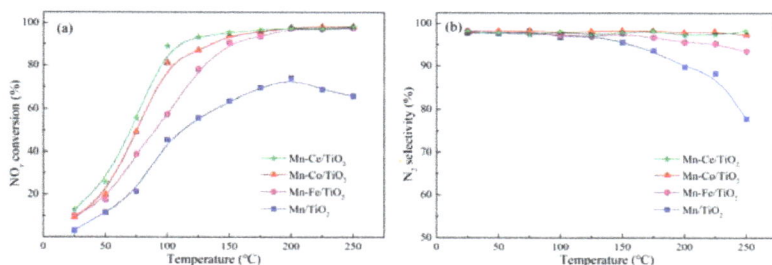

Figure 1. Catalytic performance of catalysts without plasma. Gas mixture composition: 300 ppm NO, 300 ppm NH_3, 8% O_2, ~0.1% H_2O and N_2 as balance gas. Gas hourly space velocity (GHSV) 20,000 h^{-1}. (a) NO_x conversion of Mn-based nanocatalysts; and (b) N_2 selectivity of Mn-based nanocatalysts.

2.2. NO_x Conversion of Plasma-Catalyst Hybrid Catalytic Process

The NO conversion and NO_2 concentration over three typical Mn-based bimetallic nanocatalysts were compared in Figure 2. The performance of all prepared nanocatalysts were measured in terms of various discharge energies to reveal the interaction of Mn−O−Fe, Mn−O−Co, and Mn−O−Ce. As shown in Figure 2a, both Mn−Co/TiO_2 and Mn−Ce/TiO_2 nanocatalysts could reach NO_x conversion maximum >99% within the discharge energy range of 18~24 W·h/m^3. While the start discharge energy of Mn−Ce/TiO_2 nanocatalyst with superior SCR activities was much lower than that of Mn−Co/TiO$_2$nanocatalyst. The Mn−Ce/TiO$_2$ bimetallic nanocatalyst raised the optimal NO$_x$ conversion to 93.3% with the relatively low discharge energy of 12 W·h/m^3. For the other Mn-based bimetallic nanocatalysts, a lower NO$_x$ elimination efficiency was achieved with NO$_x$ conversion less than 85% at 15 W·h/m^3 and the maximum obtained at 24 W·h/m^3, which meant that the higher discharge energy was required to induce the plasma-catalyst catalytic process, and the narrower discharge energy window was limited to the hybrid catalytic reaction.

The N$_2$ and O$_2$ contained in the gas mixture were motivated to form N and O atoms via the collision of active electrons in the plasma-catalyst hybrid system. Compared to the chemical-bond dissociation energies of N$_2$ (945.33 kJ/mol), the O$_2$ was much easier to react with the energetic electrons for its lower chemical-bond dissociation energies of 498.36 kJ/mol. As a result, a high concentration of O radicals was produced in the plasma-catalyst system. The generated dominating O radicals and subordinate N radicals could react with NO/O$_2$/N$_2$/NH$_3$ gas mixture in the following reactions (3)~(8) [19]. The oxidation reactions (5) and (6) occurred between the radicals of O and O$_3$ and the NO molecules to generate NO$_2$ were regarded as the positive main steps to enhance NO conversion [7,9].

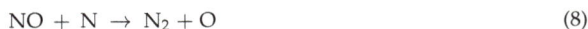

$$O + O_2 \rightarrow O_3 \tag{3}$$

$$O_3 + NO \rightarrow NO_2 + O_2 \tag{4}$$

$$O + NO \rightarrow NO_2 \tag{5}$$

$$O + NO_2 \rightarrow NO + O_2 \tag{6}$$

$$O + N \rightarrow NO \tag{7}$$

$$NO + N \rightarrow N_2 + O \tag{8}$$

Thus, in the plasma-catalyst hybrid catalytic process, the catalytic reactions (2) and (3) become the predominant paths for NO elimination [4]. It had been testified that the reaction rate of fast SCR reaction (2) was more than 10 times larger than that of standard SCR reaction (1) below 200 °C [17].

Meanwhile, the NO$_2$ concentrations over Mn-based bimetallic nanocatalysts were relatively lower compared to that of plasma without catalyst assistance. Under discharge energy of 24 W·h/m^3, more than 120 ppm NO$_2$ generated in the plasma-only catalytic process. However, the NO$_2$ concentration in Mn-based bimetallic nanocatalysts combining with plasma was no more than 20 ppm, which indicated that almost 100 ppm NO$_2$ took part in the catalytic reaction probably via the fast SCR reaction or the catalytic oxidation, as shown in Figure 2b. Therefore, it was believed that both the fast SCR and the standard SCR reactions occurred in the plasma-catalyst hybrid system simultaneity and the proportion of NO$_x$ conversion via the fast SCR reaction improved with the discharge energy increasing. The N$_2$ selectivity over the Mn-based nanocatalysts was displayed in Figure 2c. The N$_2$ selectivity of the plasma-catalyst hybrid catalytic process was obviously larger than that of plasma-only process within discharge energy range of 0~24 W·h/m^3, which was owing to the possibility of higher NO conversion and lower NO$_2$ formation, discussed above in Figure 2a,b. All test results presented a decreasing trend of N$_2$ selectivity with the discharge energy rising, which resulted from a great deal of N$_2$O produced in this reaction operation. It was proposed that the pivotal disadvantages of catalyzing NO$_x$ by plasma were the low selectivity and the complex chemical productions that formed via diverse reaction pathways [10]. In order to verify the actual reactions during the plasma-catalyst hybrid process over the Mn-based bimetallic nanocatalysts, the NO$_x$ conversion and the N$_2$ selectivity over Mn−Ce/TiO$_2$ sample in the balance gas of N$_2$ and Ar were tested, as shown in Figure 2d. It was obvious that the variation tendency of NO$_x$ conversion obtained in the balance gas of N$_2$ and Ar were quite similar. While within the whole discharge energy range of 0~24 W·h/m^3, the NO$_x$ conversion in Ar was slightly higher. According to a previous report, under abundant O radicals or O$_2$, the N species is ten times more likely to react with O$_2$ than with NO [20]. Hence, almost N atoms produced from N$_2$ in the plasma transformed to NO via reaction (7), which was further oxidized into NO$_2$ and eliminated via fast SCR reactions immediately [7]. Therefore, in the balance gas of N_2, the NO concentration formed from N and O radicals was relatively small compare to the initial NO_x concentration, which caused little influence on the NO_x conversion during the plasma-catalyst hybrid process. Meanwhile, there was no obvious difference between the N_2 selectivity obtained in the balance gas of N_2 and Ar. The NO_x conversions over $Mn-Ce/TiO_2$ nanocatalyst with and without O_2 were analyzed as exhibited in Figure 2e. The NO_x conversion decreased drastically from 99.1% to 43% with the O_2 concentration dropping from 8% to 0%, which demonstrated the oxidation pathway for NO reduction by O species via reactions (3), (4), and (5) was dominant during the plasma-catalyst hybrid process. The NO_x conversions under O_2 8% and 4% were almost the same, indicating the amount of oxygen excessive for NO_x redox reactions. Due to the dissociation energy of O_2 much smaller than that of N_2, the rate for dissociation of O_2 was much higher compared to the dissociation of N_2, which was the main reason for the remarkable promotion of O_2 on NO_x conversion [21].

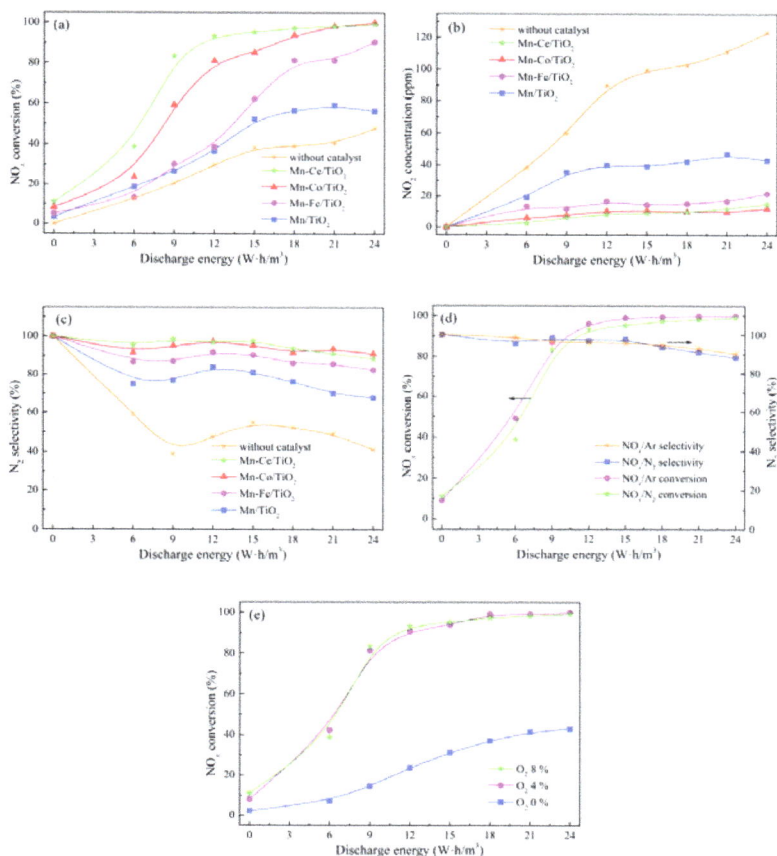

Figure 2. Catalytic performance of plasma-catalyst hybrid catalytic process at ambient temperature. Gas mixture composition: 300 ppm NO, 300 ppm NH_3, 8% O_2, ~0.1% H_2O, and N_2/Ar as balance gas. Gas hourly space velocity (GHSV) 20,000 h^{-1}. (**a**) NO_x conversion; (**b**) NO_2 concentration; (**c**) N_2 selectivity; (**d**) NO_x conversion and N_2 selectivity in balance gas of N_2 and Ar over $Mn-Ce/TiO_2$; and (**e**) NO_x conversion over $Mn-Ce/TiO_2$ with and without O_2.

The interaction effects of discharge energy and temperature on NO_x conversion in plasma with and without Mn-based bimetallic nanocatalysts reaction process were shown in Figure 3. The variation tendency of NO_x conversion in plasma-only process was opposite to that in plasma-catalyst hybrid process with reaction temperature increasing. As displayed in Figure 3a, high reaction temperature led to significant reduction in NO_x elimination, with the maximum catalytic conversion of 49.5% at 21 W·h/m^3, 25 °C declining to 20.5% at 12 W·h/m^3, 200 °C. In the plasma catalytic process, the NO_2 generation via interaction between the radicals of O and O_3 and the NO molecules was conducive to deNO$_x$ as analyzed above. While with the formed NO_2 accumulation in the plasma-only process, the inhibition of reaction (6) on NO elimination progressively intensified. The concentration of O radical could be improved under high temperature, which could further promote reaction (5), (6), and (7). As a result, the temperature increase spurred the formation of NO and impeded NO oxidation into NO_2 [19]. Considering the energy consume during the plasma-catalyst process, the reaction temperature in the catalyst bed could be higher than the outside of nanocatalysts. In order to clearly realize the relation between reaction temperatures and plasma, an infrared thermometer was introduced to detect the specific temperature of discharge area during the plasma process. The test results were shown in Figure 3b. The plasma energy caused the temperature of the discharge area improved at different degrees and the largest temperature increase could reach 47 °C under the discharge energy of 24 W·h/m^3. While the reaction temperature of gas mixture influenced by the plasma energy was relatively smaller with the Maximum temperature rise no larger than 13 °C, due to the short residence time of the gas mixture in the discharge area. Therefore, under the experiment conditions of this research, the plasma effects on NO conversion could be primarily analyzed by the discharge energy based on the gas mixture temperature.

It was apparent that the trends of NO conversion of these three Mn-based bimetallic nanocatalysts were consistent, as exhibited in Figure 3c–e). The NO conversion under different reaction temperatures and various discharge energies could be divided into three zones. In zone I, the NO_x conversion >90% only depended on the discharge energy and not affected by the reaction temperature. In zone II, the satisfied NO_x conversion (>90%) was achieved and both depended on the discharge energy and the reaction temperature. In zone III, it was impossibility to acquire a desired NO_x conversion. Mn−Ce/TiO$_2$ nanocatalyst presented superior catalytic property than Mn−Co/TiO$_2$ and Mn−Fe/TiO$_2$ samples with much broader zone I, which signified high NO_x conversion obtained with lower reaction temperature and the less discharge energy. A variety of previous works had revealed the optimal NO_x conversions obtained with the specific input energy varying from 4.7 to 40.3 W·h/m^3 and the temperature changing from 25 to 350 °C, as shown in Table 1. In this study, Mn−Ce/TiO$_2$ sample exhibited the superior performance with NO_x conversion of 99.5% under 15 W·h/m^3 at 25 °C, respectively, which was believed to be a potential excellent catalyst for the NO removal via the plasma-catalyst process.

Table 1. Plasma-catalyst performance in previous researches.

Samples	Specific Input Energy (W·h/m^3)	NO$_x$ Conversion (%)	Temperature (°C)	Reductant	GHSV (h^{-1})	Gas Flow Rate (m^3/h)	Ref
V$_2$O$_5$-WO$_3$/TiO$_2$	4.7 [a]	~76.5	170	NH$_3$	–	31.8	[11]
H-mordenite	5	76	160	NH$_3$	20,000	31	[7]
Ag/Al$_2$O$_3$	16.7 [a]	~91	350	C$_3$H$_6$	10,000	1.2	[12]
BaTiO$_3$-Al$_2$O$_3$	~40.3 [a]	~61.5	150	CH$_3$OH	11,000	–	[22]
Cu-ZSM-5	37.5 [a]	~90	25 [b]	C$_2$H$_4$	–	0.12	[13]
Co-ZSM-5	8.3 [a]	~70.6	150	C$_2$H$_4$+NH$_3$	1000	0.12	[22]
Co-HZSM-5	38.3	~92	300	C$_2$H$_2$	12,000	0.03	[23]
Mn−Ce/ZSM5−MWCNTs	16.7 [a]	~85	25	NH$_3$	60,000	0.12	[4]
Mn−Ce/TiO$_2$	15	99.5	25	NH$_3$	20,000	0.1	This study

[a] calculated according to the data in the report (1 W·h/m^3 = 3.6 J/L); [b] room temperature.

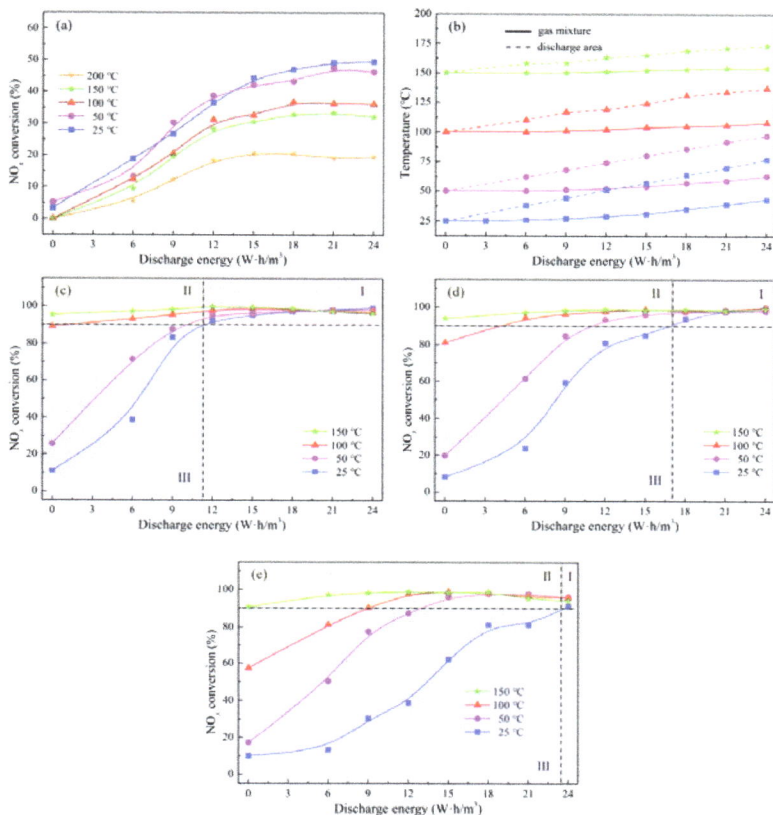

Figure 3. Effect of temperature on NO conversion. Gas mixture composition: 300 ppm NO, 300 ppm NH$_3$, 8% O$_2$, ~0.1% H$_2$O, and N$_2$ as balance gas. Gas hourly space velocity (GHSV) 20,000 h^{-1}. (**a**) plasma-only process; (**b**) plasma cooperate with Mn−Ce/TiO$_2$ nanocatalysts; (**c**) plasma cooperate with Mn−Co/TiO$_2$ nanocatalysts; and (**d**) plasma cooperate with Mn−Fe/TiO$_2$ nanocatalysts.

2.3. Morphological Characterization

2.3.1. BET Measurements

In order to achieve the physical properties of these three typical Mn-based bimetallic nanocatalysts, the results of specific surface areas (S$_{BET}$), total pore volumes (V$_{total}$), and average pore diameters (D$_p$) were summarized in Table 2. It was evident that Mn−Ce/TiO$_2$ nanocatalyst obtained larger specific surface areas than Mn−Co/TiO$_2$ nanocatalyst and was more than twice as much as Mn−Fe/TiO$_2$ nanocatalyst, which was probable, owing to the Mn−Ce−O$_x$ species better dispersed on the nanocatalyst surface. Meanwhile, there were noticeable changes of D$_p$, increasing from 17.57 nm in Mn−Ce/TiO$_2$ to 33.06 nm in Mn−Co/TiO$_2$ and further rising to 54.85 nm in Mn−Fe/TiO$_2$. It was proposed that the Mn−Ce−O$_x$ species were more likely to promote nanocatalyst to form micropores compared with Mn−Co−O$_x$ and Mn−Fe−O$_x$ species [24,25]. However, the difference of total pore volumes among these three Mn-based bimetallic nanocatalysts was not obvious. The total pore volumes of Mn−Ce/TiO$_2$ and Mn−Co/TiO$_2$ samples centered on 0.53 cm^3·g^{-1}, approximately. While the total pore volume of Mn−Fe/TiO$_2$ decreased to 0.424 cm^3·g^{-1} slightly, which was probable due to the mesoporosity formation that suppressed the micropore generation, resulting in the total pore volume reduced a little. Thereby, it was believed that Mn−Ce/TiO$_2$ nanocatalyst had superior physical properties than the other two samples with larger specific surface area, more micropores structure and satisfied total pore volumes, which coincided with catalytic performance of catalysts without plasma, as shown in Figure 1.

Table 2. Physical properties of Mn-based bimetallic nanocatalysts.

Samples	S$_{BET}$ (m^2·g^{-1})	V$_{total}$ (cm^3·g^{-1})	D$_p$ (nm)
Mn−Ce/TiO$_2$	239.7	0.527	17.57
Mn−Co/TiO$_2$	189.9	0.531	33.06
Mn−Fe/TiO$_2$	104.6	0.424	54.85

2.3.2. TEM Analysis

The morphological characterization and grain structure of these three typical Mn-based bimetallic nanocatalysts were collected by TEM analysis. From Figure 4a, it could be observed that Mn−Ce/TiO$_2$ nanocatalyst was constituted of fine uniform nanoparticles with narrow size distribution, smooth elliptic surfaces, and without evident agglomeration. The distinct and unbroken mesh structure of micropore was formed in the Mn−Ce/TiO$_2$ sample. According to the TEM images of Mn−Co/TiO$_2$ nanocatalyst, as shown in Figure 4b, there were some tightly aggregated metal oxide nanoparticles interfused into the smaller regular particles, which increased the average pore diameters and reduced the specific surface areas to some extent. However, from Figure 4c, a noticeable augment in the particle size was observed over Mn−Fe/TiO$_2$ nanocatalyst, which was consistent with Barrett–Joyner–Halenda (BJH) results. The nanoparticles were irregular, lots of which stacked on the catalyst surface with an abundant micropore structure collapsing and regional accumulations.

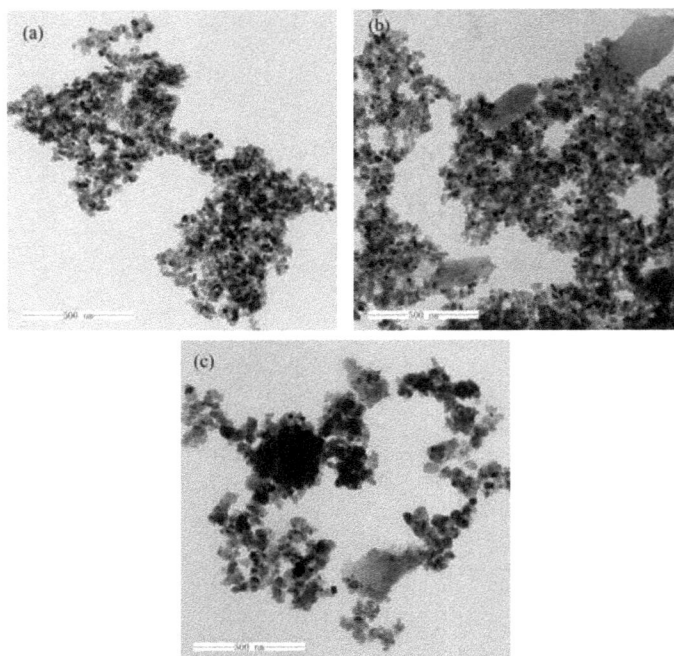

Figure 4. TEM of Mn-based bimetallic nanocatalysts. temperature on NO conversion. (**a**) Mn−Ce/TiO$_2$; (**b**) Mn−Co/TiO$_2$; and (**c**) Mn−Fe/TiO$_2$.

2.4. Structural Characterization

2.4.1. Textural Properties

Figure 5a exhibited the XRD spectra of Mn-based bimetallic nanocatalysts and the phases contained in the nanocatalyst samples were identified by the software of MDI Jade 6.5. Among all these three nanocatalysts, there were strong and distinguished diffraction peaks at about 2θ values of 25.3°, 37.8°, 48.0°, 53.9°, 62.7°, 68.8°, 70.3°, 75.1°, and 82.7° well matched the XRD pattern of anatase TiO_2 (ICDD PDF card # 71-1166) [26]. While the diffraction peaks for the structure of TiO_2 support were reserved completely, the diffraction angles of the matching peaks shifted at different degrees. In Mn−Ce/TiO_2 nanocatalyst, the anatase TiO_2 presented the lowest diffraction angle for every corresponding peak, which probably verified the interaction between $MnCeO_x$ and anatase TiO_2 was stronger than that between $MnCoO_x$ or $MnFeO_x$ and anatase TiO_2. Comparing these three nanocatalysts, it could be found that the diffraction peaks of anatase TiO_2 in Mn−Ce/TiO_2 nanocatalyst were broader and weaker than that in the other two nanocatalysts, indicating the crystalline of TiO_2 reduced by the $MnCeO_x$ loading. Meanwhile, there was no obvious characterization reflections for MnO_x or CeO_x in Mn−Ce/TiO_2 nanocatalyst that manifested the active species were finely dispersed on the nanocatalyst surface or the active species of MnO_x and CeO_x incorporated into TiO_2 lattice [27].

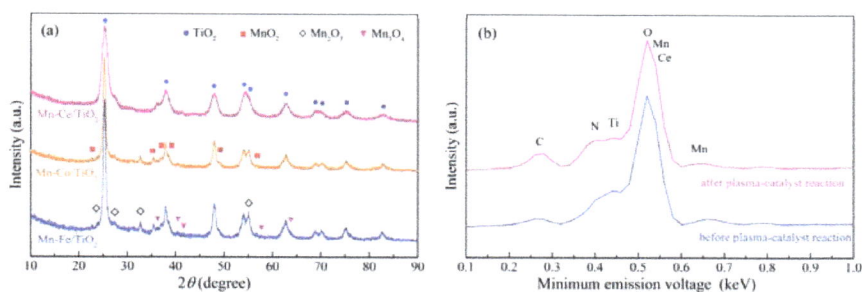

Figure 5. XRD patterns of Mn-based bimetallic nanocatalysts and element contents of Mn−Ce/TiO_2 before and after plasma-catalyst reaction. (**a**) XRD patterns of Mn−Ce/TiO_2, Mn−Co/TiO_2 and Mn−Fe/TiO_2 nanocatalysts; and (**b**) EDS patterns of element contents on the surface of Mn−Ce/TiO_2.

In the XRD patterns of Mn−Co/TiO_2 and Mn−Fe/TiO_2 nanocatalysts, the diffraction peaks accord with MnO_x were very complex due to the transformation among MnO_2, Mn_2O_3, Mn_3O_4, and MnO in the incomplete crystallization of manganese oxides. The diffraction peaks matched with MnO_2 exactly at 2θ = 22.10°, 35.19°, 36.96°, 38.72°, 47.86°, and 57.166°, corresponding to the crystallographic plane reflections of (110), (310), (201), (111), (311), and (420), respectively (ICDD PDF card # 82-2169) [28]. At the same time, the diffraction peaks of Mn_2O_3 and Mn_3O_4 were evident in Mn−Co/TiO_2 and Mn−Fe/TiO_2 nanocatalysts. The intensive and sharp characteristic peaks at 2θ values of 23.08°, 26.72°, 32.87°, and 56.89° could be primarily ascribed to Mn_2O_3 matching with the crystallographic plane reflections of (211), (220), (222), and (433), correspondingly (ICDD PDF card # 78-0390), and the distinct signals at 36.28°, 40.67°, 41.80°, 57.73°, and 64.17° could be assigned to Mn_3O_4 corresponding to the crystallographic plane reflections of (112), (130), (131), (115), and (063), respectively (ICDD PDF card # 75-0765) [28,29]. Comparing the pattern of Mn−Co/TiO_2 nanocatalyst, it could be noticed that the diffraction peaks of both Mn_2O_3 and Mn_3O_4 were remarkably decreased in Mn−Fe/TiO_2 nanocatalyst, simultaneously, the diffraction peaks matched anatase TiO_2 were also visibly weakened. These possibly suggested the addition of cobalt into manganese oxides had better effects than iron on diminishing the crystallization of MnO_x and TiO_2 at the same time. Furthermore, there were no obvious distinct diffraction peaks of CoO_x were observed in Mn−Co/TiO_2 nanocatalysts, which indicated the addition ratios of cobalt not only enhanced the dispersion of MnO_x, but also promoted the dispersion of CoO_x entirely on the nanocatalyst surface. A similar proposal could be obtained over $Mn-Fe/TiO_2$ nanocatalysts. Generally, among $Mn-Ce/TiO_2$, $Mn-Co/TiO_2$, and $Mn-Fe/TiO_2$ nanocatalysts, the $MnCeO_x$ loading on anatase TiO_2 performed the superior properties with smaller the nanoparticle sizes, reducing the chemical compounds crystallinities and increasing the active species distributions, which were facilitated to the SCR reactions [30]. In order to confirm the presence of nitrates in the mixtures during the plasma-catalyst process, the Energy Dispersive Spectrometer (EDS) test was introduced to qualitatively analyze the elements changes, as exhibited in Figure 5b. It was apparent that the variation of nitrogen contents on the $Mn-Ce/TiO_2$ sample before and after the plasma-catalyst reaction was tiny, which indicated little deposition of nitrates on the catalyst surface.

2.4.2. Reducibility Properties

In order to explore the oxidation states and the reduction potentials of the active species contained in the Mn-based bimetallic nanocatalysts, H_2-TPR analysis was performed with the reduction peaks fitted by Gaussian functions, as exhibited in Figure 6. The H_2 consumptions together with all reduction temperature values were summarized in Table 3. On account of the support of anatase TiO_2 induced no noteworthy reduction peaks in the test temperature region, all the H_2 consumption peaks displayed in Figure 6 could be ascribed to the reduction reactions of diverse active species of MnO_x, CeO_x, CoO_x, and FeO_x. For Mn-based catalysts, the typical reduction peaks were regard as following the order of $MnO_2 \rightarrow Mn_2O_3$ (Mn_3O_4) $\rightarrow MnO$ [31]. For $Mn-Ce/TiO_2$ nanocatalyst, as shown in Figure 6a, there were five main H_2 consumption peaks appearing within the temperature range of 50~850 °C. The initial dominating reduction peak (R1) at around 261 °C was mainly caused by the reduction reaction of the high oxidation state of Mn^{4+} reducing to Mn^{3+} [32]. The subsequent asymmetrical reduction peak from 260 °C to 410 °C could be further divided into two reduction peaks (R2 and R3), according to the two processes of Mn_2O_3 reducing to Mn_3O_4 and Mn_2O_3 reducing to MnO reported in previous literatures [28,31]. The converting from Mn_2O_3 to Mn_3O_4 preferred to occur on the primal amorphous Mn_2O_3 [33], which was consistent with appearance of R2 peak. While the transformation from Mn_2O_3 to MnO was apt to happen at higher reaction temperatures [34], well coinciding with the temperature value of R3 peak. For Ce-containing sample, the typical CeO_x reduction process usually presented two separated peaks, the one of $CeO_2{}^s$ converting to $Ce_2O_3{}^s$ on the catalyst surface occurred at about 450 °C, the other one of $CeO_2{}^b$ transforming to $Ce_2O_3{}^b$ in the catalyst bulk came up at 730 °C approximately [35]. Therefore, the fourth wide reduction peak (R4) in the $Mn-Ce/TiO_2$ nanocatalyst was related to the reduction processes of Mn_3O_4 to MnO and $CeO_2{}^s$ to $Ce_2O_3{}^s$ simultaneously, and the fifth peak (R5) at around 717 °C was potentially associated with the $CeO_2{}^b$ reduction reaction. Among these three Mn-based bimetallic nanocatalysts, $Mn-Ce/TiO_2$ nanocatalyst displayed the highest low-temperature reducibility and exhibited a noticeable lack of high-temperature reduction peaks at the same time, which manifested the higher oxidation states of manganese ion (Mn^{4+} and Mn^{3+}) constituted the dominating phase [34].

Comparing with $Mn-Ce/TiO_2$ nanocatalyst, the H_2-TPR curve of $Mn-Co/TiO_2$ nanocatalyst was conspicuously different in both the reduction temperatures and the peak intensions. For $Mn-Co/TiO_2$ nanocatalyst, the reduction peak of MnO_2 to Mn_2O_3 shifted toward lower temperature (218 °C) and weakened significantly. Meanwhile, the reduction peaks of Mn_2O_3 to Mn_3O_4 and Mn_2O_3 to MnO moved to higher temperatures and strengthened noticeably. The two reduction processes of Mn_2O_3 presented as a whole peak centered at about 418 °C. The reduction reaction of cobalt oxides exhibited two peaks at around 327 (R2) and 517 °C (R4), which could be ascribed to the transformation of Co^{3+} $\rightarrow Co^{2+}$ and $Co^{2+} \rightarrow Co^0$, respectively [15]. However, these two reduction peaks were overlapped with the MnO_x peaks in whole or partly. For $Mn-Fe/TiO_2$ nanocatalyst, considering the coexistence of FeO_x and MnO_x, the joint peaks (R2 and R3) from 330 °C to 530 °C were mainly attributed to the conversion of Mn_2O_3 to Mn_3O_4 combining with the transformation of Fe_2O_3 to Fe_3O_4. According to previous report [36], the majority of Fe_2O_3 ($Fe_2O_3{}^m$) was in the form of nanoparticles, oligomeric clusters or isolated ions locating at effortlessly reducible sites. After the $Fe_2O_3{}^m$ reduction reaction, the reduction of residual Fe_2O_3 ($Fe_2O_3{}^r$) to Fe_3O_4 accomplished at the higher temperature [37]. The remarkable strong peak (R4) at about 501 °C was ascribed to the overlapped peaks of Mn_3O_4 to MnO and Fe_3O_4 to FeO.

As exhibited in Table 3, the total H_2 consumptions of $Mn-Ce/TiO_2$ and $Mn-Co/TiO_2$ nanocatalysts were 4.86 mmol·g^{-1} and 4.43 mmol·g^{-1}, respectively, much larger than that of $Mn-Fe/TiO_2$ nanocatalysts. It was proposed that the peaks appearing at lower temperatures demonstrated superior catalytic activity in low temperature region [2]. While the starting reduction peak temperature of $Mn-Co/TiO_2$ nanocatalyst was the lowest at 218 °C, its total H_2 consumption was obvious smaller than that of $Mn-Ce/TiO_2$ nanocatalyst, which was regarded as a more important factor affecting the reducing capacity. Based on the H_2 consumption as a vital factor to the redox property of catalyst, it was reasonable that $Mn-Ce/TiO_2$ nanocatalyst presented the higher NO_x conversion with and without plasma than $Mn-Co/TiO_2$ and $Mn-Fe/TiO_2$ nanocatalysts.

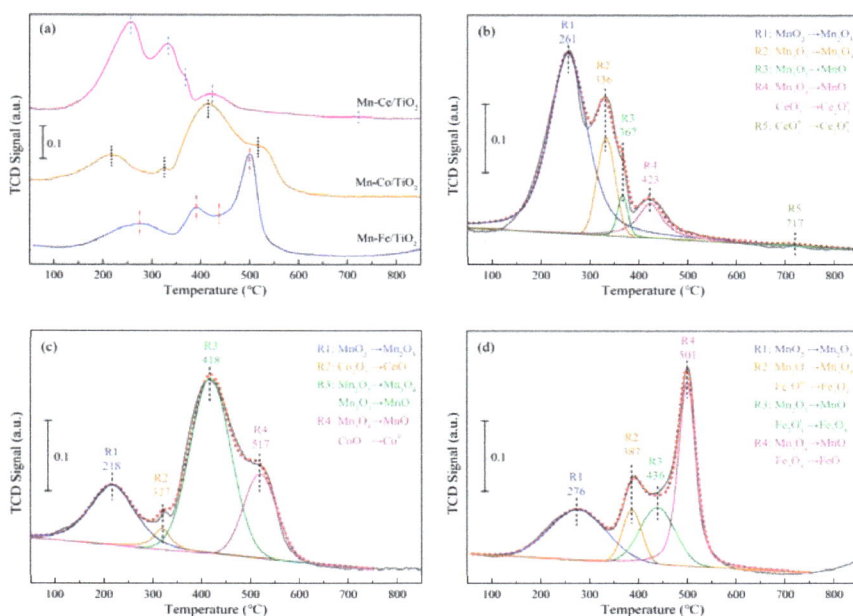

Figure 6. H_2-TPR profiles of Mn-based bimetallic nanocatalysts: (a) Total H_2-TPR curves; (b) multi-peaks Gaussian fitting for $Mn-Ce/TiO_2$ nanocatalyst; (c) multi-peaks Gaussian fitting for $Mn-Co/TiO_2$ nanocatalyst; and (d) multi-peaks Gaussian fitting for $Mn-Fe/TiO_2$ nanocatalyst.

Table 3. H_2-TPR quantitative analysis of Mn-based bimetallic nanocatalysts.

Samples	Temperature (°C)					H_2 Consumption (mmol·g^{-1})					
	R 1	R 2	R 3	R 4	R 5	R 1	R 2	R 3	R 4	R 5	Total
$Mn-Ce/TiO_2$	261	336	367	423	717	2.52	1.24	0.36	0.67	0.07	4.86
$Mn-Co/TiO_2$	218	327	418	517	–	1.14	0.21	2.12	0.96	–	4.43
$Mn-Fe/TiO_2$	276	387	436	501	–	0.83	0.41	0.54	1.59	–	3.37

2.4.3. Ammonia Adsorption Properties

Besides the redox property, the acid capacity on the catalyst surface was another crucial factor influencing the catalytic performance in SCR reactions [14,38]. NH_3-TPD and NO-TPD tests were introduced in order to establish the connection between the surface acidities and the SCR activities for the Mn-based bimetallic nanocatalysts. The test results were analyzed and compared in Figure 7 and Table 4, respectively. The NH_3-TPD curves for these three typical Mn-based bimetallic nanocatalysts were attributed to four desorption peaks of chemisorbed NH_3 within the temperature range of 150~750 °C. It was obvious that the first weak NH_3 desorption peak (P1) of Mn−Fe/TiO_2 nanocatalyst appeared at about 209 °C ascribed to the NH_3 desorption from the weak acid sites, which was deemed too weak to be stable bound NH_3 in the gas mix during the SCR reactions [14]. The second and the third successive desorption peaks (P2 and P3) located from 398 °C to 585 °C indicated the distribution of medium strong acid sites. Additionally, the final and relatively stronger peak (P4) at 649 approximately were attributed to strong acid sites, which could be regard as plenty of Lewis acid sites generated on the nanocatalyst surface with adsorbing a large amount of strongly bound ammonia [39]. For Mn−Co/TiO_2 nanocatalyst, the NH_3 desorption result demonstrated superior acidity capacity at medium and high temperatures, but an undesirable temperature shift towards higher temperature regions appeared at the same time. The desorption peak temperature value of weak acid sites (P1), medium strong acid sites (P2 and P3), and strong acid sites (P4) reached 276 °C, 402 °C, 587 °C, and 653 °C, respectively, which signified that it is more difficult for the chemisorbed NH_3 to desorb from the acid sites and participate in SCR reactions [40]. Compared with the NH_3-TPD curve of Mn−Co/TiO_2 nanocatalyst, the four desorption peaks of Mn−Ce/TiO_2 nanocatalyst shifted towards lower temperatures slightly and became much stronger. The enrichments of the weak acid sites, the medium strong acid sites and the strong acid sites were all positive to ammonia adsorption, which could undoubtedly form more abundant Brønsted acid sites and Lewis acid sites promoting NH_3 adsorption on the nanocatalyst surface [41,42]. This change was regard as an important reason for the outstanding catalytic performance of Mn−Ce/TiO_2 nanocatalyst with and without plasma.

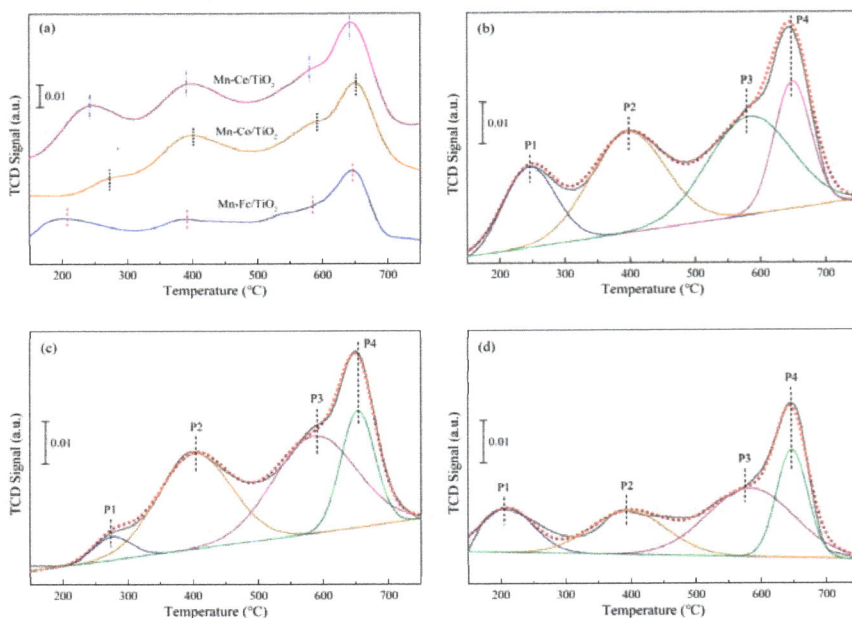

Figure 7. NH_3-TPD profiles of Mn-based bimetallic nanocatalysts: (**a**) Total NH_3-TPD curves; (**b**) multi-peaks Gaussian fitting for Mn−Ce/TiO_2 nanocatalyst; (**c**) multi-peaks Gaussian fitting for Mn−Co/TiO_2 nanocatalyst; (**d**) multi-peaks Gaussian fitting for Mn−Fe/TiO_2 nanocatalyst.

Table 4. Quantitative analysis of NH$_3$-TPD profiles.

Samples	Temperature (°C)				NH$_3$ Composition (mmol·g^{-1})				
	Peak 1	Peak 2	Peak 3	Peak 4	Peak 1	Peak 2	Peak 3	Peak 4	Total
Mn−Ce/TiO$_2$	245	396	583	647	0.27	0.52	0.58	0.39	1.76
Mn−Co/TiO$_2$	276	402	587	653	0.11	0.49	0.52	0.32	1.44
Mn−Fe/TiO$_2$	209	398	585	649	0.14	0.24	0.39	0.20	0.97

In order to exactly confirm the total acid capacity, the quantitative analysis of the desorption peaks of Mn−Ce/TiO$_2$, Mn−Co/TiO$_2$ and Mn−Fe/TiO$_2$ nanocatalyst was performed and summarized in Table 4. Among these three nanocatalysts, the total NH$_3$ desorption of Mn−Ce/TiO$_2$ sample achieved the maximum value of 1.76 mmol·g^{-1}, which further verified the better promotion effects of MnCeO$_x$ on the surface acidity than that of MnCoO$_x$ and MnFeO$_x$. It was noteworthy that, although Mn−Fe/TiO$_2$ nanocatalyst exhibited the weak acid sites at the lowest temperature, the acid capacity of its weak acid sites was only 0.14 mmol·g^{-1} too small to satisfy the NH$_3$ adsorption requirements during SCR recations. Therefore, Mn−Ce/TiO$_2$ nanocatalyst with NH$_3$ desorption of 0.27 mmol·g^{-1} at around 245 °C was qualified for the best acidity properties at the low temperature. However, it had been revealed that NH$_3$ could block NO adsorption and activation onto the metal active sites on the catalyst surface via the undesired electronic contact between the adsorbed NH$_3$ and the metal sites [43]. As a result, NH$_3$ presented inhibiting effects on SCR reactions at low temperature. Hence, the acidity properties on the surface of the Mn-based bimetallic nanocatalysts were closely related to the redox performance. It was required to achieve an adequate equilibrium between the oxidation states of active species and the acidity properties of active metal compounds in order to develop the optimal catalyst.

2.4.4. Oxidation States of Active Species

The elemental valence states and the atomic concentrations on the surface of Mn-based bimetallic nanocatalysts were explored by XPS analysis for the purpose of a better insight into the metal oxidation states and the surface compositions. The XPS spectra of Mn 2p, O 1s, Ti 2p, Fe 2p, Co 2p, and Ce 3d in the nanocatalysts were exhibited in Figure 8. The valence state of every element was determined numerically according to Gaussian fitting, respectively. The specific binding energy and the individual element concentration in various valence states were summarized in Table 5. Figure 8a displayed the XPS spectrum for Ti 2p of catalyst support, which comprised two peaks of Ti 2$p_{1/2}$ locating at about 464.3 eV and Ti 2$p_{3/2}$ situating at around 458.7 eV, respectively [44]. Comparing the three XPS spectrum of Ti, it could be easily found that, although the Mn-based bimetallic nanocatalysts were doped with different active elements of iron, cobalt, and manganese, no significant changes were observed in the Ti peaks. The +4 valence state of titanium on the catalyst surface was stabilized and dominating in Mn−Fe/TiO$_2$, Mn−Co/TiO$_2$, and Mn−Ce/TiO$_2$ nanocatalysts.

The XPS spectra of Mn 2p in these three typical Mn-based bimetallic nanocatalysts consisted of two characteristic peaks, assigned to Mn 2$p_{1/2}$ peak at around 653 eV and Mn 2$p_{3/2}$ peak at about 642 eV [45], as shown in Figure 8b. The asymmetrical Mn 2$p_{3/2}$ peak further verified the complicated manganese species in divers valence states coexisting in the nanocatalysts. The curve of Mn 2$p_{3/2}$ peak could be split into three peaks via multi-peaks gaussian fitting, the first peak at around 641.2 ± 0.3 eV was assigned to Mn^{2+}, the second one at 642.6 ± 0.2 eV associated with Mn^{3+}, and the third one at 644.1 ± 0.5 eV consistent with Mn^{4+}, respectively [46]. The complex MnO$_x$ including three valence states were apparently difficult to distinguish with the binding energy difference no more than 3.7 eV. In order to make an accurate identification of the atomic composition and Mn^{n+} concentration on the nanocatalyst surface, a quantitative analysis was introduced based on the area covered under each separated peak, as listed in Table 4. According to previous studies [28,33,47], NO$_x$ conversion over pure MnO$_x$ could been ranked as MnO$_2$ > Mn$_2$O$_3$ > Mn$_3$O$_4$. Hence, the improved concentration of MnO$_2$ on the nanocatalyst surface was advantageous to SCR reactions [48]. Among the three Mn-based bimetallic nanocatalysts, the major amount of manganese primarily presenting +4 valence state on the catalyst surface dispersedly, as shown in Table 4. While $Mn-Ce/TiO_2$ sample presented the highest atomic composition of Mn^{4+}/Mn^{n+} (56.5%), which was much larger than 46.8% and 40.2% in $Mn-Co/TiO_2$ and $Mn-Fe/TiO_2$ nanocatalysts, respectively. It was regard as the reason for the higher catalytic activity of $Mn-Ce/TiO_2$ nanocatalyst, as exhibited in the above results.

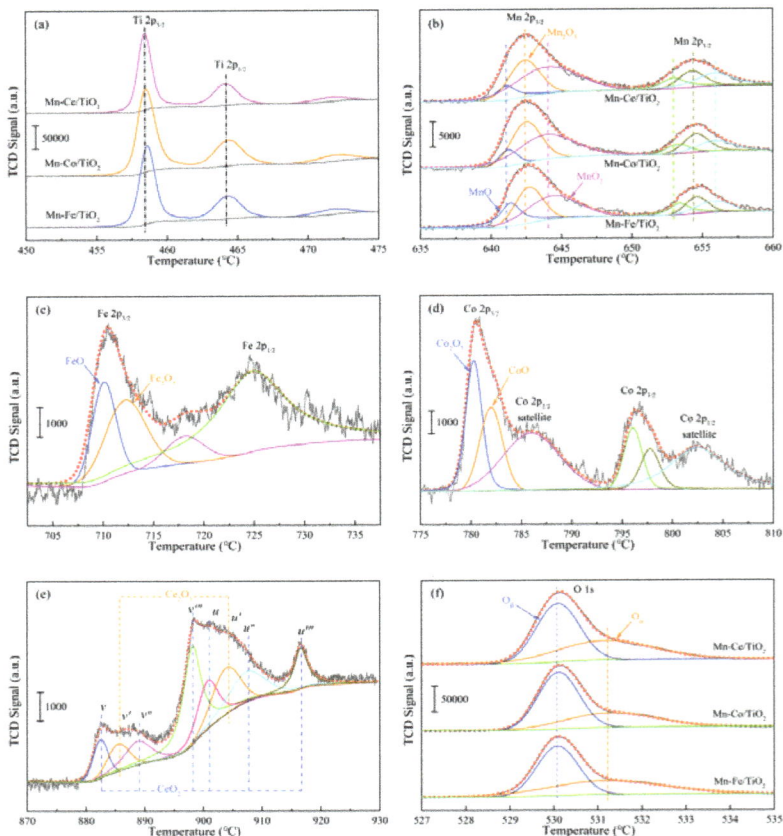

Figure 8. XPS analysis of Mn-based bimetallic nanocatsalysts: (**a**) XPS spectra for Ti 2*p*; (**b**) XPS spectra for Mn 2*p*; (**c**) XPS spectra for Fe 2*p*; (**d**) XPS spectra for Co 2*p*; (**e**) XPS spectra for Ce 3*d*; and (**f**) XPS spectra for O 1*s*.

Table 5. Surface atomic compositions of the catalysts determined by XPS.

Samples	Binding Energy (eV)/Atomic Composition (%)										
	Mn			Fe		Co		Ce		O	
	Mn^{2+}	Mn^{3+}	Mn^{4+}	Fe^{2+}	Fe^{3+}	Co^{2+}	Co^{3+}	Ce^{3+}	Ce^{4+}	O_α	O_β
$Mn-Ce/TiO_2$	640.9/ 9.3	642.4/ 34.2	644.0/ 56.5	-/-	-/-	-/-	-/-	885.6/ 40.4	882.4/ 59.6	531.2/ 40.4	530.2/ 59.6
$Mn-Co/TiO_2$	641.2/ 13.8	642.6/ 39.4	644.1/ 46.8	-/-	-/-	779.6 39.3	782.1/ 60.7	-/-	-/-	531.4/ 33.7	530.3/ 66.3
$Mn-Fe/TiO_2$	641.4/ 21.3	642.8/ 38.5	644.6/ 40.2	709.6/ 43.4	711.7/ 56.6	-/-	-/-	-/-	-/-	531.6/ 28.2	530.3/ 71.8

For Mn−Fe/TiO$_2$ nanocatalyst, the Fe 2p XPS spectra was was presented in Figure 8c with two individual peaks attributed to Fe $2p_{3/2}$ at about 710 eV, Fe $2p_{1/2}$ at around 724 eV and a satellite peak of Fe^{3+} in Fe$_2$O$_3$ exhibited at 718.3 eV [49]. The broad peak of Fe $2p_{3/2}$ was composed of two overlapped peaks, the one ascribed to Fe^{2+} locating at around 709.6 eV and the other one attributed to Fe^{3+} seated at about 711.6 eV. These two peaks confirmed the co-occurrence of iron in +2 and +3 valence states on Mn−Fe/TiO$_2$ nanocatalyst surface, as quantified in Table 5. The promotion effect of iron on Mn-based bimetallic nanocatalysts was ascribed to the interaction happening in the redox circulation: Fe^{3+} + Mn^{3+} ↔ Fe^{2+} + Mn^{4+} [50]. For Mn−Co/TiO$_2$ nanocatalyst, there were two main peaks in the Co 2p spectrum ascribed to Co $2p_{3/2}$ at about 780.6 eV and Co $2p_{1/2}$ at around 796.5 eV. Each of these two peaks was accompanied by an adjacent satellite peak at 786.8 eV and 803.1 eV correspondingly, as depicted in Figure 8d. The two broader and gentler satellite structures at the relatively higher binding energy region were caused by the metal-to-ligand charge transfer, also known as the shakeup process of cobalt in its high spin state. While this process can only be observed with the high spin state of Co^{2+} ion, but cannot be observed with the diamagnetic low-spin Co^{3+} ion [51]. The XPS spectra of Co $2p_{3/2}$ scope could be further divided into Co^{3+} spectrum at binding energy of 780.0 eV and Co^{2+} spectrum at 781.6 eV. This test result showed the ions of Co^{2+} and Co^{3+} were co-existed on Mn−Co/TiO$_2$ nanocatalyst surface and the Co^{3+} exhibited a comparatively higher atomic composition of 60.7%. The Co^{3+} species existed in a relatively high valence state and gave rise to more anionic defects, generating abundant surface oxygen to enhance the process of adsorption and oxidation during SCR reactions [52]. For Mn−Ce/TiO$_2$ nanocatalyst, the Ce 3d spectra result was depicted in Figure 8e. The Ce 3d pattern was composed of u and v multi-peaks matching to the spin orbit split $3d_{5/2}$ and $3d_{3/2}$ core holes [53]. According to the binding energies of different peaks, the Ce 3d spectra could be elaborately separated into eight peaks, labeled as u, u′, u″, u‴ and v, v′, v″, v‴, respectively [54]. The u′ and v′ peaks were matched with the $3d^{10}4f^1$ electronic state of Ce^{3+}, and the u, u″, u‴, v, v″, and v‴ peaks were ascribed to the $3d^{10}4f^0$ electronic state of Ce^{4+} [55]. These distinctive peaks verified the coexistence of of Ce^{3+} and Ce^{4+} species on Mn−Co/TiO$_2$ nanocatalyst surface. The Ce^{3+} species were important incentives for the formation of unsaturated chemical bonds and the generation of electric charge balance. [56]. In the active compounds of manganese and cerium, the negative charge transferred from Mn^{2+} or Mn^{3+} to Ce^{4+} strengthening the interaction between manganese and cerium [1,2]. The oxygen circle of storing and releasing was easier for the Mn−Ce/TiO$_2$ nanocatalyst with the redox couple of Ce^{3+}/Ce^{4+} to form more surface oxygen vacancies that were advantageous to oxygen adsorption and chemisorbed oxygen generation [57].

The O 1s spectra for Mn-based bimetallic nanocatalysts were displayed in Figure 8f. On the base of curve-fitting results, the O 1s spectra was divided into two peaks: The O$_\alpha$ peak ascribed to chemisorbed oxygen centered at binding energy of 531.2 ~ 531.6 eV, the O$_\beta$ peak attributed to lattice oxygen appeared at binding energy of 530.2 ~ 530.3 eV. Compared with the O 1s spectra of nanocatalysts, it could be found that the binding energies of O$_\alpha$ shifted to lower values from 531.6 eV in Mn−Fe/TiO$_2$ to 531.4 eV in Mn−Co/TiO$_2$ and to 531.2 eV in Mn−Ce/TiO$_2$, and similar variation tendency occurred on the binding energies of O$_\beta$. Meanwhile, as shown in Table 4, the surface atomic composition of chemisorbed oxygen over Mn−Ce/TiO$_2$ nanocatalyst reached the maximum 40.4%, much higher than 33.7% on Mn−Co/TiO$_2$ and 28.2% on Mn−Fe/TiO$_2$ samples. The chemisorbed oxygen was the most energetic oxygen species due to its high mobility [58]. Therefore, these surface atomic composition of Mn−Ce/TiO$_2$ nanocatalyst were regard as another reason for its superior catalytic performance with and without plasma.

2.5. Reaction Mechanism Analysis

According to the catalytic performance of NO$_x$ conversion over Mn-based bimetallic nanocatalysts with and without plasma and the physicochemical properties of these nanocatalysts presented above, the complex bimetallic oxides of MnFeO$_x$, MnCoO$_x$, and MnCeO$_x$ affected the hybrid catalyst-plasma catalytic process obviously with the different redox characteristics of active chemisorbed sites. All the three bimetallic nanocatalysts enhanced the catalytic ability of manganese species by increasing the ratio of Mn^{4+}/Mn^{n+}, generating more lattice oxygen and plenty of oxygen vacancy on the catalyst surface [2]. In the catalyst-plasma hybrid catalytic system, the plasma derivatives reformed the chemical compositions of the gas mix and modified the electronic states on the nanocatalyst surface. For $Mn-Fe/TiO_2$, $Mn-Co/TiO_2$ and $Mn-Ce/TiO_2$ nanocatalysts, a dynamic equilibrium was sustained on their surfaces with the electron transfer between Mn and Fe (Co or Ce) ions during the catalytic oxidation process, which could be expressed as $Fe^{3+} + Mn^{3+} \leftrightarrow Fe^{2+} + Mn^{4+}$, $Co^{3+} + Mn^{3+} \leftrightarrow Co^{2+} + Mn^{4+}$, $Ce^{3+} + Mn^{3+} \leftrightarrow Ce^{2+} + Mn^{4+}$ and $Ce^{4+} + Mn^{3+} \leftrightarrow Ce^{3+} + Mn^{4+}$, respectively.

Besides originally partial NO oxidation into NO_2 over the lattice oxygen of MnO_x, FeO_x, CoO_x and CeO_x, more NO was oxidized to NO_2 via the reaction (4) and (5) under the energetic particles. The valence state part of manganese cations increased, which was caused by the electron transition from Mn^{3+} to Mn^{4+} via lattice oxygen [59]. Mn^{4+} was more desirable for the oxidation of NO to NO_2 over Mn-based catalysts [16] and it was reduced to Mn^{3+} during the SCR reactions [58]. Under the plasma derived species, such as O_3 or O radicals, the Mn^{3+} could be fast re-oxidation into Mn^{4+}, thus accelerating the catalytic oxidation process and the fast SCR reaction. Furthermore, the concentration of chemisorbed oxygen on the nanocatalyst surface was also improved in the catalyst-plasma hybrid catalytic system. More surface oxygen species could form via the direct interaction of MnO_x, FeO_x, CoO_x, and CeO_x with plasma excited oxygen species. Considering the inhibiting effect of NO on O_3 formation, the surface oxygen species were more likely to generate from O radicals. The adsorbed oxygen reacted with NO to form NO_2 according to the following reaction steps (9)~(11):

$$O + M \rightarrow M-O_{ads} \tag{9}$$

$$NO + M-O_{ads} \rightarrow M-O-NO_{ads} \tag{10}$$

$$M-O-NO_{ads} \rightarrow M + NO_2 \tag{11}$$

where M represented the active sites on the nanocatalyst surface, O_{ads} and NO_{ads} represented adsorbed NO and oxygen on the nanocatalyst surface, respectively. During this process, the NO_{ads} liberated electron to Mn^{4+} and the O_{ads} trapped electron from Fe^{2+} or Co^{2+} or Ce^{3+}, respectively, which transform into absorbed NO^+ and O^-. Then formed NO^+ further reacted with O^- to generate NO_2. Simultaneously, a part of NO was oxidized to NO_2 directly by the active oxygen produced from O_2 activation on the surface oxygen vacancies. The possible catalyst-plasma hybrid catalytic process of SCR reaction over Mn-based bimetallic nanocatalysts was exhibited in Figure 9.

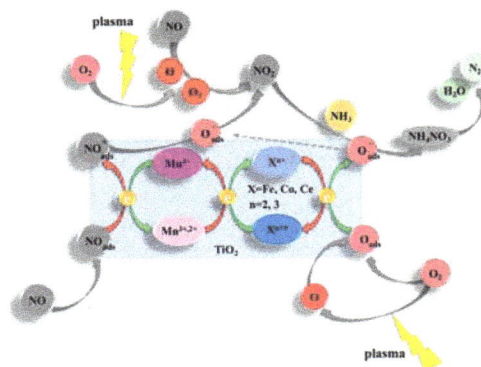

Figure 9. The possible catalyst-plasma hybrid catalytic process of SCR reaction over Mn-based bimetallic nanocatalysts.

3. Materials and Methods

3.1. Catalysts Preparation

The three typical Mn-based bimetallic nanocatalysts were prepared by hydrothermal method. $Mn(NO_3)_2$ (analytical pure 50%, Sinopharm, Shanghai, China), $Fe(NO_3)_3 \cdot 9H_2O$ (analytical pure 99.9%, Sinopharm, Shanghai, China), $Co(CH_3COO)_2 \cdot 4H_2O$ (analytical pure 99.9%, Kermel, Tianjin, China), and $Ce(NO_3)_3 \cdot 6H_2O$ (analytical pure 99.9%, Nanjing-reagent, Nanjing, China) were introduced as the precursors of MnO_x, FeO_x, CoO_x, and CeO_x, respectively. The tetrabutyl titanate was used as the precursors of TiO_2 for supporting the active metallic oxides. $Mn(NO_3)_2$ was added into deionized water at room temperature and then $Fe(NO_3)_3 \cdot 9H_2O$ was dissolved in the solution. Amount of glycol was added into the above mixture with magnetic stirring continuously. A Teflon-lined stainless steel autoclave was introduced to heat the homogeneous solution at 180 °C for 8 h. After the autoclave cooling down to the ambient temperature, tetrabutyl titanate was added into this solution and aged in the autoclave again at 180°C for 3 h. The mixture was collected by reduplicative centrifugation and wash. Finally, the precipitate was dried at 150 °C for 12 h and calcined in air at 500 °C for 4 h. The produced Mn-based bimetallic nanocatalysts were triturated and filtered with 60−80 mesh for activity tests and characterization analysis. The nanocatalyst was denoted as $Mn-Fe/TiO_2$ with the molar ratios of Mn:Fe:Ti = 2:1:7. The $Mn-Co/TiO_2$ and $Mn-Ce/TiO_2$ nanocatalysts were prepared under the same process with $Co(CH_3COO)_2 \cdot 4H_2O$ and $Ce(NO_3)_3 \cdot 6H_2O$ replacing $Fe(NO_3)_3 \cdot 9H_2O$, respectively.

3.2. Catalysts Characterization

The Maxon Tristar II 3020 micropore-size analyzer (Maxon, Chicago, IL, USA) was used for testing N_2 adsorption isotherms of the prepared nanocatalysts at -196 °C. The surface areas and the pore-size distributions of the nanocatalysts were measured after the nanocatalysts degassing in vacuum at 350 °C for 10 h. BET plot linear portion was used to determine the nanocatalysts specific surface areas, and the desorption branch with Barrett–Joyner–Halenda (BJH) formula was introduced to calculate the pore-size distributions. The XRD data was captured by a Bruker D8 advance analyzer (Bruker, Frankfurt, Germany) with Mo K_α radiation, diffraction intensity from 10° to 90°, point counting time of 1s and point counting step of 0.02°. The element phases contained in the nanocatalys were distinguished by comparing characteristic peaks presented in the XRD patterns with the International Center for Diffraction Data (ICDD). The advanced microstructural image data and the surface element contents of the nanocatalysts were achieved by a high resolution transmission electron microscope JEOL JEM-2010 combined with EDS ((Japan electronics corporation, Tokyo, Japan). H_2-TPR and NH_3-TPD tests were performed with a Micromeritics Autochem II 2920 chemical adsorption instrument (Micromeritics, Houston, TX, USA). During H_2-TPR experiment, nanocatalysts were pretreatment in He at 400 °C for 1 h, and then cooled to environment temperature in H_2 and He gas mixture at 30 mL/min. The test temperature range of H_2 consumptions was from 50 °C to 850 °C with the heating rate of 10 °C/min. The operating process of NH_3-TPD test was similar to that of H_2-TPR test with NH_3 replacing H_2. XPS analysis was performed on a Thermo ESCALAB 250XI (Thermo Fisher, Boston, MA, USA) with pass energy 46.95 eV, Al K_α radiation 1486.6 eV, X-ray source 150 W and binding energy precision ± 0.3 eV. The C 1s line at 284.6 eV was measured as a reference.

3.3. Catalytic Performance Tests

The catalytic performance of Mn-based bimetallic nanocatalysts was explored in a catalyst-plasma hybrid system as shown in Figure 10. The dielectric barrier discharge (DBD) plasma reactor was comprised of two electrodes and a quartz tube. The high voltage electrode was a stainless-steel rod with diameter of 3 mm, installed inside the quartz tube coaxially. The ground electrode was a copper wire mesh wrapped outside the quartz tube tightly. The discharge energy was produced by an AC power transverter with a digital controller of voltage, electricity, and frequency. The quartz tube was in the height of 800 mm, outer diameter of 12 mm and thicknesses of 0.8 mm. 5 mL of nanocatalyst was filled in the discharge zone of plasma reactor. A resistance furnace was introduced to maintain the desired reaction temperature located upstream plasma reactor, connected to the temperature controller. The concentration of gas mixture was measured by German MRU MGA-5 analyzer (MRU, Berlin, Germany) joint with an external special detector for N_2O and NH_3. An Infrared Thermometer (HCJYET, HT-8872, Hongcheng, Shanghai, China) was introduced to detect the specific temperature of discharge area during the plasma process. During plasma-catalyst catalytic activity experiment, the inlet mixed gas included 300 ppm NO, 300 ppm NH_3, 8% O_2, ~0.1% H_2O and N_2 as balance gas. The gas hourly space velocity (GHSV) was about 20 000 h^{-1}. The NO_x conversion rate was calculated according to Equation (12), where $[NO_x] = [NO] + [NO_2]$. The N_2 selectivity was calculated by the concentrations of N_2O and NO_x, as shown in Equation (13). Each experiment was repeated three times to assure the results accuracy. The discharge energy density was defined as discharge power divided by the inlet gas flow rate [9], which was calculated using Equation (14) [60], where E (W·h/m^3) was energy density, P (W) was discharge power, and Q (m^3/h) was the gas flow rate. More basic data relating to the discharge energy was listed in Table 1.

$$NO_x \text{ conversion rate} = \left(\frac{[NO_x]_{in} - [NO_x]_{out}}{[NO_x]_{in}} \right) \times 100\% \tag{12}$$

$$N_2 \text{ selectivity} = 1 - \frac{2[N_2O]_{out}}{[NO_x]_{in} - [NO_x]_{out}} \times 100\% \tag{13}$$

$$E \text{ (W·h/m}^3\text{)} = \frac{P \text{ (W)}}{Q \text{ (m}^3/h\text{)}} \tag{14}$$

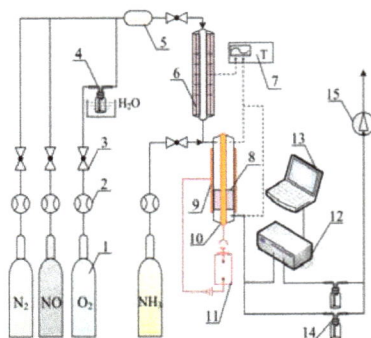

Figure 10. The schematic diagram about the catalyst-plasma hybrid system. 1, standard gas; 2, mass flowmeter; 3, shutdown valve; 4, water carrier; 5, gas mixer; 6, resistance furnace; 7, temperature controller; 8, nanocatalysts; 9, ground electrode; 10, high voltage electrode; 11, AC power transverter; 12, flue gas analyzer; 13, record system; 14, gas washing bottle; and 15, induced draft fan.

4. Conclusions

The Mn-based bimetallic nanocatalysts of Mn$-$Fe/TiO$_2$, Mn$-$Co/TiO$_2$, Mn$-$Ce/TiO$_2$, synthesized by hydrothermal method, presented obvious synergistic effects on NO_x catalytic conversion via the plasma-catalyst hybrid catalytic process. In the catalytic process with catalyst alone, the NO_x conversions of all tested catalysts were lower than 20% at ambient temperature. While in the plasma-catalyst hybrid catalytic process, the catalytic activities for NO_x elimination improved significantly with discharge energy enlarging. The maximum NO_x conversion of about 99.5% achieved on Mn$-$Ce/TiO$_2$ with discharge energy of 15 W·h/m^3 at ambient temperature. The reaction temperature had an inhibiting effect on plasma-catalyst hybrid catalysis.

Among these three Mn-based bimetallic nanocatalysts, Mn−Ce/TiO$_2$ displayed the optimal catalytic property with higher catalytic activity and superior selectivity in the plasma-catalyst hybrid catalytic process. Furthermore, based on the multiple characterizations performed on the Mn-based bimetallic nanocatalysts, it could be confirmed that the catalytic property of plasma-catalyst hybrid catalytic process was highly dependent on the phase composition of the catalyst. Mn−Ce/TiO$_2$ nanocatalyst presented the optimal structure characteristic among all tested samples, with the largest surface area, the increased active components distributions, the reduced crystallinity and the minished particle sizes. In the meantime, the ratios of Mn^{4+}/(Mn^{2+} + Mn^{3+} + Mn^{4+}) in the Mn−Ce/TiO$_2$ sample was the highest, which was beneficial to plasma-catalyst hybrid catalysis. Generally, it was believed that the plasma-catalyst hybrid catalytic process with the Mn-based bimetallic nanocatalyst was an effective approach for high-efficiency catalytic conversion of NO$_x$, especially at ambient temperature.

Author Contributions: Conceptualization, Y.G.; Funding acquisition, Y.G., T.L., W.Z.; Methodology, Y.G., T.L.; Project administration, Y.G.; Writing–original draft, Y.G.; Writing–review & editing, H.L. and W.Z.; Data curation, Y.G., W.J., W.F., H.J.

Funding: This work was supported by National Natural Science Foundation of China (Project No. 51708336), Shandong Provincial Natural Science Foundation (ZR2016EEB28), Shandong Provincial Science and Technology Development Plan (2011GSF11716), Shandong Jianzhu university open experimental project (2018yzkf023, 2018wzkf013,), and the Shandong electric power engineering consulting institute science and technology project (37-K2014-33).

Conflicts of Interest: The authors declare no conflict of interest.

References

1. Wang, B.; Chi, C.; Xu, M.; Wang, C.; Meng, D. Plasma-catalytic removal of toluene over CeO$_2$-MnOx catalysts in an atmosphere dielectric barrier discharge. *Chem. Eng. J.* **2017**, *322*, 679–692. [CrossRef]

2. France, L.J.; Yang, Q.; Li, W.; Chen, Z.; Guang, J.; Guo, D.; Wang, L.; Li, X. Ceria modified FeMnOx—Enhanced performance and sulphur resistance for low-temperature SCR of NOx. *Appl. Catal. B Environ.* **2017**, *206*, 203–215. [CrossRef]

3. Gao, Y.; Luan, T.; Lu, T.; Cheng, K.; Xu, H. Performance of V$_2$O$_5$-WO$_3$-MoO$_3$/TiO$_2$ catalyst for selective catalytic reduction of NOx by NH$_3$. *Chin. J. Chem. Eng.* **2013**, *21*, 1–7. [CrossRef]

4. Wang, T.; Liu, H.; Zhang, X.; Liu, J.; Zhang, Y.; Guo, Y.; Sun, B. Catalytic conversion of NO assisted by plasma over Mn-Ce/ZSM5-multi-walled carbon nanotubes composites: Investigation of acidity, activity and stability of catalyst in the synergic system. *Appl. Surf. Sci.* **2018**, *457*, 187–199. [CrossRef]

5. Zhang, R.; Yang, W.; Luo, N.; Li, P.; Lei, Z.; Chen, B. Low-temperature NH$_3$-SCR of NO by lanthanum manganite perovskites: Effect of A-/B-site substitution and TiO$_2$/CeO$_2$ support. *Appl. Catal. B Environ.* **2014**, *146*, 94–104. [CrossRef]

6. Gao, Y.; Luan, T.; Zhang, M.; Zhang, W.; Feng, W. Structure–Activity Relationship Study of Mn/Fe Ratio Effects on Mn−Fe−Ce−Ox/γ-Al$_2$O$_3$ Nanocatalyst for NO Oxidation and Fast SCR Reaction. *Catalysts* **2018**, *8*, 642. [CrossRef]

7. Miessner, H.; Francke, K.P.; Rudolph, R.; Hammer, T. NOx removal in excess oxygen by plasma-enhanced selective catalytic reduction. *Catal. Today* **2002**, *75*, 325–330. [CrossRef]

8. Kim, H.-H.; Teramoto, Y.; Ogata, A.; Takagi, H.; Nanba, T. Plasma Catalysis for Environmental Treatment and Energy Applications. *Plasma Chem. Plasma Process.* **2016**, *36*, 45–72. [CrossRef]

9. Penetrante, B.M.; Brusasco, R.M.; Merritt, B.T.; Pitz, W.J.; Vogtlin, G.E.; Kung, M.C.; Kung, H.H.; Wan, C.Z.; Voss, K.E. Plasma-Assisted Catalytic Reduction of NOx. In Proceedings of the International Fall Fuels and Lubricants Meeting and Exposition, San Francisco, CA, USA, 19–22 October 1998. [CrossRef]

10. Patil, B.S.; Cherkasov, N.; Lang, J.; Ibhadon, A.O.; Hessel, V.; Wang, Q. Low temperature plasma-catalytic NOx synthesis in a packed DBD reactor: Effect of support materials and supported active metal oxides. *Appl. Catal. B Environ.* **2016**, *194*, 123–133. [CrossRef]

11. Hammer, T.; Kishimoto, T.; Miessner, H.; Rudolph, R. Plasma Enhanced Selective Catalytic Reduction: Kinetics of NOx-Removal and Byproduct Formation. In Proceedings of the International Fuels & Lubricants Meeting & Exposition, Dearborn, MI, USA, 25 October 1999; Volume 1, pp. 3632–3641.

12. McAdams, R.; Beech, P.; Shawcross, J.T. Low Temperature Plasma Assisted Catalytic Reduction of NOx in Simulated Marine Diesel Exhaust. *Plasma Chem. Plasma Process.* **2008**, *28*, 159–171. [CrossRef]

13. Oda, T.; Kato, T.; Takahashi, T.; Shimizu, K. Nitric oxide decomposition in air by using non-thermal plasma processing—With additives and catalyst. *J. Electrost.* **1997**, *42*, 151–157. [CrossRef]

14. Zhang, C.; Chen, T.; Liu, H.; Chen, D.; Xu, B.; Qing, C. Low temperature SCR reaction over Nano-Structured Fe-Mn Oxides: Characterization, performance, and kinetic study. *Appl. Surf. Sci.* **2018**, *457*, 1116–1125. [CrossRef]

15. Qiu, L.; Pang, D.; Zhang, C.; Meng, J.; Zhu, R.; Ouyang, F. In situ IR studies of Co and Ce doped Mn/TiO_2 catalyst for low-temperature selective catalytic reduction of NO with NH_3. *Appl. Surf. Sci.* **2015**, *357*, 189–196. [CrossRef]

16. You, X.; Sheng, Z.; Yu, D.; Yang, L.; Xiao, X.; Wang, S. Influence of Mn/Ce ratio on the physicochemical properties and catalytic performance of graphene supported $MnOx-CeO_2$ oxides for NH_3-SCR at low temperature. *Appl. Surf. Sci.* **2017**, *423*, 845–854. [CrossRef]

17. Chen, Y.; Wang, J.; Yan, Z.; Liu, L.; Zhang, Z.; Wang, X. Promoting effect of Nd on the reduction of NO with NH_3 over CeO_2 supported by activated semi-coke: An in situ DRIFTS study. *Catal. Sci. Technol.* **2015**, *5*, 2251–2259. [CrossRef]

18. Liu, Z.; Zhu, J.; Li, J.; Ma, L.; Woo, S.I. Novel Mn–Ce–Ti Mixed-Oxide Catalyst for the Selective Catalytic Reduction of NOx with NH_3. *ACS Appl. Mater. Interfaces* **2014**, *6*, 14500–14508. [CrossRef] [PubMed]

19. Wang, T.; Zhang, X.; Liu, J.; Liu, H.; Wang, Y.; Sun, B. Effects of temperature on NOx removal with Mn-Cu/ZSM5 catalysts assisted by plasma. *Appl. Therm. Eng.* **2018**, *130*, 1224–1232. [CrossRef]

20. Penetrante, B.M.; Bardsley, J.N.; Hsiao, M.C. Kinetic Analysis of Non-Thermal Plasmas Used for Pollution Control. *Jpn. J. Appl. Phys.* **1997**, *36*, 5007. [CrossRef]

21. Heck, R.M. Catalytic abatement of nitrogen oxides–stationary applications. *Catal. Today* **1999**, *53*, 519–523. [CrossRef]

22. Kim, H.H.; Takashima, K.; Katsura, S.; Mizuno, A. Low-temperature NOx reduction processes using combined systems of pulsed corona discharge and catalysts. *J. Phys. D Appl. Phys.* **2001**, *34*, 604. [CrossRef]

23. Niu, J.; Yang, X.; Zhu, A.; Shi, L.; Sun, Q.; Xu, Y.; Shi, C. Plasma-assisted selective catalytic reduction of NOx by C_2H_2 over Co-HZSM-5 catalyst. *Catal. Commun.* **2006**, *7*, 297–301. [CrossRef]

24. Liu, Z.; Yi, Y.; Zhang, S.; Zhu, T.; Zhu, J.; Wang, J. Selective catalytic reduction of NOx with NH_3 over Mn-Ce mixed oxide catalyst at low temperatures. *Catal. Today* **2013**, *216*, 76–81. [CrossRef]

25. Li, X.; Zhang, S.; Jia, Y.; Liu, X.; Zhong, Q. Selective catalytic oxidation of NO with O_2 over Ce-doped $MnOx/TiO_2$ catalysts. *J. Nat. Gas Chem.* **2012**, *21*, 17–24. [CrossRef]

26. Ma, Z.; Yang, H.; Li, Q.; Zheng, J.; Zhang, X. Catalytic reduction of NO by NH_3 over Fe–Cu–Ox/CNTs-TiO_2 composites at low temperature. *Appl. Catal. A Gen.* **2012**, *427–428*, 43–48. [CrossRef]

27. Shi, Y.; Chen, S.; Sun, H.; Shu, Y.; Quan, X. Low-temperature selective catalytic reduction of NOx with NH_3 over hierarchically macro-mesoporous Mn/TiO_2. *Catal. Commun.* **2013**, *42*, 10–13. [CrossRef]

28. Gao, F.; Tang, X.; Yi, H.; Chu, C.; Li, N.; Li, J.; Zhao, S. In-situ DRIFTS for the mechanistic studies of NO oxidation over α-MnO_2, β-MnO_2 and γ-MnO_2 catalysts. *Chem. Eng. J.* **2017**, *322*, 525–537. [CrossRef]

29. Gao, Y.; Luan, T.; Zhang, W.; Li, H. The promotional effects of cerium on the catalytic properties of Al_2O_3-supported MnFeOx for NO oxidation and fast SCR reaction. *Res. Chem. Intermed.* **2018**, in press. [CrossRef]

30. Pérez Vélez, R.; Ellmers, I.; Huang, H.; Bentrup, U.; Schünemann, V.; Grünert, W.; Brückner, A. Identifying active sites for fast NH_3-SCR of NO/NO_2 mixtures over Fe-ZSM-5 by operando EPR and UV–vis spectroscopy. *J. Catal.* **2014**, *316*, 103–111. [CrossRef]

31. Gong, P.; Xie, J.; Fang, D.; Han, D.; He, F.; Li, F.; Qi, K. Effects of surface physicochemical properties on NH_3-SCR activity of MnO_2 catalysts with different crystal structures. *Chin. J. Catal.* **2017**, *38*, 1925–1934. [CrossRef]

32. Boningari, T.; Pappas, D.K.; Ettireddy, P.R.; Kotrba, A.; Smirniotis, P.G. Influence of SiO_2 on M/TiO_2 (M = Cu, Mn, and Ce) Formulations for Low-Temperature Selective Catalytic Reduction of NOx with NH_3: Surface Properties and Key Components in Relation to the Activity of NOx Reduction. *Ind. Eng. Chem. Res.* **2015**, *54*, 2261–2273. [CrossRef]

33. Luo, S.; Zhou, W.; Xie, A.; Wu, F.; Yao, C.; Li, X.; Zuo, S.; Liu, T. Effect of MnO_2 polymorphs structure on the selective catalytic reduction of NOx with NH_3 over TiO_2–Palygorskite. *Chem. Eng. J.* **2016**, *286*, 291–299. [CrossRef]

34. Jiang, H.; Wang, C.; Wang, H.; Zhang, M. Synthesis of highly efficient MnOx catalyst for low-temperature NH_3-SCR prepared from Mn-MOF-74 template. *Mater. Lett.* **2016**, *168*, 17–19. [CrossRef]

35. Cheng, X.; Zhang, X.; Su, D.; Wang, Z.; Chang, J.; Ma, C. NO reduction by CO over copper catalyst supported on mixed CeO_2 and Fe_2O_3: Catalyst design and activity test. *Appl. Catal. B Environ.* **2018**, *239*, 485–501. [CrossRef]

36. Stanciulescu, M.; Caravaggio, G.; Dobri, A.; Moir, J.; Burich, R.; Charland, J.P.; Bulsink, P. Low-temperature selective catalytic reduction of NOx with NH_3 over Mn-containing catalysts. *Appl. Catal. B Environ.* **2012**, *123–124*, 229–240. [CrossRef]

37. Kong, F.; Qiu, j.; Liu, H.; Zhao, R.; Zeng, H. Effect of NO/SO_2 on Elemental Mercury Adsorption by Nano-Fe_2O_3. *Proc. CSEE (China)* **2010**, *30*, 43–48.

38. Cao, L.; Chen, L.; Wu, X.; Ran, R.; Xu, T.; Chen, Z.; Weng, D. TRA and DRIFTS studies of the fast SCR reaction over CeO_2/TiO_2 catalyst at low temperatures. *Appl. Catal. A Gen.* **2018**, *557*, 46–54. [CrossRef]

39. Jin, R.; Liu, Y.; Wu, Z.; Wang, H.; Gu, T. Low-temperature selective catalytic reduction of NO with NH_3 over MnCe oxides supported on TiO_2 and Al_2O_3: A comparative study. *Chemosphere* **2010**, *78*, 1160–1166. [CrossRef] [PubMed]

40. Ma, L.; Seo, C.Y.; Nahata, M.; Chen, X.; Li, J.; Schwank, J.W. Shape dependence and sulfate promotion of CeO_2 for selective catalytic reduction of NOx with NH_3. *Appl. Catal. B Environ.* **2018**, *232*, 246–259. [CrossRef]

41. Xiong, Z.-B.; Liu, J.; Zhou, F.; Liu, D.-Y.; Lu, W.; Jin, J.; Ding, S.-F. Selective catalytic reduction of Nox with NH_3 over iron-cerium-tungsten mixed oxide catalyst prepared by different methods. *Appl. Surf. Sci.* **2017**, *406*, 218–225. [CrossRef]

42. Liu, F.; He, H.; Zhang, C.; Shan, W.; Shi, X. Mechanism of the selective catalytic reduction of NOx with NH_3 over environmental-friendly iron titanate catalyst. *Catal. Today* **2011**, *175*, 18–25. [CrossRef]

43. Fan, X.; Qiu, F.; Yang, H.; Tian, W.; Hou, T.; Zhang, X. Selective catalytic reduction of NOX with ammonia over Mn–Ce–Ox/TiO_2-carbon nanotube composites. *Catal. Commun.* **2011**, *12*, 1298–1301. [CrossRef]

44. Guan, D.S.; Wang, Y. Synthesis and growth mechanism of multilayer TiO_2 nanotube arrays. *Nanoscale* **2012**, *4*, 2968–2977. [CrossRef] [PubMed]

45. Lu, X.; Shen, C.; Zhang, Z.; Barrios, E.; Zhai, L. Core–Shell Composite Fibers for High-Performance Flexible Supercapacitor Electrodes. *ACS Appl. Mater. Interfaces* **2018**, *10*, 4041–4049. [CrossRef] [PubMed]

46. Wang, C.; Yu, F.; Zhu, M.; Wang, X.; Dan, J.; Zhang, J.; Cao, P.; Dai, B. Microspherical MnO_2-CeO_2-Al_2O_3 mixed oxide for monolithic honeycomb catalyst and application in selective catalytic reduction of NOx with NH_3 at 50–150 °C. *Chem. Eng. J.* **2018**, *346*, 182–192. [CrossRef]

47. Kapteijn, F.; Singoredjo, L.; Andreini, A.; Moulijn, J.A. Activity and selectivity of pure manganese oxides in the selective catalytic reduction of nitric oxide with ammonia. *Appl. Catal. B Environ.* **1994**, *3*, 173–189. [CrossRef]

48. Gao, G.; Shi, J.-W.; Liu, C.; Gao, C.; Fan, Z.; Niu, C. Mn/CeO_2 catalysts for SCR of NOx with NH_3: Comparative study on the effect of supports on low-temperature catalytic activity. *Appl. Surf. Sci.* **2017**, *411*, 338–346. [CrossRef]

49. Zhang, W.; Wang, F.; Li, X.; Liu, Y.; Liu, Y.; Ma, J. Fabrication of hollow carbon nanospheres introduced with Fe and N species immobilized palladium nanoparticles as catalysts for the semihydrogenation of phenylacetylene under mild reaction conditions. *Appl. Surf. Sci.* **2017**, *404*, 398–408. [CrossRef]

50. Wang, T.; Wan, Z.; Yang, X.; Zhang, X.; Niu, X.; Sun, B. Promotional effect of iron modification on the catalytic properties of Mn-Fe/ZSM-5 catalysts in the Fast SCR reaction. *Fuel Process. Technol.* **2018**, *169*, 112–121. [CrossRef]

51. Hu, H.; Cai, S.; Li, H.; Huang, L.; Shi, L.; Zhang, D. Mechanistic Aspects of deNOx Processing over TiO_2 Supported Co–Mn Oxide Catalysts: Structure–Activity Relationships and In Situ DRIFTs Analysis. *ACS Catal.* **2015**, *5*, 6069–6077. [CrossRef]

52. Li, K.; Tang, X.; Yi, H.; Ning, P.; Kang, D.; Wang, C. Low-temperature catalytic oxidation of NO over Mn–Co–Ce–Ox catalyst. *Chem. Eng. J.* **2012**, *192*, 99–104. [CrossRef]

53. Qi, G.; Yang, R.T. Performance and kinetics study for low-temperature SCR of NO with NH$_3$ over MnOx–CeO$_2$ catalyst. *J. Catal.* **2003**, *217*, 434–441. [CrossRef]
54. Huang, B.; Yu, D.; Sheng, Z.; Yang, L. Novel CeO$_2$@TiO$_2$ core–shell nanostructure catalyst for selective catalytic reduction of NOx with NH$_3$. *J. Environ. Sci.* **2017**, *55*, 129–136. [CrossRef] [PubMed]
55. Boningari, T.; Somogyvari, A.; Smirniotis, P.G. Ce-Based Catalysts for the Selective Catalytic Reduction of NOx in the Presence of Excess Oxygen and Simulated Diesel Engine Exhaust Conditions. *Ind. Eng. Chem. Res.* **2017**, *56*, 5483–5494. [CrossRef]
56. Ruggeri, M.P.; Grossale, A.; Nova, I.; Tronconi, E.; Jirglova, H.; Sobalik, Z. FTIR in situ mechanistic study of the NH$_3$NO/NO$_2$ "Fast SCR" reaction over a commercial Fe-ZSM-5 catalyst. *Catal. Today* **2012**, *184*, 107–114. [CrossRef]
57. Huang, T.-J.; Zhang, Y.-P.; Zhuang, K.; Lu, B.; Zhu, Y.-W.; Shen, K. Preparation of honeycombed holmium-modified Fe-Mn/TiO$_2$ catalyst and its performance in the low temperature selective catalytic reduction of NOx. *J. Fuel Chem. Technol.* **2018**, *46*, 319–327. [CrossRef]
58. Marbán, G.; Valdés-Solís, T.; Fuertes, A.B. Mechanism of low-temperature selective catalytic reduction of NO with NH$_3$ over carbon-supported Mn$_3$O$_4$: Role of surface NH$_3$ species: SCR mechanism. *J. Catal.* **2004**, *226*, 138–155. [CrossRef]
59. Wang, J.; Yi, H.; Tang, X.; Zhao, S.; Gao, F.; Yang, Z. Oxygen plasma-catalytic conversion of NO over MnOx: Formation and reactivity of adsorbed oxygen. *Catal. Commun.* **2017**, *100*, 227–231. [CrossRef]
60. Wang, T.; Zhang, X.; Liu, J.; Liu, H.; Guo, Y.; Sun, B. Plasma-assisted catalytic conversion of NO over Cu-Fe catalysts supported on ZSM-5 and carbon nanotubes at low temperature. *Fuel Process. Technol.* **2018**, *178*, 53–61. [CrossRef]

catalysts

MDPI

Article

Plasma Oxidation of H_2S over Non-stoichiometric La_xMnO_3 Perovskite Catalysts in a Dielectric Barrier Discharge Reactor

Kejie Xuan [1], Xinbo Zhu [1,2,3,*], Yuxiang Cai [2] and Xin Tu [3,*]

[1] Faculty of Maritime and Transportation, Ningbo University, Ningbo 315211, China; xuankejie@sina.com
[2] State Key Laboratory of Clean Energy Utilization, Zhejiang University, Hangzhou 310027, China; yxcai01@sina.com
[3] Department of Electrical Engineering and Electronics, University of Liverpool, Liverpool L69 3GJ, UK
* Correspondence: zhuxinbo@nbu.edu.cn (X.Z.); xin.tu@liv.ac.uk (X.T.); Tel.: +86-574-8760-0505 (X.Z.); +44-151-794-4513 (X.T.)

Received: 12 July 2018; Accepted: 23 July 2018; Published: 2 August 2018

Abstract: In this work, plasma-catalytic removal of H_2S over La_xMnO_3 ($x = 0.90, 0.95, 1, 1.05$ and 1.10) has been studied in a coaxial dielectric barrier discharge (DBD) reactor. The non-stoichiometric effect of the La_xMnO_3 catalysts on the removal of H_2S and sulfur balance in the plasma-catalytic process has been investigated as a function of specific energy density (SED). The integration of the plasma with the La_xMnO_3 catalysts significantly enhanced the reaction performance compared to the process using plasma alone. The highest H_2S removal of 96.4% and sulfur balance of 90.5% were achieved over the $La_{0.90}MnO_3$ catalyst, while the major products included SO_2 and SO_3. The missing sulfur could be ascribed to the sulfur deposited on the catalyst surfaces. The non-stoichiometric La_xMnO_3 catalyst exhibited larger specific surface areas and smaller crystallite sizes compared to the $LaMnO_3$ catalyst. The non-stoichiometric effect changed their redox properties as the decreased La/Mn ratio favored the transformation of Mn^{3+} to Mn^{4+}, which contributed to the generation of oxygen vacancies on the catalyst surfaces. The XPS and H_2-TPR results confirmed that the Mn-rich catalysts showed the higher relative concentration of surface adsorbed oxygen (O_{ads}) and lower reduction temperature compared to $LaMnO_3$ catalyst. The reaction performance of the plasma-catalytic oxidation of H_2S is closely related to the relative concentration of O_{ads} formed on the catalyst surfaces and the reducibility of the catalysts.

Keywords: plasmas-catalysis; non-thermal plasmas; perovskite catalysts; nonstoichiometry; H_2S oxidation

1. Introduction

The emission of odors from various sources including wastewater treatment and municipal solid waste (MSW) treatment facilities have become a public concern due to their negative effect on air quality and human health, especially on sensitive or sick people [1]. As a result, air quality control in the waste treatment facilities is important to ensure a comfortable environment for the workers and local residents. Great efforts have been devoted to the research and development of odor abatement technologies including wet scrubbing, active carbon adsorption, incineration and biofiltration, etc. [2–4]. However, these technologies are not cost-effective when dealing with low concentrations of odors in high-volume waste gas streams. For example, biological treatment is not flexible for the variation of odor loadings and volume of waste gas streams [5].

Recently, non-thermal plasma (NTP) has been regarded as a promising alternative for deodorization due to its unique characteristics of fast reaction, compact system and adaptability towards complex working conditions [6,7]. Typically, air plasma could generate electrons and various chemically reactive species, which could react with the odor molecules, leading to the purification of the odor-containing waste gas streams. Anderson et al. performed an on-site test of a pilot scale plasma reactor to purify a 138 $m^3 \cdot h^{-1}$ odor-containing waste gas stream with the main compositions of acetic acid, propanoic acid, trimethylamine and indole, etc. The average odor removal efficiency of 97% could be achieved at an input power of 140.8 W [8]. Kuwahara et al. developed a dielectric barrier discharge (DBD) reactor with a laminated film-electrode for odor control, while the complete removal of 100 ppm odor was achieved at a discharge power of 10 W and a flow rate of 5 $L \cdot min^{-1}$ [9]. Lu et al. reported the decomposition of ammonia and hydrogen sulfide with self-designed gliding arc plasma reactor, while a removal efficiency of 100% was achieved at the applied voltage of 11 kV and the velocity of 4.72 $m \cdot s^{-1}$ [10]. Some researcher also performed scale-up studies of plasma technologies in odor control [11]. Dobslaw et al. performed a 35-day-long scale-up test of plasma odor removal in two industrial sites in Germany, while the odor concentration was significantly reduced by 95.9–98.3% [12]. They also used the scale-up plasma units to enhance the performance of a biotrickling filter, the odor removal efficiency of using plasma alone reached up to 93.9% [13]. However, the relative low selectivity towards the desired final products in the NTP process remains the main challenge for the use of NTP-based technologies in environmental clean-up.

In the last three decades, the combination of non-thermal plasma and heterogeneous catalysis, known as "plasma-catalysis" has been demonstrated as a promising emerging process for the removal of low concentration gas pollutants including odors with reduced formation of by-products and enhanced process performance [14–17]. Although $LaMnO_3$ perovskite type catalysts have been used in thermal catalytic reactions due to their relatively low cost, comparable activity, and thermal stability [18–20], the use of these catalysts in plasma chemical reactions for environmental clean-up or the synthesis of fuels and chemicals has been very limited. Hueso et al. found that the combination of lanthanum based perovskite catalysts and a microwave discharge plasma enhanced the conversion of low concentration methane (3600 ppm) by around 30% compared to the plasma process in the absence of a catalyst at a same energy density of 10 $W \cdot m^{-2}$ [21,22]. Vandenbroucke et al. reported that the removal of trichloroethylene was increased by 13.9% when packing a $Pd/LaMnO_3$ catalyst into a negative DC corona discharge reactor at a specific energy density (SED) of 460 $J \cdot L^{-1}$ compared to the plasma reaction without a catalyst. They also found that the coupling of the plasma-catalyst coupling significantly reduced the formation of major by-product $CHCl_3$ [23]. It is well recognized that the substitution of A and B sites of the perovskite catalysts could result in a structural non-stoichiometry, which might generate excess oxygen species on the catalyst surfaces and consequently affect the reaction performance [24]. To the best of our knowledge, the use of perovskite catalysts in plasma-catalytic odor control has not been reported yet, while the knowledge about the underlying mechanisms of the non-stoichiometric effect of the perovskite catalyst on the plasma-catalytic oxidation processes are still missing.

In this work, hydrogen sulfide (H_2S) is chosen as a model pollutant since it accounts for over 90% of the total mass concentration in the odor emissions [25]. The non-stoichiometric effect of La_xMnO_3 catalysts on the plasma-catalytic removal of low concentration H_2S was investigated in a DBD plasma reactor in terms of the removal of H_2S and sulfur balance. The physicochemical properties of the catalysts were determined using various characterization techniques, including Brunauer-Emmett-Teller (BET) surface measurement, X-ray diffraction (XRD), X-ray photoelectron spectroscopy (XPS) and temperature programmed reduction of H_2 (H_2-TPR) to understand the structure-activity relationships between the La_xMnO_3 catalysts and the plasma-catalytic process and the role of these catalysts in the plasma-catalytic process.

2. Results and Discussions

2.1. Physicochemical Properties of the Catalysts

Table 1 show the results of N_2 adsorption-desorption experiments. The non-stoichiometric effect of the La_xMnO_3 catalysts enlarged the specific surface areas (S_{BET}) by 5.7% to 23.5% compared to that of the $LaMnO_3$ catalyst (12.3 $m^2 \cdot g^{-1}$), while the $La_{0.9}MnO_3$ catalyst had the highest S_{BET}. Figure 1 shows the XRD patterns of the La_xMnO_3 catalysts. The formation of the perovskite phase was observed in all the catalysts and no obvious segregated phases of La_2O_3 and MnO_x were found. The stoichiometric $LaMnO_3$ catalyst exhibited a typical cubic perovskite structure (JCPDS 75-0440) of $LaMnO_3$. The cubic structure of the La_xMnO_3 catalysts remained unchanged when the La doping amount was decreased, while the orthorhombic structure of $LaMnO_3$ (JCPDS 89-2470) was observed for the La-rich catalysts ($x > 1$). According to Table 1, the cell volume of the La_xMnO_3 catalysts increased with the increase of La/Mn molar ratio. The expansion of unit cell could be attributed to the transformation of Mn^{4+} into Mn^{3+} and the incorporation of La^{3+} into the perovskite lattice of La_xMnO_3 considering the larger ion radius of La^{3+} (1.06 Å) and Mn^{3+} (0.65 Å) compared to that of Mn^{4+} (0.54 Å). The characteristic diffraction peak of $LaMnO_3$ (1 2 1) located at around 32.6° was slightly shifted to a lower 2θ value for the La-rich catalysts, while this peak in the XRD pattern of the Mn-rich catalysts ($x < 1$) was shifted to a higher 2θ value [26]. The non-stoichiometric catalysts had a smaller crystallite size compared to the $LaMnO_3$ catalyst (16.1 nm), while the $La_{0.90}MnO_3$ catalyst had the smallest crystallite size of 15.4 nm.

Table 1. Physicochemical properties of La_xMnO_3 catalysts.

Sample	S_{BET} $(m^2 \cdot g^{-1})$	[1] Crystalline Size (nm)	Cell Volume (Å^3)	$Mn^{4+}/(Mn^{3+} + Mn^{4+})$ (%)	$O_{ads}/(O_{ads} + O_{lat})$ (%)
$La_{0.90}MnO_3$	15.2	15.4	58.6	43.2	60.7
$La_{0.95}MnO_3$	13.0	15.8	58.7	42.2	59.9
$LaMnO_3$	12.6	16.1	58.9	41.5	58.1
$La_{1.05}MnO_3$	13.7	15.9	233.2	40.8	57.6
$La_{1.10}MnO_3$	14.6	15.7	234.4	38.2	56.6

[1] Calculated from the diffraction peak of La_xMnO_3 located at $2\theta = 32.6°$.

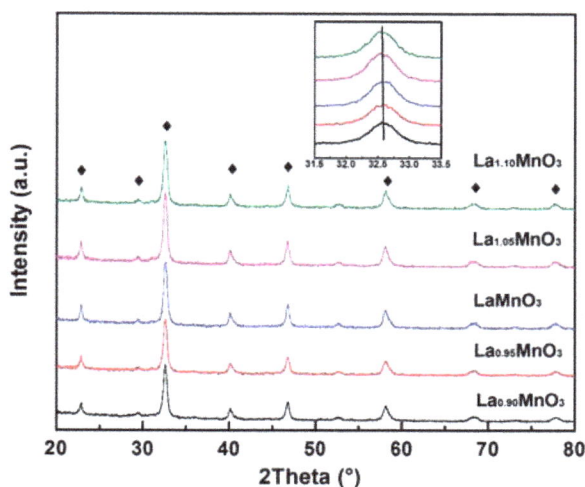

Figure 1. XRD patterns of La_xMnO_3 catalysts.

2.2. Redox Properties of the Catalysts

The chemical states of major elements (Mn 2p and O 1s) in the La_xMnO_3 catalysts were examined using XPS, as shown in Figure 2. The binding energies of La $3d_{5/2}$ (838.1 eV and 855.0 eV) were observed for all the catalysts (not shown), while those of La $3d_{3/2}$ were located at around 834.8 eV and 851.5 eV. The binding energies and the spin-orbit splitting of La 3d were close to those of pure La_2O_3, indicating that the lanthanum ions were in trivalent state of the La_xMnO_3 catalysts [27].

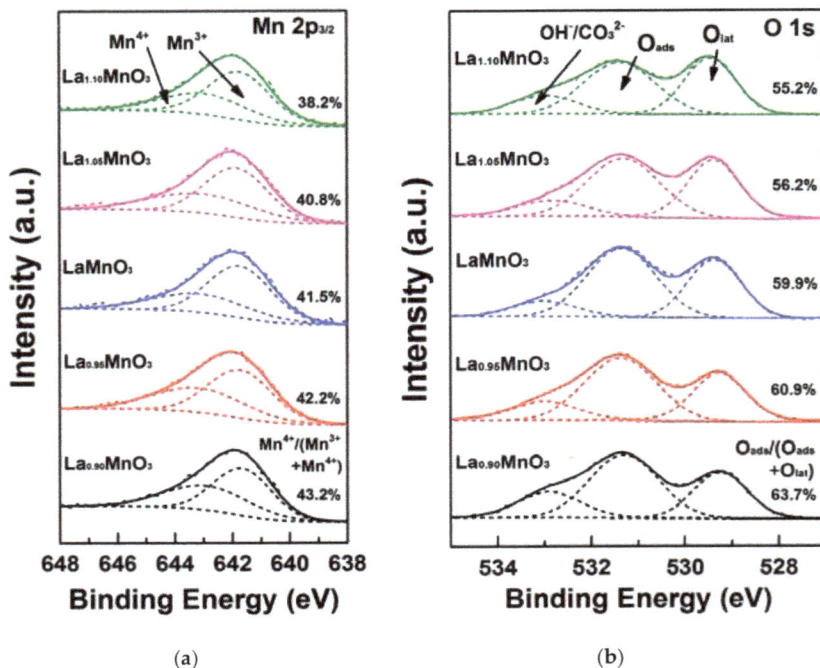

Figure 2. XPS spectra of La_xMnO_3 catalysts: (**a**) Mn $2p_{3/2}$; and (**b**) O 1s.

Figure 2a shows the XPS spectra of Mn $2p_{3/2}$ of the La_xMnO_3 catalysts. The co-existence of Mn^{2+}, Mn^{3+} and Mn^{4+} species in the $LaMnO_3$-based perovskite catalysts has been reported [28]. However, in this work, Mn^{2+} species were unlikely existed due to the missing satellite peaks at around 648.8 eV [29]. The XPS signals of Mn $2p_{3/2}$ can be deconvoluted into two major peaks. The XPS peaks at 641.6 eV were assigned to the generation of Mn^{3+} cations, while the peaks at around 643.0 eV was associated to the formation of Mn^{4+} cations. The relative concentration of Mn^{4+} in the La_xMnO_3 catalysts, defined as $Mn^{4+}/(Mn^{3+}+Mn^{4+})$, was calculated based on the deconvoluted peaks (Table 1). The relative concentration of Mn^{4+} of these catalysts varied from 38.2% to 43.2%, and the $La_{0.90}MnO_3$ catalyst showed the highest value of 43.2%. The relative Mn^{4+} concentration of the catalysts decreased with the increase of nominal La/Mn molar ratio.

The deconvoluted O 1s spectra of the La_xMnO_3 perovskites suggested the coexistence of various oxygen species on the catalyst surface (Figure 2b). The peaks located at around 532.9 eV, 531.2 eV and 529.4 eV could be attributed to the formation of oxygen-containing groups: hydroxyl/carbonate species, surface adsorbed oxygen (O_{ads}) and lattice oxygen (O_{lat}), respectively [30]. The non-stoichiometric effect did not significantly change the binding energy of these three major oxygen species. Table 1 presents the relative concentration of O_{ads} of the catalysts, defined as $O_{ads}/(O_{ads} + O_{lat})$. The $La_{0.90}MnO_3$ catalyst showed the highest relative O_{ads} concentration of 60.7%, while the relative concentration of O_{ads} of the catalysts decreased with the increase of nominal La/Mn molar ratio. Note that the variation of $O_{ads}/(O_{ads} + O_{lat})$ with the La/Mn molar ratio followed the same trend as $Mn^{4+}/(Mn^{3+} + Mn^{4+})$.

The reducibility of the La_xMnO_3 catalysts was analyzed using H_2-TPR experiment (Figure 3). For all the La_xMnO_3 catalysts, two major reduction peaks were observed. The first broad peak located between 200 °C and 500 °C could be divided into three sub-peaks. The shoulder peak centered between 200 °C and 300 °C could be ascribed to the reduction of weakly adsorbed oxygen species, while the second peak locate at around 350 °C was associated to the reduction of Mn^{4+} to Mn^{3+} with concomitant of surface adsorbed oxygen species. The third peak indicated the reduction of Mn^{3+} with coordination-unsaturated compose, while the peaks at higher temperature (above 600 °C) can be attributed to further reduction of Mn^{3+} to Mn^{2+} [27,31]. The lowest reduction temperature of the first broad peaks (335 °C and 392 °C) was observed over the $La_{0.90}MnO_3$ catalyst. Increasing the nominal La/Mn molar ratio slightly shifted the reduction peaks to higher temperatures. For example, the reduction temperatures of the $LaMnO_3$ catalyst were 342 °C and 393 °C, while those for $La_{1.10}MnO_3$ were 351 °C and 401 °C, respectively. This phenomenon indicated stronger interactions between La and Mn species, which inhibited the oxygen mobility on the La_xMnO_3 catalysts, and consequently decreased the reducibility of the catalysts.

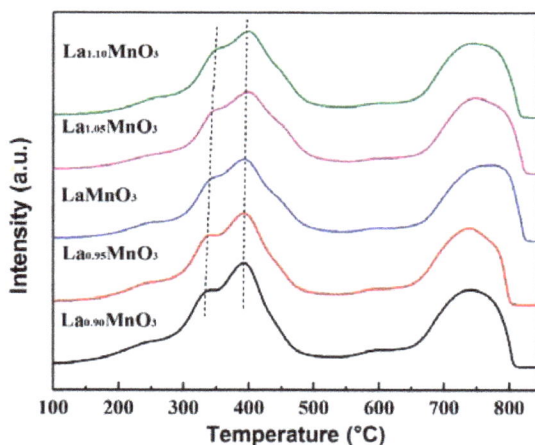

Figure 3. H_2-TPR profiles of La_xMnO_3 catalysts.

2.3. Plasma-Catalytic Oxidation of H_2S

The effect of the La_xMnO_3 catalysts on the removal of H_2S and the sulfur balance of the plasma-catalytic process was shown in Figure 4. The removal of H_2S increased monotonically with the increasing SED regardless of the presence of the catalysts. In the plasma reaction without a catalyst, the H_2S removal increased significantly from 10.7% to 41.5% when increasing the SED from 304.8 J·L^{-1} to 604.0 J·L^{-1}, while the sulfur balance of the plasma process decreased from 98.2% to 74.3%. The input energy in a typical air plasma is mainly used to generate highly energetic electrons and chemically reactive species, including O, OH·, and nitrogen excited states $N_2(A)$, which contributed to the decomposition of the H_2S through direct electron-impact dissociation of H_2S and radical attack [6]. The dissociation of H_2S molecules by electrons with sufficient energy forms H· and SH· radicals [32], while the collision of H_2S with radicals (e.g., O and OH·) and excited nitrogen species also takes place in the plasma reaction [33]:

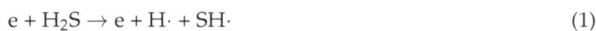

$$e + H_2S \rightarrow e + H\cdot + SH\cdot \tag{1}$$

$$N_2(A) + H_2S \rightarrow N_2 + H\cdot + SH\cdot \tag{2}$$

$$O + H_2S \rightarrow OH\cdot + SH\cdot \tag{3}$$

$$OH\cdot + H_2S \rightarrow SH\cdot + H_2O \tag{4}$$

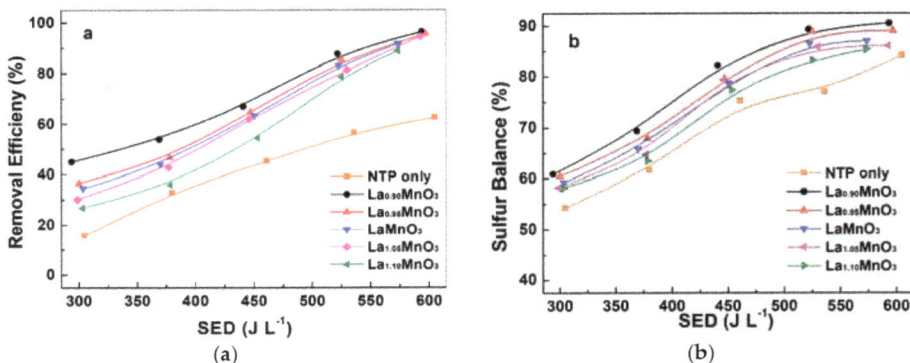

Figure 4. Effect of SED on plasma-catalytic removal of H_2S over La_xMnO_3 catalysts: (**a**) removal efficiency; and (**b**) sulfur balance of the plasma-catalytic process.

The generated $H\cdot$ radicals could be oxidized to form $OH\cdot$ by the oxidative species and participate the consequent plasma-induced reactions [6]. The $SH\cdot$ radicals were unstable in the oxidative environment and could be reacted with O and $OH\cdot$ radicals to form the major reaction product of SO_2:

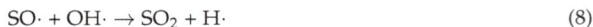

$$SH\cdot + O\cdot \rightarrow SO\cdot + H\cdot \tag{5}$$

$$SH\cdot + OH\cdot \rightarrow SO\cdot + H\cdot \tag{6}$$

$$SO\cdot + O\cdot \rightarrow SO_2 \tag{7}$$

$$SO\cdot + OH\cdot \rightarrow SO_2 + H\cdot \tag{8}$$

The inner energy of the rotational and vibrational excited species, which were insufficient to break chemical bonds of H_2S molecules, could be transferred to the electronically excited species and accelerate the plasma-catalytic reactions [34]. NTP generated by the dielectric barrier discharge reactor consisted of numerous micro-discharges in the plasma region. The number of micro-discharges in each discharge period was increased at a higher SED with fixed gas flow rate, which creates more reaction channels for the decomposition of H_2S and intermediates, resulting in the enhanced removal of H_2S [35–37]. It is worth noting that no ozone was detected in this work, which might be consumed by local heating, catalytic effect, or plasma oxidation reactions [38]. The decreased sulfur balance at a higher SED can be ascribed to the formation of sulfur via: (1) electron impact dissociation of $SH\cdot$ radicals, and (2) disproportionation of $SH\cdot$ radical and recombination between SO_2 and H_2S [14]:

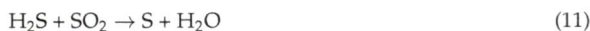

$$e + SH\cdot \rightarrow e + S + H\cdot \tag{9}$$

$$SH\cdot + SH\cdot \rightarrow S + H_2S \tag{10}$$

$$H_2S + SO_2 \rightarrow S + H_2O \tag{11}$$

The oxidation of SO_2 to SO_3 by highly oxidative species present in a plasma was reported by Jarrige et al. [39]:

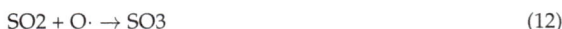

$$SO2 + O\cdot \rightarrow SO3 \tag{12}$$

$$SO2 + OH \cdot \rightarrow SO3 + H \cdot \tag{13}$$

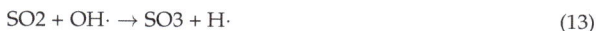

The presence of the La_xMnO_3 catalysts in the plasma region significantly enhanced the removal of H_2S and the sulfur balance compared to the plasma processing of H_2S without a catalyst. Figure 4 shows that the removal of H_2S increased from 34.3% to 91.6% in the plasma reaction over the stoichiometric $LaMnO_3$ when increasing the SED from 303.6 $J \cdot L^{-1}$ to 573.6 $J \cdot L^{-1}$, while the sulfur balance of the plasma-catalytic process increased from 59.2% to 87.0% under the same operating conditions. The increase of sulfur balance could be attributed to the enhanced oxidation of solid state sulfur to SO_2 at a higher SED, whilst the formation of a small amount of SO_3 was also observed. However, the formation of SO_3 in the plasma process was relatively low in this work (less than 8%), which was comparable to the results reported by Chang [40] and Kim et al. [41]. The non-stoichiometric effect of the La_xMnO_3 catalysts imposes a distinct effect on the plasma-catalytic removal of H_2S. Coupling the plasma with the $La_{0.90}MnO_3$ catalyst showed the highest H_2S removal of 96.4% and highest sulfur balance of 90.5% at a SED of 593.7 $J \cdot L^{-1}$. The increase of nominal La/Mn molar ratio inhibited the performance of the plasma-catalytic system for H_2S removal. The removal of H_2S and the sulfur balance follows the order of $La_{0.90}MnO_3 > La_{0.95}MnO_3 > LaMnO_3 > La_{1.05}MnO_3 > La_{1.10}MnO_3$ > NTP in the tested SED range, which is the opposite to the order of La/Mn molar ratio. Part of the missing sulfur element was found on the surface of the spent catalysts by XPS, which was also confirmed by a previous work [14].

Clearly, the La_xMnO_3 catalysts played a crucial role in the enhancement of the performance of the plasma-catalytic system. The non-stoichiometric effect slightly increased the specific surface area of the La_xMnO_3 catalysts by 3.2 to 20.6% compared to pure $LaMnO_3$ catalyst (12.6 $m^2 \cdot g^{-1}$), while the crystalline size of the La_xMnO_3 catalysts was decreased by 0.9–9.8% (16.1 nm for the $LaMnO_3$ catalyst). The changes in the specific surface area and crystalline size of the catalysts indicated that the non-stoichiometric effect contributed to the strong interactions between the La and Mn species of the catalysts. Larger specific surface area could offer more adsorption sites and active sites in the plasma-catalytic reaction, which prolonged the residence time of H_2S and intermediates in the plasma reaction [42]. Previous study also confirmed that smaller crystallite sizes of the catalysts favored the exposures of active sites on the catalyst surface [43]. To sum up, the changes in the physicochemical properties of the La_xMnO_3 catalysts would increase the possibilities of effective collisions between the reactive species and the pollutant including H_2S and intermediates, which benefited the reaction performance of the plasma-catalysis system.

The XPS spectra confirmed the coexistence of Mn^{3+} and Mn^{4+} cations in the La_xMnO_3 catalysts. Charge imbalance may occur in the non-stoichiometric La_xMnO_3 regarding the change of the nominal La/Mn molar ratio in the perovskite structure [18,44]. Taking the $La_{0.90}MnO_3$ catalyst as an example, the decrease of La^{3+} content could form La vacancies in the perovskite lattice. The form of La vacancies could be compensated by the conversion of Mn^{3+} to Mn^{4+} to maintain the neutrality, which was evidenced by the production of highest relative concentration of Mn^{4+} of 43.2% in the $La_{0.90}MnO_3$ catalyst. Similar results were reported for the La-rich samples as more Mn^{3+} can be generated to maintain the electron neutrality in the presence of excess La cations [45]. The redox cycle between Mn^{3+} and Mn^{4+} was closely associated with the redox properties of the catalysts. The value of $Mn^{4+}/(Mn^{3+} + Mn^{4+})$ decreased from 43.2% to 38.2% when the La/Mn molar ratio increased from 0.90 to 1.10 in this work. The presence of Mn^{4+} species was also correlated with the formation of oxygen vacancies on the surface of the La_xMnO_3 catalysts [46]. Oxygen vacancies acted as the adsorption-desorption centers for the generation of O_{ads} from gas-phase oxygen species including O_2 molecules and O radicals in the plasma process. The oxygen species adsorbed on the oxygen vacancies (O_{ads}) played an important role in the plasma-induced surface reactions for the oxidation of H_2S and the intermediates due to the higher mobility of O_{ads} compared to the lattice oxygen (O_{lat}) [47]. Thus, perovskite catalysts with abundant oxygen vacancies were more active for the oxidation of pollutants on the catalyst surface in the plasma region. The $La_{0.90}MnO_3$ catalyst possessed the highest $O_{ads}/(O_{ads} + O_{lat})$ value of 60.7%, while increasing the nominal La/Mn molar ratio reduced the value to 59.9% for LaMnO$_3$ and 55.2% for La$_{1.10}$MnO$_3$, respectively. Note the decrease of O$_{ads}$/(O$_{ads}$ + O$_{lat}$) at higher La/Mn molar ratio was in line with the decreased Mn^{4+}/(Mn^{3+} + Mn^{4+}).

The reducibility of the La$_x$MnO$_3$ catalysts was also significantly affected by the non-stoichiometric effect. Figure 3 shows that the reduction temperatures of the Mn-rich catalysts were much lower compared to the La-rich catalysts and LaMnO$_3$, indicating that on the Mn-rich catalysts had a higher oxygen mobility. As mentioned before, the compensation effect of the insufficient or excess La species could promote or inhibit the formation of oxygen vacancies, which could contribute to the formation of surface adsorbed oxygen (O$_{ads}$) with higher oxygen mobility and consequently contributes to the enhanced oxidation of the adsorbed H$_2$S and intermediates to the final product of H$_2$O, SO$_2$, sulfur and possibly SO$_3$ in the plasma-catalytic system. It is noteworthy that the order of the reduction temperature of these catalysts was in consistent with the sequence of O$_{ads}$/(O$_{ads}$ + O$_{lat}$) and Mn^{4+}/(Mn^{3+} + Mn^{4+}). Combining the results of XPS and H$_2$-TPR of these catalysts, it is clear that the redox properties of the catalysts play a determining role in the reaction performances of the plasma-catalytic removal of H$_2$S, while the non-stoichiometric effect at higher nominal La/Mn ratio inhibited the removal of H$_2$S in the plasma-catalytic system. The possible reaction mechanisms in the plasma-catalytic oxidation of H$_2$S over the La$_x$MnO$_3$ catalysts have been summarized in Figure 5.

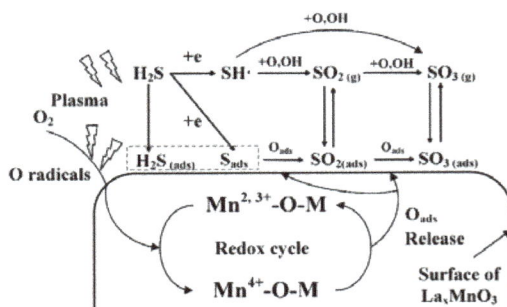

Figure 5. Reaction mechanisms of plasma-catalytic removal of H$_2$S over the La$_x$MnO$_3$ catalysts.

3. Experimental Setup

3.1. Catalyst Preparation

The non-stoichiometric La$_x$MnO$_3$ (x = 0.9, 0.95, 1, 1.05 and 1.1) perovskite-type catalysts were synthesized using a citric acid method, while lanthanum nitrate hexahydrate and 50 wt % manganese nitrate solution were used as the precursors [1]. All chemicals were analytic reagent and purchased from Aladdin Co. Ltd. (Shanghai, China) The synthesis procedures of the La$_x$MnO$_3$ catalysts were: (1) weighted amounts of all precursors and deionized water were mixed to obtain 0.1 M solutions; (2) a 50% excess amount of citric acid (comparing to the total amount of the metal cations) was added to the solution obtained in step 1 as the ligand; (3) the solution was thoroughly stirred for 3 h and evaporated in an 80 °C water bath to form viscous gel; (4) the gel was dried in an oven overnight at 110 °C and then calcined at 700 °C for 5 h; (5) the samples was then pressed and sieved to 35–60 meshes for testing.

3.2. Catalyst Characterizations

The specific surface area, average pore size and pore volume of the La$_x$MnO$_3$ catalysts were measured using N$_2$ adsorption-desorption experiments (Quantachrome Autosorb-1, Boynton Beach, FL, USA) at 77 K. The La$_x$MnO$_3$ catalyst samples were degassed at 200 °C for 5 h before each test. The specific surface area (S$_{BET}$) of the La$_x$MnO$_3$ catalyst was obtained by the Brunauer-Emmett-Teller (BET) equation.

The X-ray diffraction (XRD) measurements were carried out on a Rikagu D/max-2000 X-ray diffractometer (Tokyo, Japan) using Cu-Kα as the radiation source.

The X-ray photoelectron spectroscopy (XPS) experiments were performed with a Thermo ESCALAB 250 spectrometer (Waltham, MA, USA) using an Al Kα X-ray source (hν = 1486.6 eV). The XPS results were calibrated using the C 1s spectra at a binding energy (B.E.) of 284.6 eV.

The hydrogen temperature-programmed reduction (H_2-TPR) were performed using a Micrometrics Autochem II 2920 instrument (Ottawa, ON, Canada). For each test, 50 mg La_xMnO_3 catalyst samples were loaded and pretreated at 250 °C in a flowing N_2 for 1 h to remove the weakly adsorbed impurities and then cooled down to 50 °C. The samples were then heated from room temperature to 800 °C in a 5 vol % H_2/Ar flow (40 mL·min^{-1}). The heating rate was constant at 10 °C·min^{-1}. The H_2 consumption was calculated based on the H_2-TPR profiles.

3.3. Experimental Set-Up

The schematic experimental set-up is shown in Figure 6. The details of the DBD reactor were described elsewhere [26]. The reactor was co-axial type with the discharge length of 60 mm. The inner diameter of the discharge tube was 8 mm, while the outer diameter was 10 mm. The diameter of the stainless high voltage rod was 2 mm. Thus, the total plasma discharge volume was 2.26 mL. The carrier gas (high-purity air) and H_2S was generated from gas cylinders purchased from Jingong, Hangzhou, China. All gas streams were regulated by mass flow controllers and mixed prior to the plasma reactor. The total flow rate was 2 L·min^{-1}, while the initial H_2S concentration was 100 ppm. In this work, 100 mg La_xMnO_3 catalyst was used in each test. The catalysts was held in place by quartz wool, while only quartz wool was packed in the plasma region in the cases using plasma alone. The reactor was energized by an AC power supply (Suman CTP2000-K, Nanjing, China) and the discharge frequency was fixed at 10 kHz. The temperature at the outer wall of the reactor was measured using an infrared thermometer (Omega OS540, Norwalk, CT, USA). During the experiment process, it gradually increased to approximately 80 °C, while the outlet gas temperature was about 30 °C.

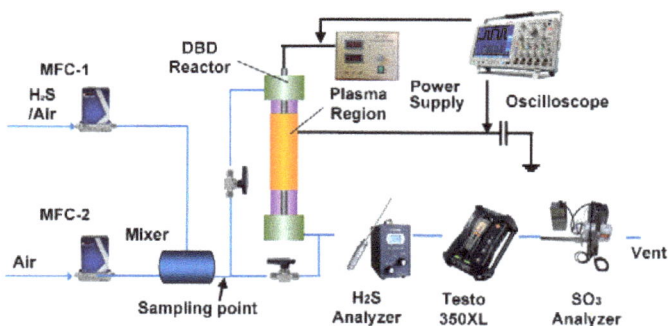

Figure 6. Schematic diagram of the experimental setup.

The applied voltage across the plasma-catalytic reactor was measured using a Tektronix 4034B digital oscilloscope connected to a Tektronix 6015A voltage probe (1000:1). An external capacitor was connected in series with the grounded electrode across which the voltage was measured using a TPP500 voltage probe (Tektronix, Beaverton, OR, USA). The average discharge power was calculated by integrating the voltage waveforms by Q-U Lissajous method. The specific energy density (SED) of the plasma-catalysis process can be obtained as:

$$SED(J \cdot L^{-1}) = \frac{P(W)}{Q\left(L \cdot min^{-1}\right)} \times 60 \tag{14}$$

where P is the discharge power of plasma-catalytic process and Q is the total gas flow rate.

H$_2$S was measured by using a PTM-400 H$_2$S analyzer (Yiyuntian, Shenzhen, China) with the measuring range of 0–200 ppm and an accuracy of $\pm 2\%$, while the concentration of formed SO$_2$ was measured by using a Testo 350XL gas analyzer (Lenzkirch, Germany) with an accuracy of $\pm 5\%$. The concentration of outlet SO$_3$ was measured using a SO$_3$ analyzer (RJ-SO3-M) (Ruijing Tech., Shenzhen, China) based on the US EPA method 8 (specific IPA absorption method for SO$_3$ measurement). The detection range of the SO$_3$ analyzer was 0–200 ppm, while the average measuring error was less than 10% [48]. All data given in this work was the average value of three repeated measurements. The removal of H$_2$S (η_{H_2S}) and sulfur balance of the plasma-catalytic oxidation of H$_2$S can be defined as:

$$\eta_{H_2S}(\%) = \frac{c_{in} - c_{out}}{c_{in}} \times 100\% \tag{15}$$

$$Sulfur\ balance(\%) = \frac{c_{SO_2} + c_{SO_3}}{c_{in} - c_{out}} \times 100\% \tag{16}$$

where c_{in} and c_{out} are the H$_2$S concentration at the inlet and outlet of the reactor, respectively, while c_{SO_2} and c_{SO_3} are the outlet SO$_2$ and SO$_3$ concentrations.

4. Conclusions

A series of non-stoichiometric La$_x$MnO$_3$ catalysts were evaluated in the plasma-catalytic oxidative removal of H$_2$S using a non-thermal plasma DBD reactor in terms of H$_2$S removal and sulfur balance of the plasma process. The coupling of the DBD plasma with the La$_x$MnO$_3$ catalysts enhanced both H$_2$S removal and sulfur balance compared to the plasma process without using a catalyst. The highest H$_2$S removal of 96.4% was achieved at a SED of 593.7 J·L^{-1} when placing the La$_{0.90}$MnO$_3$ catalyst in the DBD plasma reactor. In addition, the combination of the plasma with the La$_{0.90}$MnO$_3$ catalyst also showed the highest sulfur balance in the plasma-catalytic process. The non-stoichiometric effect of the La$_x$MnO$_3$ catalysts significantly affected the removal of H$_2$S since the reaction performance of the plasma-catalytic process decreased with the increasing nominal La/Mn molar ratio. Compared to pure LaMnO$_3$ catalyst, the non-stoichiometric La$_x$MnO$_3$ catalyst showed enhanced specific surface area and smaller crystallite size. The decreased La/Mn molar ratio in the La$_x$MnO$_3$ catalysts favored the transformation of Mn^{3+} cations to Mn^{4+}, which contributed to the formation of oxygen vacancies on the catalyst surfaces. The results of XPS and H$_2$-TPR demonstrated that the Mn-rich catalysts possessed higher relative concentration of surface adsorbed oxygen (O$_{ads}$) and better reducibility (lower reduction temperature) compared to pure LaMnO$_3$, which suggests that the relative concentration of O$_{ads}$ and reducibility of the catalysts played a key role in determining the reaction performance of the plasma-catalytic removal of H$_2$S.

Author Contributions: Conceptualization: X.Z. and X.T.; data curation: K.X., X.Z., and Y.C.; formal analysis: Y.C.; investigation: K.X. and X.Z.; methodology: X.Z. and Y.C.; writing—original draft: X.Z.; writing—review and editing: X.Z. and X.T.

Funding: This research was funded by National Natural Science Foundation of China (no. 51606166) and K.C. Wong Magna Fund in Ningbo University, China.

Conflicts of Interest: The authors declare no conflict of interest.

References

1. Bogner, J.E.; Chanton, J.P.; Blake, D.; Abichou, T.; Powelson, D. Effectiveness of a florida landfill biocover for reduction of CH_4 and nmhc emissions. *Environ. Sci. Technol.* **2010**, *44*, 1197–1203. [CrossRef] [PubMed]
2. Burgess, J.E.; Parsons, S.A.; Stuetz, R.M. Developments in odour control and waste gas treatment biotechnology: A review. *Biotechnol. Adv.* **2001**, *19*, 35–63. [CrossRef]
3. Schlegelmilch, M.; Streese, J.; Stegmann, R. Odour management and treatment technologies: An overview. *Waste Manag.* **2005**, *25*, 928–939. [CrossRef] [PubMed]
4. Zhang, L.; De Schryver, P.; De Gusseme, B.; De Muynck, W.; Boon, N.; Verstraete, W. Chemical and biological technologies for hydrogen sulfide emission control in sewer systems: A review. *Water Res.* **2008**, *42*, 1–12. [CrossRef] [PubMed]
5. Chen, L.; Hoff, S.J. Mitigating odors from agricultural facilities: A review of literature concerning biofilters. *Appl. Eng. Agric.* **2009**, *25*, 751–766. [CrossRef]
6. Vandenbroucke, A.M.; Morent, R.; De Geyter, N.; Leys, C. Non-thermal plasmas for non-catalytic and catalytic VOC abatement. *J. Hazard. Mater.* **2011**, *195*, 30–54. [CrossRef] [PubMed]
7. Van Durme, J.; Dewulf, J.; Leys, C.; Van Langenhove, H. Combining non-thermal plasma with heterogeneous catalysis in waste gas treatment: A review. *Appl. Catal. B Environ.* **2008**, *78*, 324–333. [CrossRef]
8. Andersen, K.B.; Feilberg, A.; Beukes, J.A. Abating odour nuisance from pig production units by the use of a non-thermal plasma system. In Proceedings of the International conference on environmental odour monitoring and control (NOSE), Florence, Italy, 22–24 September 2010; pp. 351–356.
9. Kuwahara, T.; Okubo, M.; Kuroki, T.; Kametaka, H.; Yamamoto, T. Odor removal characteristics of a laminated film-electrode packed-bed nonthermal plasma reactor. *Sensors* **2011**, *11*, 5529–5542. [CrossRef] [PubMed]
10. Lu, S.Y.; Chen, L.; Huang, Q.X.; Yang, L.Q.; Du, C.M.; Li, X.D.; Yan, J.H. Decomposition of ammonia and hydrogen sulfide in simulated sludge drying waste gas by a novel non-thermal plasma. *Chemosphere* **2014**, *117*, 781–785. [CrossRef] [PubMed]
11. Andersen, K.B.; Feilberg, A.; Beukes, J.A. Use of non-thermal plasma and UV-light for removal of odour from sludge treatment. *Water Sci. Technol.* **2012**, *66*, 1656. [CrossRef] [PubMed]
12. Dobslaw, D.; Ortlinghaus, O.; Dobslaw, C. A combined process of non-thermal plasma and a low-cost mineral adsorber for VOC removal and odor abatement in emissions of organic waste treatment plants. *J. Environ. Chem. Eng.* **2018**, *6*, 2281–2289. [CrossRef]
13. Dobslaw, D.; Schulz, A.; Helbich, S.; Dobslaw, C.; Engesser, K.-H. VOC removal and odor abatement by a low-cost plasma enhanced biotrickling filter process. *J. Environ. Chem. Eng.* **2017**, *5*, 5501–5511. [CrossRef]
14. Maxime, G.; Amine, A.A.; Abdelkrim, B.; Dominique, W. Removal of gas-phase ammonia and hydrogen sulfide using photocatalysis, nonthermal plasma, and combined plasma and photocatalysis at pilot scale. *Environ. Sci. Pollut. Res.* **2014**, *21*, 13127–13137. [CrossRef] [PubMed]
15. Ragazzi, M.; Tosi, P.; Rada, E.C.; Torretta, V.; Schiavon, M. Effluents from MBT plants: Plasma techniques for the treatment of VOCs. *Waste Manag.* **2014**, *34*, 2400–2406. [CrossRef] [PubMed]
16. Ochiai, T.; Ichihashi, E.; Nishida, N.; Machida, T.; Uchida, Y.; Hayashi, Y.; Morito, Y.; Fujishima, A. Field performance test of an air-cleaner with photocatalysis-plasma synergistic reactors for practical and long-term use. *Molecules* **2014**, *19*, 17424–17434. [CrossRef] [PubMed]
17. Almarcha, D.; Almarcha, M.; Jimenez-Coloma, E.; Vidal, L.; Puigcercós, M.; Barrutiabengoa, I. Treatment efficiency by means of a nonthermal plasma combined with heterogeneous catalysis of odoriferous volatile organic compounds emissions from the thermal drying of landfill leachates. *J. Eng.* **2014**, *2014*, 831584. [CrossRef]
18. Dinh, M.T.N.; Giraudon, J.M.; Lamonier, J.F.; Vandenbroucke, A.; De Geyter, N.; Leys, C.; Morent, R. Plasma-catalysis of low tce concentration in air using $LaMnO_{3+\delta}$ as catalyst. *Appl. Catal. B Environ.* **2014**, *147*, 904–911. [CrossRef]
19. Nuns, N.; Beaurain, A.; Dinh, M.T.N.; Vandenbroucke, A.; De Geyter, N.; Morent, R.; Leys, C.; Giraudon, J.M.; Lamonier, J.F. A combined tof-sims and Xps study for the elucidation of the role of water in the performances of a post-plasma process using $LaMnO_{3+\delta}$ as catalyst in the total oxidation of trichloroethylene. *Appl. Surf. Sci.* **2014**, *320*, 154–160. [CrossRef]

20. Shi, C.; Zhang, Z.S.; Crocker, M.; Xu, L.; Wang, C.Y.; Au, C.T.; Zhu, A.M. Non-thermal plasma-assisted NO_x storage and reduction on a $LaMn_{0.9}Fe_{0.1}O_3$ perovskite catalyst. *Catal. Today* **2013**, *211*, 96–103. [CrossRef]

21. Hueso, J.; Cotrino, J.; Caballero, A.; Espinos, J.; Gonzalezelipe, A. Plasma catalysis with perovskite-type catalysts for the removal of NO and CH_4 from combustion exhausts. *J. Catal.* **2007**, *247*, 288–297. [CrossRef]

22. Hueso, J.L.; Caballero, A.; Cotrino, J.; González-Elipe, A.R. Plasma catalysis over lanthanum substituted perovskites. *Catal. Commun.* **2007**, *8*, 1739–1742. [CrossRef]

23. Vandenbroucke, A.M.; Nguyen Dinh, M.T.; Nuns, N.; Giraudon, J.M.; De Geyter, N.; Leys, C.; Lamonier, J.F.; Morent, R. Combination of non-thermal plasma and $Pd/LaMnO_3$ for dilute trichloroethylene abatement. *Chem. Eng. J.* **2016**, *283*, 668–675. [CrossRef]

24. Choi, J.J.; Billinge, S.J. Perovskites at the nanoscale: From fundamentals to applications. *Nanoscale* **2016**, *8*, 6206–6208. [CrossRef] [PubMed]

25. Kim, K.-H.; Choi, Y.J.; Jeon, E.C.; Sunwoo, Y. Characterization of malodorous sulfur compounds in landfill gas. *Atmos. Environ.* **2005**, *39*, 1103–1112. [CrossRef]

26. Zhu, X.; Gao, X.; Yu, X.; Zheng, C.; Tu, X. Catalyst screening for acetone removal in a single-stage plasma-catalysis system. *Catal. Today* **2015**, *256*, 108–114. [CrossRef]

27. Zhang, C.; Wang, C.; Zhan, W.; Guo, Y.; Guo, Y.; Lu, G.; Baylet, A.; Giroir-Fendler, A. Catalytic oxidation of vinyl chloride emission over $LaMnO_3$ and $LaB_{0.2}Mn_{0.8}O_3$ (B = Co, Ni, Fe) catalysts. *Appl. Catal. B Environ.* **2013**, *129*, 509–516. [CrossRef]

28. Zhu, X.; Liu, S.; Cai, Y.; Gao, X.; Zhou, J.; Zheng, C.; Tu, X. Post-plasma catalytic removal of methanol over Mn-Ce catalysts in an atmospheric dielectric barrier discharge. *Appl. Catal. B Environ.* **2016**, *183*, 124–132. [CrossRef]

29. Hou, Y.-C.; Ding, M.-W.; Liu, S.-K.; Wu, S.-K.; Lin, Y.-C. Ni-substituted $LaMnO_3$ perovskites for ethanol oxidation. *RSC Adv.* **2014**, *4*, 5329. [CrossRef]

30. Shen, M.Q.; Zhao, Z.; Chen, J.H.; Su, Y.G.; Wang, J.; Wang, X.Q. Effects of calcium substitute in $LaMnO_3$ perovskites for NO catalytic oxidation. *J. Rare Earth.* **2013**, *31*, 119–123. [CrossRef]

31. Quiroz, J.; Giraudon, J.-M.; Gervasini, A.; Dujardin, C.; Lancelot, C.; Trentesaux, M.; Lamonier, J.-F. Total oxidation of formaldehyde over MnO_x-CeO_2 catalysts: The effect of acid treatment. *ACS Catal.* **2015**, *5*, 2260–2269. [CrossRef]

32. Continetti, R.E.; Balko, B.A.; Lee, Y.T. Photodissociation of H_2S and the HS radical at 193.3 nm. *Chem. Phys. Lett.* **1991**, *182*, 400–405. [CrossRef]

33. Liang, W.-J.; Fang, H.-P.; Li, J.; Zheng, F.; Li, J.-X.; Jin, Y.-Q. Performance of non-thermal DBD plasma reactor during the removal of hydrogen sulfide. *J. Electrostat.* **2011**, *69*, 206–213. [CrossRef]

34. Aerts, R.; Martens, T.; Bogaerts, A. Influence of vibrational states on CO_2 splitting by dielectric barrier discharges. *J. Phys. Chem. C* **2012**, *116*, 23257–23273. [CrossRef]

35. Zheng, C.; Zhu, X.; Gao, X.; Liu, L.; Chang, Q.; Luo, Z.; Cen, K. Experimental study of acetone removal by packed-bed dielectric barrier discharge reactor. *J. Ind. Eng. Chem.* **2014**, *20*, 2761–2768. [CrossRef]

36. Aerts, R.; Tu, X.; De Bie, C.; Whitehead, J.C.; Bogaerts, A. An investigation into the dominant reactions for ethylene destruction in non-thermal atmospheric plasmas. *Plasma Process. Polym.* **2012**, *9*, 994–1000. [CrossRef]

37. Zhu, X.; Gao, X.; Qin, R.; Zeng, Y.; Qu, R.; Zheng, C.; Tu, X. Plasma-catalytic removal of formaldehyde over Cu-Ce catalysts in a dielectric barrier discharge reactor. *Appl. Catal. B Environ.* **2015**, *170–171*, 293–300. [CrossRef]

38. Holzer, F.; Kopinke, F.D.; Roland, U. Influence of ferroelectric materials and catalysts on the performance of non-thermal plasma (NTP) for the removal of air pollutants. *Plasma Chem. Plasma Process.* **2005**, *25*, 595–611. [CrossRef]

39. Jarrige, J.; Vervisch, P. Decomposition of gaseous sulfide compounds in air by pulsed corona discharge. *Plasma Chem. Plasma Process.* **2007**, *27*, 241–255. [CrossRef]

40. Chang, M.B.; Balbach, J.H.; Rood, M.J.; Kushner, M.J. Removal of SO_2 from gas streams using a dielectric barrier discharge and combined plasma photolysis. *J. Appl. Phys.* **1991**, *69*, 4409–4417. [CrossRef]

41. Hyun Ha, K.; Wu, C.; Takashima, K.; Mizuno, A. The influence of reaction conditions on SO_2 oxidation in a discharge plasma reactor. *IEEE Trans. Ind. Appl.* **1999**, *3*, 1478–1482.

42. Neyts, E.C. Plasma-surface interactions in plasma catalysis. *Plasma Chem. Plasma Process.* **2015**, *36*, 185–212. [CrossRef]

43. Tu, X.; Whitehead, J.C. Plasma-catalytic dry reforming of methane in an atmospheric dielectric barrier discharge: Understanding the synergistic effect at low temperature. *Appl. Catal. B Environ.* **2012**, *125*, 439–448. [CrossRef]

44. Hosseini, S.A.; Salari, D.; Niaei, A.; Oskoui, S.A. Physical-chemical property and activity evaluation of $LaB_{0.5}Co_{0.5}O_3$ (B = Cr, Mn, Cu) and $LaMn_xCo_{1-x}O3$ (x = 0.1, 0.25, 0.5) nano perovskites in VOC combustion. *J. Ind. Eng. Chem.* **2013**, *19*, 1903–1909. [CrossRef]

45. Chen, J.; Shen, M.; Wang, X.; Qi, G.; Wang, J.; Li, W. The influence of nonstoichiometry on $LaMnO_3$ perovskite for catalytic NO oxidation. *Appl. Catal. B Environ.* **2013**, *134–135*, 251–257. [CrossRef]

46. Dai, Y.; Wang, X.Y.; Li, D.; Dai, Q.G. Catalytic combustion of chlorobenzene over Mn-Ce-La-O mixed oxide catalysts. *J. Hazard. Mater.* **2011**, *188*, 132–139.

47. Zhang, Z.; Jiang, Z.; Shangguan, W. Low-temperature catalysis for VOCs removal in technology and application: A state-of-the-art review. *Catal. Today* **2016**, *264*, 270–278. [CrossRef]

48. Yang, Z.; Zheng, C.; Zhang, X.; Zhou, H.; Silva, A.A.; Liu, C.; Snyder, B.; Wang, Y.; Gao, X. Challenge of SO_3 removal by wet electrostatic precipitator under simulated flue gas with high SO_3 concentration. *Fuel* **2018**, *217*, 597–604. [CrossRef]

catalysts

MDPI

Article

Ammonia Plasma-Catalytic Synthesis Using Low Melting Point Alloys

Javishk R. Shah, Joshua M. Harrison and Maria L. Carreon *

Russell School of Chemical Engineering, The University of Tulsa, 800 S Tucker Dr, Tulsa, OK 74104, USA;
jrs943@utulsa.edu (J.R.S.); jmh1756@utulsa.edu (J.M.H.)
* Correspondence: maria-carreon@utulsa.edu; Tel.: +1-918-631-2424

Received: 19 September 2018; Accepted: 30 September 2018; Published: 3 October 2018

Abstract: The Haber-Bosch process has been the commercial benchmark process for ammonia synthesis for more than a century. Plasma-catalytic synthesis for ammonia production is theorized to have a great potential for being a greener alternative to the Haber-Bosch process. However, the underlying reactions for ammonia synthesis still require some detailed study especially for radiofrequency plasmas. Herein, the use of inductively coupled radiofrequency plasma for the synthesis of ammonia when employing Ga, In and their alloys as catalysts is presented. The plasma is characterized using emission spectroscopy and the surface of catalysts using Scanning Electron Microscope. A maximum energy yield of 0.31 g-NH$_3$/kWh and energy cost of 196 MJ/mol is achieved with Ga-In (0.6:0.4 and 0.2:0.8) alloy at 50 W plasma power. Granular nodes are observed on the surface of catalysts indicating the formation of the intermediate GaN.

Keywords: plasma catalysis; gallium; indium; Ga–In alloys; radiofrequency plasma; ammonia synthesis

1. Introduction

The alternatives to the Haber–Bosch process have been explored for decades. The global production of ammonia will exceed 176 million tons by the end of 2018. Such huge production also accounts for 1–2% of the global greenhouse gas emissions [1,2]. With the growing consciousness as well as the threat of global warming, alternatives to the Haber–Bosch process for ammonia synthesis are not only required for scientific advancement but also as an effort to reduce the effects of global warming.

Ammonia was discovered by Fritz Haber at laboratory scale but the scale-up of the process to an economical scale was performed by Carl Bosch [3]. The early catalysts osmium and uranium used by Haber were replaced with a pure iron metal catalyst by Bosch [4]. The nitrogen is obtained from air and hydrogen is obtained via methane reformation with oxygen from air. The N$_2$ and H$_2$ are separated from other gases by refrigeration process and sent to the reactor. The iron catalyst in the reactor helps with the breaking of the dinitrogen triple bond. The ammonia is removed from the exit gas stream via refrigeration [5]. In the terms of formation mechanism, the major energy consumption step for the Haber–Bosch process is the dinitrogen triple bond breaking. Due to the high stability of the dinitrogen triple bond, either high pressure or high temperature is needed. The commercial plants running on the Haber–Bosch process operate at 400–500 °C and 150–250 bar to keep the production economically feasible [5]. This limits the thermodynamic yield for ammonia synthesis to 15%, since at these process conditions the reaction of ammonia formation is reversible. At higher pressure and higher temperature, the decomposition reaction becomes more dominant. The yield can be improved at lower temperatures, but the exothermic nature of the reaction increases the process temperature thus imposing a thermodynamic limit on the ammonia yield [6]. To curb the energy consumption issue, iron catalysts with promoters such as potassium, calcium, and aluminum are employed to break the dinitrogen bond [7].

There have been some advances in the field of alternative pathways for ammonia synthesis. Plasma catalysis has gained attention as an alternative. The discharge is created by exciting the nitrogen and hydrogen molecules, which further break down into charged species, ions, and excited atoms. The already excited species help in shifting the rate-limiting step from breaking of dinitrogen to a surface reaction [8]. In recent years, there have been several reports of ammonia synthesis using dielectric barrier discharge, microwave, glowing arc discharge, and radiofrequency (RF) plasma [9]. Among the most notable reports are Patil et al. (Ru/Tubular Alumina) [10], Iwamoto et al. (metal wool) [11,12], and Kim et al. (Ru/Al$_2$O$_3$) [13], all of them performed in a dielectric barrier discharge (DBD). These reports indicated the highest energy yield of ammonia in the range of 5–35 g-NH$_3$/kWh. The reaction mechanism in a DBD plasma reactor is reasonably understood without a catalyst, but with the introduction of a catalyst it becomes extremely complex [14]. Herein, we focus on tailoring the catalyst for radiofrequency plasma catalytic ammonia synthesis. The catalysts used by other researchers for plasma catalytic ammonia synthesis include Pt,Ni,Fe/Alumina Membrane [15], Ni/SiO$_2$ [16], Ru/MCM-41 [17], Mo wires [18], MgO [19], CaO, WO$_3$ [20], and Ru/Carbon nanotube [18,21–24]. The only reports for RF plasma ammonia synthesis were presented by Matsumoto's group with an energy yield of 0.075 g-NH$_3$/kWh. Moreover, the only catalyst explored by this group was iron and the effect of the catalyst introduction on plasma species was not clearly described. The use of a simple metal catalyst opens the probability of understanding the complex interactions of catalyst and plasma experimentally by studying the point-emission spectra at the metal plasma interface. Pure transition metals have been quite widely studied using molecular simulation for ammonia synthesis [25,26]. However, a novel class of metals and alloys known as low-melting point alloys has been ignored for catalytic applications in experimental as well as simulation reports. Very few reports on the use of molten metals and alloys as catalysts for chemical reactions exist (discussed in the following paragraph).

Raney Ni, a very well-known catalyst, is a Ni rich alloy with Al, developed by Murray Raney in 1929 [27,28]. Since then, alloys have been used as catalysts. Molten metals are postulated to have better activity as compared to solid forms due to higher enthalpy and entropy of the liquid (or semi-liquid) state [29]. Molten metals and alloys have been used as catalysis since the 1970s. Ogino's group has been utilizing molten metal Ga, In, Ti, Pb, Cd, Bi, Sn, and Zn for this purpose. These catalysts were used for dehydrogenation in different organic compounds such as ethers, amines, and alcohols [30–33]. It was stated that a lone pair interaction from oxygen with the molten metal was responsible for the catalytic effect [31]. Another important finding was that there was little decline in catalytic activity for a catalyst used over a year, assuring extremely large catalyst life-cycles [32]. Stelmachowski reported the use of molten metals for pyrolysis of polyolefins and rubber from waste-tires using Zn, Pd, and Sn for the studies. It is noteworthy that he was able to obtain a good yield of liquid hydrocarbons in the range of gasoline and diesel [34,35]. In 2017, Upham et al. reported hydrogen production via methane pyrolysis using molten metal alloys. Interestingly, the catalyst was not poisoned even after several recycles. It was stated that molten alloys worked as bifunctional catalysts. One metal helped in breaking the C–H bond while the other helped in precipitating the carbon, mitigating the poisoning [36]. In recent years, a major application has been the growth of carbon and silicon nanostructures. Specifically, Ga and In have gained attention for this purpose [37–43].

Carreon et al. employed Ga and Ga alloys for the growth of Si nanowires. Pure Ga and Ga-Al alloy presented different reaction kinetics as compared to traditional Au catalyst while the activation energy indicated the difference in catalyst-plasma interaction with the new catalysts [44,45]. Carreon et al. also reported the interaction of hydrogen and nitrogen plasma with gallium. It is essential to note that hydrogen as well as nitrogen species were absorbed into the molten metal, which was strongly inferred by the pressure drop in the chamber. Interestingly, this only happened when the plasma was ignited and not with the neutral gas [46]. From our catalytic results for pure metals, Ga showed the best energy efficiency as well as highest conversion in gentler plasma. Indium has very similar properties to gallium. It forms a MN (metal nitride) similar to gallium and also has a low melting point. Owing to the enhanced catalytic activity of gallium and other molten metals as well as alloys, it was thought interesting to study the catalytic effect of In and Ga-In alloys. It would not only shed light on the behavior of these low-melting point alloys in plasma but might also help in understanding the driving factor for ammonia yield for pure Ga and In. Also, the melting points of these metals make the alloy preparation and loading easier.

2. Results and Discussion

2.1. Catalytic Activity for Ammonia Synthesis

The catalytic activity was tested for all catalysts at a temperature of 400 °C and pressure of 0.27 torr. As the reflected power was less than 5% of input power, the input power is assumed to be the plasma power. Plasma power is defined as the electrical energy supplied from the power supply to the RF coils. To understand the effect of plasma power, the catalytic activity was tested at 50 W, 150 W, and 300 W. The reactions were repeated twice with standard deviation being less than 2–5%. A total of seven different catalyst compositions were tried. The various alloy compositions and their melting points are represented in Table 1.

Table 1. Composition of various alloys with their melting points.

Gallium Weight %	Indium Weight %	Melting Point (°C) [1]
100	0	29.8
80	20	16.3
60	40	50.2
50	50	61
40	60	65
20	80	90
0	100	156.6

[1] Melting points interpreted from phase diagram in Anderson and Ansara [47].

The molten catalyst was spread on the outside of glass tubes and the tubes were placed in the reaction chamber. The detailed experimental setup is described in Section 3. The temperature of the gas was assumed to be the furnace temperature i.e., 400 °C. To decouple the effect of plasma and thermal energy, a reaction was run at 400 °C without any plasma. Interestingly, no ammonia was detected, which confirms that plasma is the main driving factor for the synthesis of ammonia and the temperature only aids the process. To confirm the catalytic activity of metal coated glass tubes vs. uncoated glass tubes, reactions were run with an empty reaction chamber with uncoated glass tubes in the reaction chamber. There was no difference in the ammonia yield for these sets of reactions. In the manuscript they are referred to as blank reactions. For the complete plot of catalytic activity at different powers for blank reactions, please refer to Figure S1.

The catalytic activity for various alloy compositions at different powers is presented in Figure 1. The steady state ammonia yield is reported in these plots. For ammonia yield (%) vs. time (min) plots for various alloys at different powers, please refer to Figure S2, Figure S3, Figure S4, Figure S5, Figure S6, Figure S7 and Figure S8, in Supporting Information. The order of catalytic activity for Ga and In changes with increase in plasma power. At 50 W and 150 W, pure Ga exhibits better catalytic activity as compared to In, whereas at 300 W, In leads to an ammonia yield 1.5 times higher than employing pure Ga as catalyst. At 50 W, the ammonia yield is slightly higher when a catalyst is employed as compared to blank reactions. As can be observed at 150 W and 300 W, there is a drastic increase in the ammonia yield. This can be attributed to the increase in concentration of excited species with increase in plasma power. At low powers the concentration of excited species is lower. Hence, it can be conjectured that to achieve a significant increase in yields of the final products or to achieve the full potential of a catalyst in plasma reactions, a minimum threshold concentration of excited species must be achieved.

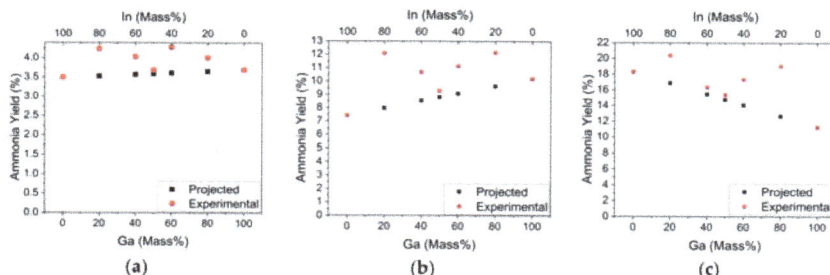

Figure 1. Ammonia yield at steady state (%) vs. alloy composition at various powers (**a**) 50 W, (**b**) 150 W, (**c**) 300 W.

Ideally, it was assumed that the catalytic activity would follow a linear trend as we changed the composition of the alloy. The formula used to determine the projected catalytic activity is represented as Equation (1)

$$X_A = (X_{Ga} * m_{Ga}) + (X_{In} * m_{In}) \tag{1}$$

where X_A = projected ammonia yield for alloy, X_{Ga} = ammonia yield using gallium as catalyst, m_{Ga} = mass fraction of Ga, X_{In} = ammonia yield using indium as catalyst and m_{In} = mass fraction of indium. From experiments, this hypothesis was proven wrong. The alloys show better activity than pure metals. Irrespective of the power used for the reactions, the ammonia yield followed the same trend with change in alloy composition. The catalytic activity follows a bow-type curve instead of a linear one. The bow-type curve (experimental yields) overlaps the linear curve (projected yields) at three points, obviously at the points of pure metals and at equal mass composition i.e., 50:50 for all powers (Figure 1). There are two maxima achieved with varying the alloy composition. The maxima have almost identical yields at a particular power. At 50 W, the maxima are achieved at alloy concentrations of 80:20 and 40:60 (In:Ga) and the maximum yield is 4.5%. The ammonia yield using pure In as catalyst is 3.5% while it is 3.6% with pure Ga. At 150 W, the maxima are achieved at alloy compositions of 80:20 and 20:80 (In:Ga) with the ammonia yield being 12%. The ammonia yields for pure Ga and In as catalysts are 9.8% and 7.3% respectively. At 300 W, the slope of the curve inverted as the ammonia yield with In as the catalyst i.e., 18.1% was higher than ammonia yield with Ga as the catalyst i.e., 11.2%. The maxima are found at alloy compositions of 80:20 and 20:80 (In:Ga) with the ammonia yield being 20%. Interestingly, as the alloy concentration approaches 50:50, the ammonia yields tend to approach the projected yields.

These experiments shed light on the fact that there is some interaction between the metals in the alloys which leads to enhanced catalytic activity as opposed to the pure metals. There is a slight dip in the melting point of the Ga-In phase diagram (see Figure S9). It was hypothesized that it might be a factor for such a catalytic activity trend. Ammonia yields were plotted against the melting points, of the alloys but no pattern emerged (see Figure S10). Daeneke et al. in their review on liquid metals and alloys described in brief the interactions between Ga-In alloys. They found that the alloy properties are highly altered by the composition of the alloy. Specifically, the metallic bonds, free electron density, and plasma frequency (frequency of oscillation of free electron cloud) were determined to be important parameters in determining of the alloy properties for application like energy storage and catalysis. Moreover, indium also acts as an atomic lubricant for free electron clouds to move freely in the alloy [48]. This leads to complex phenomena and interaction between the free electrons in the plasma phase and in the alloys. The electric potential also adds to the dynamics of electron densities and plasma oscillations (also known as Langmuir waves). Hence, it becomes extremely complex to explain the phenomena at quantum or molecular level without simulations, but it does explain the non-linear change in ammonia yield with respect to composition as well as change in catalytic activity order with respect to power. The electron densities in plasma increase as the power increases which increases the interaction between the free electrons of plasma and the free electrons in alloys. However, to explain the ammonia yield trends, a more simplistic approach using the emission spectra is discussed in the next section.

The energy yields plot for all powers is shown in Figure S11. The highest energy yield observed in this study is 0.31 g-NH$_3$/kWh with an energy cost of 196 MJ/mol. The energy cost decreased by 20% by using an alloy instead of pure metal while all other parameters were fixed. The energy cost achieved in this study is the lowest as compared to other pure metals employed as catalysts in this study and in our previous publication [49]. Earlier, we reported an energy cost of 237 MJ/mol when Au was used as catalyst in the same reaction system at the same conditions [49]. The present study materializes the potential of molten alloys as catalysts for ammonia synthesis and other plasma-catalytic reactions employing metal catalysts. It unlocks a new range of bimetallic and possible polymetallic materials as catalysts in plasma systems, with the possibility of increasing the energy yield by simply tailoring the catalyst. Moreover, it can help in decreasing the overall cost of the operation by reducing the cost of expensive catalysts like gold or silver by exchanging them with molten alloys, for example gold–gallium alloy, silver–indium alloy, etc.

2.2. Emission Spectroscopy of Plasma

The emission spectra were studied at the same point for all catalysts using a CCD detector and a fiber optic cable. Multiple species were detected in the plasma. There were four nitrogen species detected: N$_2$ second positive system $(C^3\Pi_u \rightarrow B^3\Pi_g, 337.1$ nm$)$, N$_2^+$ first negative system $(B^2\Sigma_u^+ \rightarrow X^2\Sigma_g^+, 391.4$ nm$)$, N$_2$ first positive system $(B^3\Pi_g \rightarrow A^3\Sigma_u^+, 662.3$ nm$)$, and atomic N $(2p^23p \rightarrow 2p^23s, 746.8$ $nm)$. The hydrogen lines observed at 486.1 nm and 656.3 nm represent H$_\beta$ and H$_\alpha$ species, respectively. The only NH$_x$ species detected is NH at 336 nm (a pre-shoulder to the N$_2$ second positive system). The representative spectra for different catalysts are shown in Figure S12, Figure S13 and Figure S14. In our previous work, it was found that H$_\alpha$ species played a major role in determining the ammonia yield when Cu, Pd, Ag, and Au were used as catalysts [49]. The intensities of the H$_\alpha$ species were calculated by processing the data with Origin software package. The intensities are plotted against alloy composition in Figure 2. No clear pattern is observed for 50 W but for 150 W and 300 W, it follows the same trend as ammonia yield with some shift in the maximum concentrations. It can be easily inferred that the H$_\alpha$ species concentration has a direct impact on ammonia yield, but there are also some other factors which regulate the yield. As the intensity of H$_\alpha$ species increases, the ammonia yield also increases which indicates that the pathway for ammonia formation in metals like Cu, Pd, Ag, and Au [49] is different than the pathway for metals like Ga and In. As discussed in the earlier section, there are many more quantum phenomena which involve free electron densities, Langmuir waves, free electron clouds, and the combined conductivity of the alloy at bulk and quantum levels which also have a great impact on the ammonia yield.

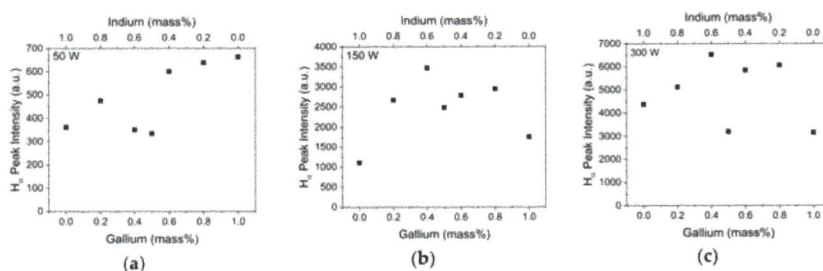

Figure 2. H$_\alpha$ Peak Intensity (a.u.) vs. Alloy Composition (mass%) for various powers (**a**) 50 W, (**b**) 150 W and (**c**) 300 W.

2.3. Characterization of Alloys

The surface of the catalysts underwent some color change which can be observed with the naked eye. Further characterization was carried out using Scanning Electron Microscopy. The SEM images of Ga-In (50;50 wt%) alloy unexposed and exposed to N_2-H_2 are shown in Figure 3. In the unexposed alloy sample (Figure 3a), it is possible to observe a clean surface with minor irregularities due to rough covering of the glass tubes. On the contrary, the alloy exposed to N_2-H_2 plasma shows two distinct phases. The left side of the image in Figure 3b is similar to the unexposed sample but the right side has granular nodes on its surface. There is a boundary differentiating both the phases. It was suspected that these granular formations were occurring due to the interaction of N_2 derived species with Ga, which at optimal processing conditions will lead to GaN formation. To validate this hypothesis, experiments were conducted at elevated temperature to support the formation of GaN only in N_2 plasma. Similar granular nodes were observed for short-time synthesis (<15 min) of GaN with gallium coated on glass substrate. At longer times (>60 min), these structures form a more uniform durian-like formation finally converting to GaN nanowires. These durian-like structures have been reported by Nabi et al. [50]. The SEM images of Ga treated in N_2 plasma for various times are shown in Figure S15. The non-uniformity in the samples is due to lower temperature and milder plasma conditions. Carreon et al. reported that hydrogen has a better interaction with Ga as compared to nitrogen [46]. Having a hydrogen-rich environment can result in non-uniformity of these granular nodes on the surface as hydrogen as well as nitrogen derived species competing to interact with the metallic surface. Also, the temperature does not provide enough activation energy for the formation of GaN nano- or micro-structures, for which elevated temperatures are required. When gallium was treated in pure H_2 plasma, no granular nodes were observed. Instead, etching was very dominant.

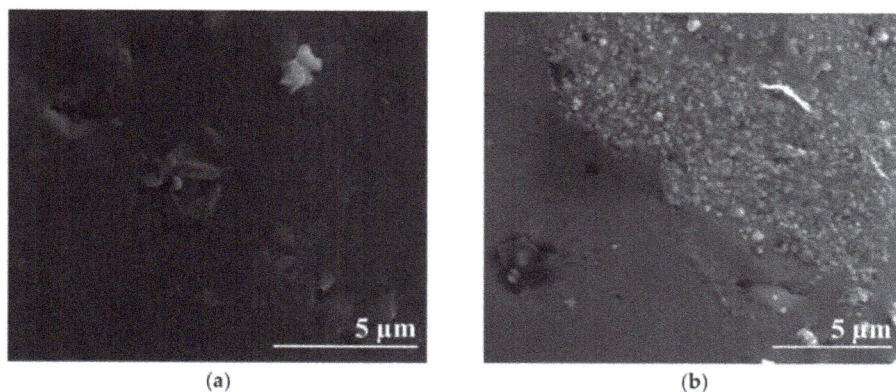

(a)   (b)

Figure 3. Scanning electron microscope (SEM) images of Ga-In (50:50 wt%), (**a**) fresh (unexposed to plasma), (**b**) spent (exposed to N_2-H_2 plasma).

3. Materials and Methods

The experiments were performed in an in-house built plasma reactor (Figure 4). The reaction was conducted by introducing nitrogen (Praxair, 99%) and hydrogen (Praxair, 99.99%) at a 1:4 N_2:H_2 ratio to the reaction chamber using mass flow controllers. The nitrogen and hydrogen flow rates were 4 and 16 sccm, respectively. The plasma was ignited using an RF Power Supply with a Matching Network from Seren IPS, Inc. (Vineland, NJ, USA). The typical reaction pressure and temperature were 0.27 torr and 400 °C, respectively. The plasma excitation was started when the furnace reached the desired temperature. The catalysts were coated on inert glass capillaries and loaded in the reactor. The mass of the catalyst loaded was 1 g for all catalysts. The reaction products were bubbled into deionized water. The reactor was uniquely designed for ammonia synthesis by adding an on-line Agilent 7820A gas chromatograph (GC) (Santa Clara, CA, USA), equipped with a gas sampling valve and HP-PlotQ column (30 m × 0.32 mm × 20 μm). The gases were analyzed every 3 min for 30 min using the GC. All experiments were repeated twice. The experiments were performed for input powers of 50 W, 150 W, and 300 W.

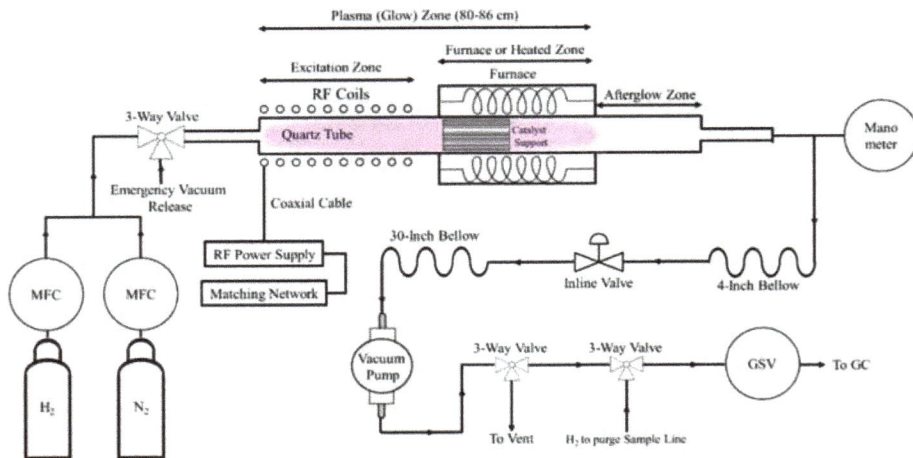

Figure 4. Schematic of the in-house built RF plasma reactor and surrounding equipment for gas inlet and outlet.

The SEM characterization was performed on an FEI Helios NanoLab Scanning Electron Microscope in field emission mode. The emission spectra of plasma were collected with a UV-VIS-NIR spectrophotometer (Avantes ULS3648 series, Louisville, CO, USA) equipped with a dual channel CCD detector. The light was transmitted to the spectrophotometer using a stainless-steel jacketed bifurcated fiber optic cable with a 400 μm fiber. The spectra were collected in the range of 200–1100 nm. The spectral resolution of the spectrophotometer 0.4 nm and grating is 600 lines/mm.

4. Conclusions

Herein, we describe the unusual catalytic behavior of Ga-In alloys for plasma catalytic ammonia synthesis. An increase of 20% is achieved in ammonia energy yield by using an alloy instead of pure metals. There are two optimum compositions of alloys for obtaining maximum ammonia yield for every power. The maximum energy yield of ammonia was obtained at 50 W with catalyst composition of 0.6:0.4 and 0.2:0.8 (Ga:In) having a value of 0.31 g-NH$_3$/KWh. The emission spectra show the direct dependence of ammonia yield on the concentration of H$_\alpha$ species in the plasma. The SEM characterization shows surface changes on the catalyst due to plasma which are indicative of GaN formation. It is extremely difficult to determine the exact quantum or molecular phenomena experimentally due to involvement of free electron densities and cloud from plasma as well as the alloys. It can be understood more clearly when employing a simulation approach.

Supplementary Materials: The following are available online at http://www.mdpi.com/2073-4344/8/10/437/s1, Figure S1: Ammonia Yield (%) vs. Time (min) for reactions run with no catalyst title, Figure S2: Ammonia Yield (%) vs. Time (min) for reactions with pure In as catalyst, Figure S3: Ammonia Yield (%) vs. Time (min) for reactions run with Ga-In Alloy (20:80), Figure S4: Ammonia Yield (%) vs. Time (min) for reactions run with Ga-In Alloy (40:60), Figure S5: Ammonia Yield (%) vs. Time (min) for reactions run with Ga–In Alloy (50:50), Figure S6: Ammonia Yield (%) vs. Time (min) for reactions run with Ga-In Alloy (60:40), Figure S7: Ammonia Yield (%) vs. Time (min) for reactions run with Ga:In Alloy (80:20), Figure S8: Ammonia Yield (%) vs. Time (min) for reactions run with pure Ga as catalyst, Figure S9: Ga-In Alloy Phase diagram, Figure S10: Ammonia Yield (%) vs. Alloy Melting Point (°C) for various plasma powers, Figure S11: Energy Yield (g-NH$_3$/kWh) vs. Composition of Alloy (mass%) for various plasma powers, Figure S12: Formation of GaN (plasma treatment time), (a) Starting of Nucleation

Process (5 min), (b) Dissolution of Nitrogen in the Gallium Droplet (15 min), (c) Durain-like GaN Nanostructures (30 min), (d) Nanowires of GaN being generated from a single droplet (120 min), Figure S13. XPS Spectra of spent catalyst (pure Ga), Figure S14: Schematic of the in-house built RF plasma reactor and surrounding equipment for gas inlet and outlet, Figure S15: Formation of GaN (plasma treatment time), (a) Starting of Nucleation Process (5 min), (b) Dissolution of Nitrogen in the Gallium Droplet (15 min), (c) Durain-like GaN Nanostructures (30 min), (d) Nanowires of GaN being generated from a single droplet (120 min).

Author Contributions: M.L.C. conceived and directed the presented research. J.R.S. performed the experiments for catalyst preparation and catalytic activity. Synthesis of catalysts and their catalytic performance evaluation was performed at The University of Tulsa. J.M.H. particularly helped in the collection of the catalytic activity data. The paper was written with the contribution of M.L.C., J.R.S., and J.M.H.

Funding: Maria L. Carreon acknowledges the University of Tulsa Faculty Start-up fund, the University of Tulsa Faculty Development Summer Fellowship Program for financial support of this work and the student research grant from the Office of Research and Sponsored Programs at the University of Tulsa.

Conflicts of Interest: The authors declare no conflict of interest

References

1. Tanabe, Y.; Nishibayashi, Y. Developing more sustainable processes for ammonia synthesis. *Coord. Chem. Rev.* **2013**, *257*, 2551–2564. [CrossRef]
2. Global Ammonia Capacity to Reach Almost 250 Million Tons per Year by 2018, says GlobalData. Available online: https://energy.globaldata.com/media-center/press-releases/oil-and-gas/global-ammonia-capacity-to-reach-almost-250-million-tons-per-year-by-2018-says-globaldata (accessed on 15 July 2018).
3. Smil, V. *Enriching the Earth: Fritz Haber, Carl Bosch, and the Transformation of World Food Production*; MIT Press: Cambridge, MA, USA, 2004.
4. Kiefer, D.M. Capturing Nitrogen Out of the Air Fritz Haber's high-pressure process for combining nitrogen with hydrogen was a major milestone. *Todays Chem. Work* **2001**, *10*, 117–122.
5. Jennings, J.R. *Catalytic Ammonia Synthesis: Fundamentals and Practice*; Springer: Berlin, Germany, 2013.
6. Modak, J.M. Haber process for ammonia synthesis. *Resonance* **2002**, *7*, 69–77. [CrossRef]
7. Emmett, P.; Brunauer, S. Accumulation of alkali promoters on surfaces of iron synthetic ammonia catalysts. *J. Am. Chem. Soc.* **1937**, *59*, 310–315. [CrossRef]
8. Hong, J.; Prawer, S.; Murphy, A.B. Plasma catalysis as an alternative route for ammonia production: Status, mechanisms, and prospects for progress. *ACS Sustain. Chem. Eng.* **2017**, *6*, 15–31. [CrossRef]
9. Bogaerts, A.; Neyts, E.C. Plasma Technology: An Emerging Technology for Energy Storage. *ACS Energy Lett.* **2018**, *3*, 1013–1027. [CrossRef]
10. Patil, B. Plasma (Catalyst)-Assisted Nitrogen Fixation: Reactor Development for Nitric Oxide and Ammonia Production. Ph.D. Thesis, Eindhoven University of Technology, Eindhoven, The Netherlands, 2017.
11. Aihara, K.; Akiyama, M.; Deguchi, T.; Tanaka, M.; Hagiwara, R.; Iwamoto, M. Remarkable catalysis of a wool-like copper electrode for NH_3 synthesis from N_2 and H_2 in non-thermal atmospheric plasma. *Chem. Commun.* **2016**, *52*, 13560–13563. [CrossRef] [PubMed]
12. Iwamoto, M.; Akiyama, M.; Aihara, K.; Deguchi, T. Ammonia synthesis on wool-like Au, Pt, Pd, Ag, or Cu electrode catalysts in nonthermal atmospheric-pressure plasma of N_2 and H_2. *ACS Catal.* **2017**, *7*, 6924–6929. [CrossRef]
13. Kim, H.H.; Teramoto, Y.; Ogata, A.; Takagi, H.; Nanba, T. Atmospheric-pressure nonthermal plasma synthesis of ammonia over ruthenium catalysts. *Plasma Process. Polym.* **2017**, *14*, 1600157. [CrossRef]
14. Hong, J.; Pancheshnyi, S.; Tam, E.; Lowke, J.J.; Prawer, S.; Murphy, A.B. Kinetic modelling of NH_3 production in N_2–H_2 non-equilibrium atmospheric-pressure plasma catalysis. *J. Phys. D Appl. Phys.* **2017**, *50*, 154005. [CrossRef]
15. Mizushima, T.; Matsumoto, K.; Ohkita, H.; Kakuta, N. Catalytic effects of metal-loaded membrane-like alumina tubes on ammonia synthesis in atmospheric pressure plasma by dielectric barrier discharge. *Plasma Chem. Plasma Process.* **2007**, *27*, 1–11. [CrossRef]
16. Akay, G.; Zhang, K. Process Intensification in Ammonia Synthesis Using Novel Coassembled Supported Microporous Catalysts Promoted by Nonthermal Plasma. *Ind. Eng. Chem. Res.* **2017**, *56*, 457–468. [CrossRef]

17. Peng, P.; Cheng, Y.; Hatzenbeller, R.; Addy, M.; Zhou, N.; Schiappacasse, C.; Chen, D.; Zhang, Y.; Anderson, E.; Liu, Y. Ru-Based Multifunctional Mesoporous Catalyst for Low-Pressure and Non-Thermal Plasma Synthesis of Ammonia. *Int. J. Hydrogen Energy* **2017**, *42*, 19056–19066. [CrossRef]

18. Tanaka, S.; Uyama, H.; Matsumoto, O. Synergistic effects of catalysts and plasmas on the synthesis of ammonia and hydrazine. *Plasma Chem. Plasma Process.* **1994**, *14*, 491–504. [CrossRef]

19. Mingdong, B.; Xiyao, B.; Zhitao, Z.; Mindi, B. Synthesis of Ammonia in a Strong Electric Field Discharge at Ambient Pressure. *Plasma Chem. Plasma Process.* **2000**, *20*, 511–520. [CrossRef]

20. Sugiyama, K.; Akazawa, K.; Oshima, M.; Miura, H.; Matsuda, T.; Nomura, O. Ammonia Synthesis by Means of Plasma over MgO Catalyst. *Plasma Chem. Plasma Process.* **1986**, *6*, 179–193. [CrossRef]

21. Peng, P.; Li, Y.; Cheng, Y.; Deng, S.; Chen, P.; Ruan, R. Atmospheric Pressure Ammonia Synthesis using Non-Thermal Plasma Assisted Catalysis. *Plasma Chem. Plasma Process.* **2016**, *36*, 1201–1210. [CrossRef]

22. Uyama, H.; Matsumoto, O. Synthesis of ammonia in high-frequency discharges. II. Synthesis of ammonia in a microwave discharge under various conditions. *Plasma Chem. Plasma Process.* **1989**, *9*, 421–432. [CrossRef]

23. Uyama, H.; Matsumoto, O. Synthesis of ammonia in high-frequency discharges. *Plasma Chem. Plasma Process.* **1989**, *9*, 13–24. [CrossRef]

24. Uyama, H.; Nakamura, T.; Tanaka, S.; Matsumoto, O. Catalytic effect of iron wires on the syntheses of ammonia and hydrazine in a radio-frequency discharge. *Plasma Chem. Plasma Process.* **1993**, *13*, 117–131. [CrossRef]

25. Mehta, P.; Barboun, P.; Herrera, F.A.; Kim, J.; Rumbach, P.; Go, D.B.; Hicks, J.C.; Schneider, W.F. Overcoming ammonia synthesis scaling relations with plasma-enabled catalysis. *Nat. Catal.* **2018**, *1*, 269–275. [CrossRef]

26. Singh, A.R.; Montoya, J.H.; Rohr, B.A.; Tsai, C.; Vojvodic, A.; Nørskov, J.K. Computational Design of Active Site Structures with Improved Transition-State Scaling for Ammonia Synthesis. *ACS Catal.* **2018**, *8*, 4017–4024. [CrossRef]

27. Murray, R. Method of Producing Finely-Divided Nickel. U.S. Patents 1,628,190, 10 May 1927.

28. Raney, M. Catalysts from alloys. *Ind. Eng. Chem.* **1940**, *32*, 1199–1203. [CrossRef]

29. Schwab, G.M. Catalysis on Liquids Metals. *Berichte der Bunsengesellschaft für physikalische Chemie* **1976**, *80*, 746–749. [CrossRef]

30. Ozawa, S.; Sasaki, K.; Ogino, Y. Reaction of benzyl phenyl ether over various molten metal catalysts. *Fuel* **1986**, *65*, 707–710. [CrossRef]

31. Okano, K.; Saito, Y.; Ogino, Y. The Dehydrogenation of Amines by Molten Metal Catalysts. *Bull. Chem. Soc. Jpn.* **1972**, *45*, 69–73. [CrossRef]

32. Saito, Y.; Hiramatsu, N.; Kawanami, N.; Ogino, Y. Dehydrogenation of Some Alcohols by the Molten Metal Catalysts. *Bull. Jpn. Pet. Inst.* **1972**, *14*, 169–173. [CrossRef]

33. Saito, Y.; Miyashita, F.; Ogino, Y. Studies on catalysis by molten metal: VI. Kinetics and the reaction scheme for the dehydrogenation of isopropyl alcohol over the liquid indium catalyst. *J. Catal.* **1975**, *36*, 67–73. [CrossRef]

34. Stelmachowski, M. Thermal conversion of waste polyolefins to the mixture of hydrocarbons in the reactor with molten metal bed. *Energy Convers. Manag.* **2010**, *51*, 2016–2024. [CrossRef]

35. Stelmachowski, M. Conversion of waste rubber to the mixture of hydrocarbons in the reactor with molten metal. *Energy Convers. Manag.* **2009**, *50*, 1739–1745. [CrossRef]

36. Upham, D.C.; Agarwal, V.; Khechfe, A.; Snodgrass, Z.R.; Gordon, M.J.; Metiu, H.; McFarland, E.W. Catalytic molten metals for the direct conversion of methane to hydrogen and separable carbon. *Science* **2017**, *358*, 917–921. [CrossRef] [PubMed]

37. Mukanova, A.; Tussupbayev, R.; Sabitov, A.; Bondarenko, I.; Nemkaeva, R.; Aldamzharov, B.; Bakenov, Z. CVD graphene growth on a surface of liquid gallium. *Mater. Today Proc.* **2017**, *4*, 4548–4554. [CrossRef]

38. Fujita, J.-I.; Hiyama, T.; Hirukawa, A.; Kondo, T.; Nakamura, J.; Ito, S.-I.; Araki, R.; Ito, Y.; Takeguchi, M.; Pai, W.W. Near room temperature chemical vapor deposition of graphene with diluted methane and molten gallium catalyst. *Sci. Rep.* **2017**, *7*, 12371. [CrossRef] [PubMed]

39. Fujita, J.-I.; Miyazawa, Y.; Ueki, R.; Sasaki, M.; Saito, T. Fabrication of large-area graphene using liquid gallium and its electrical properties. *Jpn. J. Appl. Phys.* **2010**, *49*, 06GC01. [CrossRef]

40. Lee, M.V.; Hiura, H.; Tyurnina, A.V.; Tsukagoshi, K. Controllable gallium melt-assisted interfacial graphene growth on silicon carbide. *Diam. Relat. Mater.* **2012**, *24*, 34–38. [CrossRef]

41. Lugstein, A.; Steinmair, M.; Hyun, Y.J.; Bertagnolli, E.; Pongratz, P. Ga/Au alloy catalyst for single crystal silicon-nanowire epitaxy. *Appl. Phys. Lett.* **2007**, *90*, 023109. [CrossRef]

42. Gewalt, A.; Kalkofen, B.; Lisker, M.; Burte, E.P. Gallium Assisted PECVD Synthesis of Silicon Nanowires. *ECS Trans.* **2010**, *28*, 45–56.
43. Pan, Z.W.; Dai, Z.R.; Ma, C.; Wang, Z.L. Molten gallium as a catalyst for the large-scale growth of highly aligned silica nanowires. *J. Am. Chem. Soc.* **2002**, *124*, 1817–1822. [CrossRef] [PubMed]
44. Carreon, M.L.; Jasinski, J.; Sunkara, M. Low temperature synthesis of silicon nanowire arrays. *Mater. Res. Express* **2014**, *1*, 045006. [CrossRef]
45. Carreon, M. Plasma catalysis using low melting point metals. Ph.D. Thesis, University of Louisville, Louisville, KY, USA, 2015.
46. Carreon, M.L.; Jaramillo-Cabanzo, D.F.; Chaudhuri, I.; Menon, M.; Sunkara, M.K. Synergistic interactions of H2 and N2 with molten gallium in the presence of plasma. *J. Vac. Sci. Technol. A* **2018**, *36*, 021303. [CrossRef]
47. Anderson, T.; Ansara, I. The Ga-In (gallium-indium) system. *J. Phase Equilib.* **1991**, *12*, 64–72. [CrossRef]
48. Daeneke, T.; Khoshmanesh, K.; Mahmood, N.; de Castro, I.; Esrafilzadeh, D.; Barrow, S.; Dickey, M.; Kalantar-Zadeh, K. Liquid metals: Fundamentals and applications in chemistry. *Chem. Soc. Rev.* **2018**, *47*, 4073–4111. [CrossRef] [PubMed]
49. Shah, J.; Wang, W.; Bogaerts, A.; Carreon, M.L. Ammonia synthesis by radio frequency plasma catalysis: Revealing the underlying mechanisms. *ACS Appl. Energy Mater.* **2018**, *1*, 4824–4839. [CrossRef]
50. Nabi, G.; Cao, C.; Khan, W.S.; Hussain, S.; Usman, Z.; Mahmood, T.; Khattak, N.A.D.; Zhao, S.; Xin, X.; Yu, D. Synthesis, characterization, photoluminescence and field emission properties of novel durian-like gallium nitride microstructures. *Mater. Chem. Phys.* **2012**, *133*, 793–798. [CrossRef]

catalysts

MDPI

Communication

Highly Dispersed Co Nanoparticles Prepared by an Improved Method for Plasma-Driven NH₃ Decomposition to Produce H₂

Li Wang [1,2,*], YanHui Yi [2] , HongChen Guo [2], XiaoMin Du [1], Bin Zhu [1] and YiMin Zhu [1,*]

[1] College of Environmental Science and Engineering, Dalian Maritime University, Dalian 116026, China; duxiaomin1202@163.com (X.D.); binzhu@dlmu.edu.cn (B.Z.)

[2] State Key Laboratory of Fine Chemicals, School of Chemical Engineering, Dalian University of Technology, Dalian 116024, China; yiyanhui@dlut.edu.cn (Y.Y.); hongchenguo@163.com (H.G.)

* Correspondence: liwang@dlmu.edu.cn (L.W.); ntp@dlmu.edu.cn (Y.Z.); Tel.: +86-411-84724357 (L.W.)

Received: 3 December 2018; Accepted: 17 January 2019; Published: 22 January 2019

Abstract: Previous studies reveal that combining non-thermal plasma with cheap metal catalysts achieved a significant synergy of enhancing performance of NH₃ decomposition, and this synergy strongly depended on the properties of the catalyst used. In this study, techniques of vacuum-freeze drying and plasma calcination were employed to improve the conventional preparation method of catalyst, aiming to enhance the activity of plasma-catalytic NH₃ decomposition. Compared with the activity of the catalyst prepared by a conventional method, the conversion of NH₃ significantly increased by 47% when Co/fumed SiO₂ was prepared by the improved method, and the energy efficiency of H₂ production increased from 2.3 to 5.7 mol(kW·h)$^{-1}$ as well. So far, the highest energy efficiency of H₂ formation of 15.9 mol(kW·h)$^{-1}$ was achieved on improved prepared Co/fumed SiO₂ with 98.0% ammonia conversion at the optimal conditions. The improved preparation method enables cobalt species to be highly dispersed on fumed SiO₂ support, which creates more active sites. Besides, interaction of Co with fumed SiO₂ and acidity of the catalyst were strengthened according to results of H₂-TPR and NH₃-probe experiments, respectively. These results demonstrate that employing vacuum-freeze drying and plasma calcination during catalyst preparation is an effective approach to manipulate the properties of catalyst, and enables the catalyst to display high activity towards plasma-catalytic NH₃ decomposition to produce H₂.

Keywords: plasma catalysis; catalyst preparation; NH₃ decomposition; H₂ generation

---

1. Introduction

NH₃ decomposition has been considered to be an attractive route to supply CO$_x$-free H₂ for proton exchange membrane fuel cell (PEMFC) vehicles [1–3]. Until now, the noble metal Ru, due to its high turnover frequency (TOF), is still the most active component for NH₃ decomposition, and the formation rate of H₂ reached as high as 4.0 mol/(h·g$_{cat}$)$^{-1}$ using K-Ru/MgO-CNTs catalyst with complete conversion of ammonia at 450 °C, but the scarcity and high price of Ru limits its use on a large scale [4–6]. Whereas, cheap metal catalysts show low activity towards NH₃ decomposition due to the strong adsorption of N atoms onto the surface of cheap metal catalysts [1,7–10]. As far as we know, the highest formation rate of H₂ was 2.0 mol/(h·g$_{cat}$)$^{-1}$ using CeO₂-doped Ni/Al₂O₃ catalyst with 98.3% ammonia conversion at 550 °C [7]. Recently, the combination of non-thermal plasma with cheap metal catalyst displayed a powerful ability in enhancing NH₃ decomposition [11–13]; 99.9% conversion of NH₃ was achieved in combination mode, but only 7.4% and 7.8% was obtained for Fe-based catalyst alone and plasma alone, respectively, which experienced an unexpected strong synergy between plasma and catalyst [11], and this synergy strongly depended on the properties of catalyst [12,14].

The preparation approach of catalyst could directly affect the properties of the catalyst, such as crystal size, shape, composition, acidity, and basicity [15–17]. Normally, catalyst preparation was operated in a thermodynamic equilibrium state of gas, liquid, and solid state, but faces the limitation of thermodynamic equilibrium. For example, the calcination temperature for supported metal catalysts is usually over 500 °C, and high temperature operation causes aggregation of metal particles, but low temperature operation results in incomplete decomposition of catalyst precursors. Besides, it is difficult to achieve a high dispersion of catalyst with a high metal loading above 20 wt % in a thermodynamic equilibrium state.

Non-thermal plasma is the fourth state of matter and characterized by non-equilibrium character. Typically, the overall gas temperature in a field of non-thermal plasma can be as low as room temperature, while the generated free electrons are highly energetic with a typical electron temperature of 1–10 eV, which can collide with carrier gas to produce chemically reactive species such as radicals, excited species, and ions [18]. Such a characteristic of plasma enables some thermodynamically unfavorable chemical reactions to proceed at moderate conditions, especially for inert molecule conversion, such as CO_2, CH_4, and N_2 [19–24].

Similarly, the non-equilibrium character of non-thermal plasma also benefits catalyst preparation to achieve controllable morphology and chemical property by controlling the reaction rate of nucleation and crystal growth in a non-equilibrium environment. Different from the conventional thermal process, the catalyst preparation with plasma is not based on the thermal effect, but on the inelastic collision of those energetic species (free electrons, radicals, excited species and ions) with catalyst precursors to accomplish the purpose of calcination or treatment. Catalyst preparation with plasma has attracted increasing interest since the 1990s [25–34], and a variety of plasmas, such as glow discharge, radio frequency discharge, microwave discharge, and dielectric barrier discharge, were employed for calcination and reduction of supported catalyst, which can make metal highly dispersed on a support with a narrow distribution of particle size, manipulate metal–support interaction, and shorten the time of catalyst preparation due to high reaction rates in the plasma process. Besides, with regard to the characteristic of low temperature, plasma removal of template was well developed for synthesis of microporous and mesoporous materials, instead of thermal removal that could destroy the porous structure of the materials [35]. Bogaerts and coworkers found that plasma could be formed inside pores of material with pore size above 200 μm at 20 kV by two-dimensional fluid modeling, and the possibility of discharge forming inside pores and discharge behavior strongly depended on pore size and applied voltage [36]; this observation helps to understand the process of plasma removal of template. Very recently, Wang, et al. and Di, et al. summarized the advances in preparation of catalyst with plasmas, and the mechanism of preparation was discussed as well [29,30]. Although low temperature operation of non-thermal plasma enables catalyst preparation in a more efficient and more controllable way, normally, low temperature operation leads to incomplete decomposition or removal of precursors, along with poor growth of materials with residues.

In this study, a novel combination of vacuum-freeze drying technique with atmospheric pressure dielectric barrier discharge (DBD) calcination technique was proposed for the preparation of supported Co catalyst with a high Co loading of 30 wt %. To calcine catalyst completely, DBD reactor was placed in a furnace to prevent heat dissipation of electric heat from discharge, so as to keep plasma calcination at a temperature of about 400 °C by tuning energy input of power supply (note that the furnace here is not used to heat the reactor, but used for electric heat preservation). Compared to Co/fumed SiO_2 with conventional preparation method, the improved preparation method enabled the conversion of plasma-catalytic NH_3 to increase by 47%, and greatly enhanced the formation rate of H_2. Besides, the reaction performance can be further improved through increasing specific energy input.

2. Results and Discussion

2.1. Characterization

The physicochemical properties of as-prepared Co catalysts were examined using various characterization techniques, including X-ray diffraction (XRD), X-ray fluorescence (XRF), transmission electron microcopy (TEM), H_2 temperature-programmed reduction (H_2-TPR), and NH_3 temperature-programmed desorption (NH_3-TPD). In this study, the fumed SiO_2 used as a support for Co catalyst was an amorphous material with a Brunauer–Emmett–Teller (BET) surface area of 297.8 $m^2 \cdot g^{-1}$. The theoretical Co loading was designed to be 30 wt %, but the actual Co loading through XRF analysis was 27.7 wt % and 27.4 wt % for the improved prepared catalyst and the conventional prepared catalyst, respectively (see Tables S1 and S2 in Supporting Information). Figure 1 shows the XRD patterns of as-prepared fumed SiO_2-supported Co catalysts using conventional and improved preparation methods, respectively. Besides, pure fumed SiO_2 was analyzed as a reference in Figure 1 (a). Clearly, the same diffraction peaks were observed at 2θ of 31.1, 36.7, 44.6, 59.2, and 65.2 as shown in Figure 1 (b) and (c), which matched well with the characteristic structure of Co_3O_4 (JCPDS file No: 43-1003), and those diffraction peaks represented the (220), (311), (400), (511), and (440) planes of Co_3O_4, respectively [37,38]. Namely, the difference in preparation approach did not influence the phase structure of Co catalysts, and they both finally existed in the form of Co_3O_4 over fumed SiO_2 support. However, by contrast, the intensity of diffraction peaks of Co catalyst prepared with improved method was weaker than that with conventional preparation method, suggesting that the average particle size of the former is smaller than that of the latter according to the Debye–Scherrer formula [39]; this observation is also supported by the results of TEM as follows.

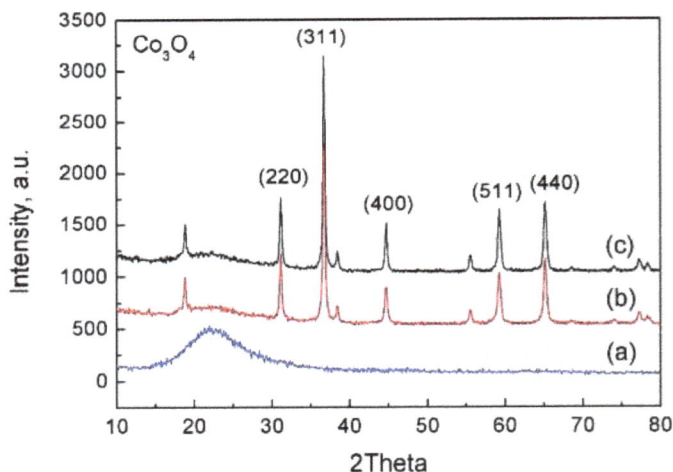

Figure 1. XRD patterns of as-prepared Co/fumed SiO_2 catalysts using different approaches (Co_3O_4, JCPDS file No: 43-1003): (**a**) pure fumed SiO_2, (**b**) improved preparation method, and (**c**) conventional preparation method.

TEM images of as-prepared Co catalyst supported on fumed SiO_2 using different approaches were shown in Figure 2. Clearly, a very poor dispersion of Co catalyst was observed on fumed SiO_2 using the conventional preparation method, and the particle size of Co was much larger than 5 nm; some particle sizes were around 50 nm, as shown in Figure 2a,b. However, the use of combining vacuum-freeze drying and plasma calcination techniques in the process of catalyst preparation enabled the Co particles to disperse highly and homogeneously onto the fumed SiO_2 support, and the average Co particle size was less than 5 nm, mostly around 2–3 nm in Figure 2c,d. Actually, it is difficult to obtain such smaller nanoparticles with a high metal loading of about 27 wt % using the conventional preparation method.

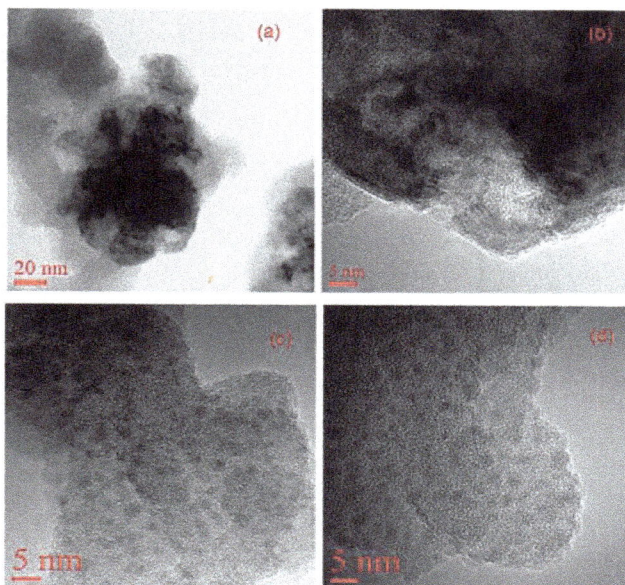

Figure 2. TEM images of as-prepared Co/fumed SiO_2 catalysts using different approaches: (**a**) and (**b**) conventional preparation method; (**c**) and (**d**) improved preparation method.

Using NH_3 as probe molecule, the influence of preparation approach on the chemical properties of catalyst was evaluated through NH_3-TPD, as displayed in Figure 3. Clearly, two major desorption peaks were observed, one at the low temperatures of 150–220 °C corresponded to the weak adsorption of NH_3 on the catalyst, and the other at the high temperatures of 220–350 °C was attributed to the strong adsorption of NH_3. It is worth noting that the desorption amount of NH_3 over Co catalyst prepared with the improved method was much higher than that with the conventional preparation method, revealing that the improved method leads to an increase in the number of active sites for NH_3 adsorption; this finding can be ascribed to the high dispersion of Co nanoparticles, as evidenced by the results of TEM in Figure 2. In addition, the desorption temperature of adsorbed NH_3 on the catalyst with improved preparation method shifted towards higher temperature, reflecting that the binding ability of NH_3 with the catalyst was stronger than that with the catalyst prepared using conventional preparation method. This inferred that the acidity of catalyst was strengthened by the improved preparation method as well and, more importantly, the increase in active site number and acid strength both facilitated the adsorption of NH_3 on the catalyst, finally promoting the dissociation of NH_3 on the catalyst.

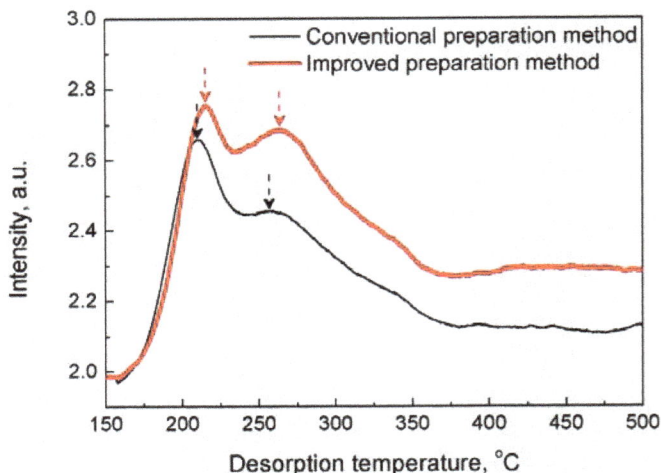

Figure 3. NH$_3$-TPD profiles of as-prepared Co/fumed SiO$_2$ catalysts using different approaches.

H$_2$-TPR technique was used to evaluate the reduction behavior of Co$_3$O$_4$/fumed SiO$_2$ prepared with different methods, and the resulting profiles are displayed in Figure 4. Clearly, the reduction of Co$_3$O$_4$ on fumed SiO$_2$ support occurred in the temperature range of 275–550 °C. Two groups of reduction peaks were observed, i.e., the low temperature reduction peaks (α) consisted of α_1 and α_2 in the range of 275–400 °C, and the high temperature reduction peaks (β) with a consecutive-broad peak consisted of β_1 and β_2 in the range of 370–550 °C. More importantly, by contrast, the reduction temperature of catalyst with improved preparation method shifted towards higher temperature, representing that the improved method strengthened the interaction of Co with fumed SiO$_2$ support. This difference in metal–support interaction can be explained by the difference in particle sizes of Co catalyst prepared by different methods (Figure 2). Actually, the reduction process of as-prepared catalyst was very complicated, since these peaks obtained were heavily overlapped. Therefore, the analysis of each peak area using peak fit function (Gaussian) of Origin software was employed to understand the H$_2$-TPR profiles obtained (see Figure S1 in Supporting Information), the area ratio of β_1/β_2 was found to be 1/3, which is quantitatively consistent with the theoretical value (1/3) of area ratio of Co$_3$O$_4$ reduction peaks [40,41]. This indicates that β_1 and β_2 corresponded to the two-step reduction of Co^{3+} $\rightarrow$ Co^{2+} $\rightarrow$ Co^0 of Co$_3$O$_4$, as do α_1 and α_2 based on 5/16 ($\approx$ 1/3) area ratio of α_1/α_2. Besides, the result of XRD in Figure 1 also supported the assignment of α and β to Co$_3$O$_4$. According to the reduction temperature of Co$_3$O$_4$, the low temperature reduction peaks (α_1 and α_2) could be due to the reduction of bulk Co$_3$O$_4$, whereas the high temperature reduction peaks (β_1 and β_2) were attributed to the reduction of Co$_3$O$_4$ that interacted with fumed SiO$_2$ [42,43].

Interestingly, the above results reveal that the application of vacuum-freeze drying and plasma calcination techniques in the preparation process of catalyst not only results in highly dispersed metal nanoparticles along with the increase of active site number, but also strengthens the acidity of catalyst and the metal–support interaction. Thus, it is feasible and crucial to manipulate the properties of catalysts through exploiting novel preparation techniques.

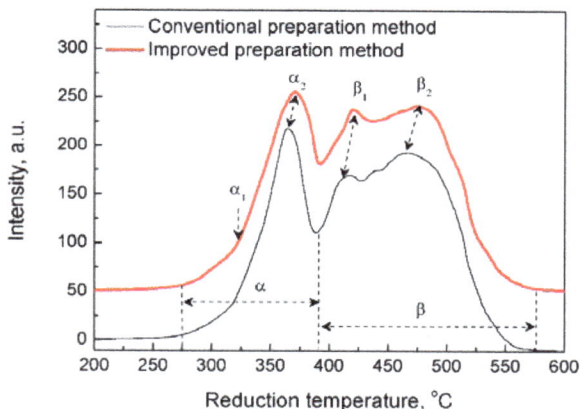

Figure 4. H_2-TPR profiles of as-prepared Co/fumed SiO_2 catalysts using different approaches.

2.2. Performance of Prepared Catalyst in Plasma-Catalytic NH_3 Decomposition

Our previous studies showed that Co-based catalyst exhibited the best activity towards NH_3 decomposition to H_2 in the presence of DBD plasma [12]. Here, the influence of catalyst preparation method on the performance of plasma-catalytic NH_3 decomposition was investigated, as shown in Figure 5. Compared to the conventional preparation method, Co/fumed SiO_2 catalyst prepared with the improved method greatly promoted the reaction performance, and the conversion of NH_3 increased from 25.8 to 72.7% at the reaction temperature of 400 °C in Figure 5a, increased by a factor of almost 3 and, correspondingly, the energy efficiency of H_2 formation increased from 2.3 to 5.7 mol(kW·h)$^{-1}$ in Figure 5b. In addition, changing the reaction temperature from 300 °C to 450 °C through increasing DBD energy input resulted in a significant increase of NH_3 conversion by 80.8% (from 16.1 to 96.9%) in the case of catalyst prepared by the improved method whereas, at the same conditions, the NH_3 conversion only increased by 47.3% (from 4% to 51.3%) over catalyst using the conventional preparation method. Note that the reaction temperature required for complete conversion of NH_3 in the case of using improved preparation method shifted towards lower temperature, at least 50 °C lower in comparison with that using conventional preparation method in Figure 5a.

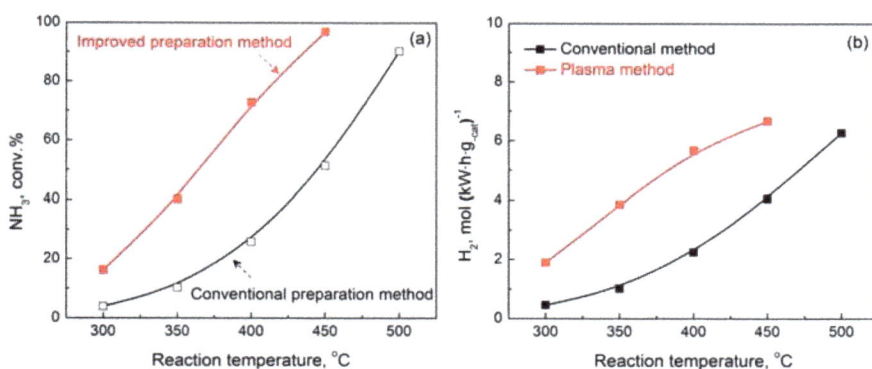

Figure 5. Plasma-catalytic NH_3 decomposition over Co/fumed SiO_2 catalyst with different preparation methods: (**a**) the conversion of NH_3; (**b**) the energy efficiency of H_2 generation (NH_3 feed rate 40 mL/min^{-1}, supported catalyst 0.88 g, discharge gap 3 mm, discharge frequency 12 kHz; The reaction temperature originated from electric heat released by discharge, and was determined using an IR camera and thermocouple tightly attached to the outer wall of the reactor [12]).

Combining the results of characterizations in Figure 1 to 4, the improved preparation method did not affect the phase composition of catalyst (Figure 1), but significantly increased the dispersion of catalyst with a narrow particle size of 2–3 nm (Figure 2), which actually creates much more active sites for NH_3 decomposition, enhancing the specific reactivity of catalyst, and this is also directly evidenced by the result of NH_3-probe experiments presented in Figure 3. Notably, the adsorption amount of NH_3 over the catalyst with improved preparation method is much larger than that with conventional preparation method (Figure 3), this directly points to the fact that enhancing the adsorption step of NH_3 decomposition is one of the reasons for the high activity of catalyst with improved preparation method. Recently, $CoPt/TiO_2$ with Co particle size of ~1 nm displayed a much higher Fischer–Tropsch reaction rate, which was also found to be due to increasing the amount of active site caused by using plasma-assisted preparation [44]. More importantly, in this study, the improved preparation method increased the acid strength of catalyst as well (Figure 3), as demonstrated by the increase in adsorption strength of NH_3 over catalyst, which can promote the dissociation step of NH_3; this is another crucial reason that explains the high activity of catalyst with the improved preparation method. Besides, the improved preparation method strengthened the interaction of Co with fumed SiO_2 (Figure 4), indicating the difference in electronic structure of catalyst with different preparation methods, and this could influence the activity of catalyst as well.

In addition, using $Co/$fumed SiO_2 catalyst prepared by the improved method, the influence of the combining mode of plasma and catalyst was investigated on the performance of plasma-catalytic NH_3 decomposition, as shown in Scheme 1 and Figure 6. About 3 g $Co/$fumed SiO_2 was packed in the reactor with a packing volume of about 3.1 mL, and the combining mode of plasma and catalyst changed through changing discharge volume "V", but the packed catalyst was fixed. Namely, changing "V" from 3.3 to 0.4 mL enabled the catalyst to be partly packed in the field of plasma, as shown in Scheme 1.

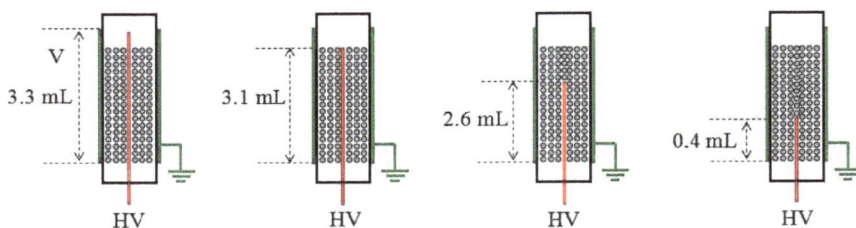

Scheme 1. Scheme of combining mode of plasma and $Co/$fumed SiO_2 catalyst (note: HV denotes high voltage; catalyst was fixed at about 3 g , but the discharge volume changes with the shortening of the length of the HV electrode, which results in the catalyst being partly packed in the field of plasma by changing the discharge volume "V" from 3.3 to 0.4 mL).

In Figure 6a, interestingly, the conversion of ammonia was greatly enhanced with discharge volume decrease, and partly packing catalyst into the discharge area was found to be better than that of full-packing mode. Among the cases studied, the discharge volume with 0.4 mL showed the best activity towards NH_3 decomposition, in this case, the reaction temperature with 98.0% NH_3 conversion was only 380 °C, which was 140 °C lower than that in the case of catalyst alone. At the reaction temperature of 380 °C, the conversion of NH_3 over $Co/$fumed SiO_2 is only 6.2% without plasma whereas, at the same conditions, the use of DBD plasma significantly enhanced the reaction performance, and the conversion of NH_3 increased by a factor of 16 (from 6.1% to 98.0%) with decreasing discharge volume from 3.3 to 0.4 mL. Correspondingly, the energy efficiency of H_2 formation increased from 11.9 to 15.9 mol$(kW \cdot h)^{-1}$; this is the highest H_2 formation rate obtained in ammonia decomposition so far, as shown in Figure 6b. In addition, Figure 6c displayed that the specific energy input (SEI) significantly increased with decreasing discharge volume, which might be the reason for the high performance shown in Figure 6a,b. To exclude the effect of heat caused by SEI increasing on the reaction performance, the reaction temperatures with different discharge volumes were all controlled at around 350 °C by adjusting energy input, then the relationship of ammonia conversion and SEI was presented in Figure 6d. Clearly, the conversion of ammonia increased with SEI increasing, demonstrating that the high performance resulting from high SEI was not due to heating of the catalyst. Furthermore, our previous studies revealed that increasing energy input of discharge can significantly facilitate the desorption of the strong-adsorbed N from catalyst surface (rate-limiting step in ammonia decomposition) [11], thus, the nature of the contribution of high SEI was to accelerate the rate-limiting step of ammonia decomposition.

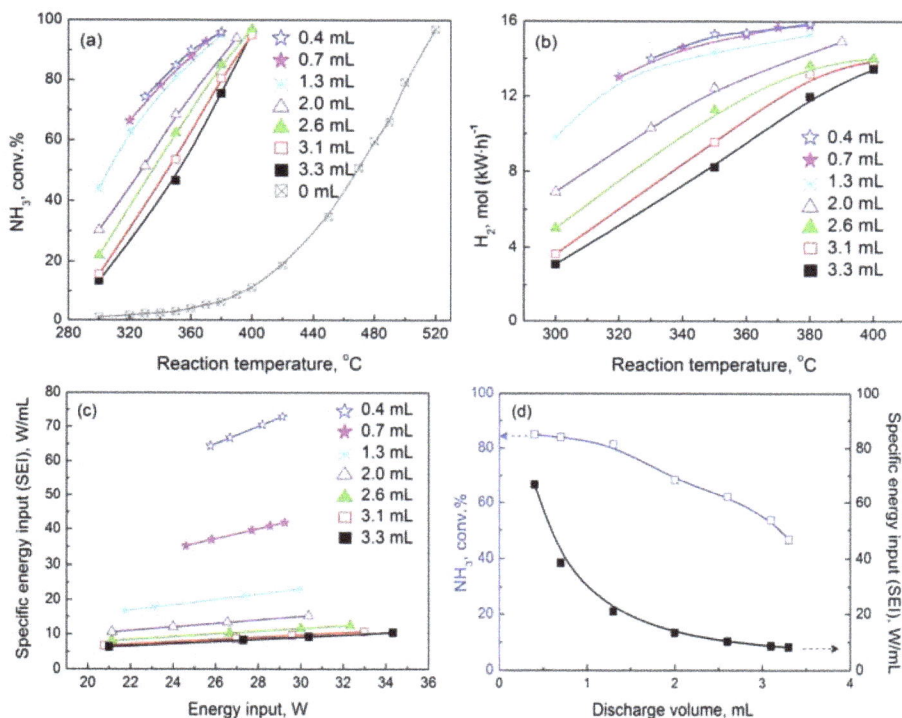

Figure 6. Influence of discharge volume on (**a**) ammonia conversion, (**b**) energy efficiency of H_2 formation, (**c**) specific energy input (SEI) and (**d**) relationship of ammonia conversion with SEI (NH_3 feed rate 40 mL/min, discharge gap 3 mm, discharge frequency 12 kHz, and the packing amount and packing volume of catalyst in the reactor was fixed at about 3 g and 3.1 mL, respectively. Changing the discharge volume "V" from 3.3 to 0.4 mL enabled the catalyst to be partly packed in the discharge area, as shown in Scheme 1, and the catalyst was fully packed in the field of plasma only when the discharge volume was over 3.1 mL; The reaction temperature originated from electric heat released by discharge, and was determined using an IR camera and thermocouple tightly attached to the outer wall of the reactor [12]).

3. Materials and Methods

3.1. DBD Plasma-Catalytic Reactor

NH_3 decomposition for H_2 generation was carried out in a DBD reactor with a catalyst bed in the discharge area at atmospheric pressure (Scheme 2). The DBD reactor was a typical cylindrical reactor using a stainless-steel rod (2 mm o.d.) as a high-voltage electrode placed along the axis of a quartz tube

(10 mm o.d. × 8 mm i.d.) which was used as a discharge dielectric. An aluminum foil sheet tightly covered the outside of the quartz cylinder and served as a ground electrode. A 3 mm of discharge gap was used, and catalyst was fully packed in the discharge area unless otherwise noted. The DBD reactor was connected to an AC high voltage power supply with a peak voltage of up to 30 kV and a variable frequency of 5–20 kHz. In this study, the discharge frequency was fixed at 12 kHz, and NH_3 with a purity of 99.999% was fed into the DBD reactor at a total flow rate of 40 ml/min. The products of NH_3 decomposition were analyzed on-line using a gas chromatograph (Shimadzu GC-2014) equipped with a thermal conductivity detector (TCD). The input power driving the reaction was determined from the product of the apparent voltage and current of AC power supply, and the discharge power was measured using a four-channel digital oscilloscope (Tektronix DPO 3012, high-voltage probe Tektronix P6015A, Tektronix Tech. Corp., Beaverton, OR, USA, current probe Pearson 6585, Pearson Electronics, Inc., San Jose, CA, USA).

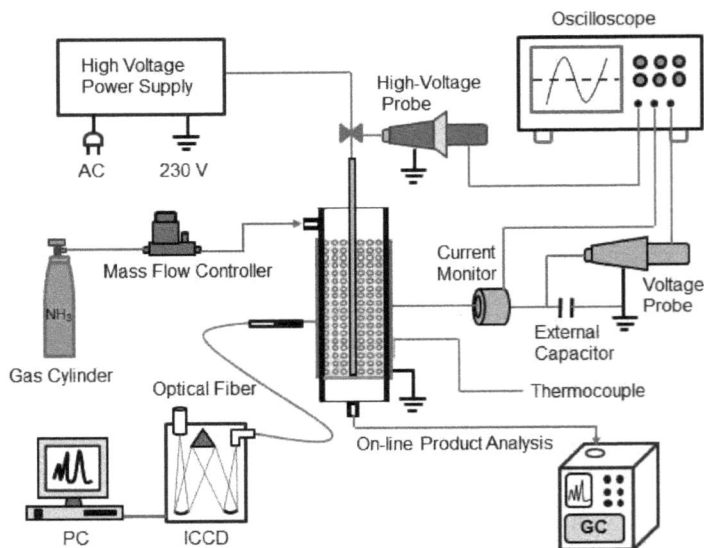

Scheme 2. Schematic diagram of experimental setup.

To evaluate the reaction performance of plasma-catalytic NH_3 decomposition to produce H_2, the conversion of NH_3 was calculated using Equation (1). The energy efficiency of H_2 formation $(mol(kW \cdot h)^{-1})$, defined as the number of moles of H_2 produced per kilowatt hour, was calculated using Equation (2). The specific energy input (SEI), defined as the energy input per discharge volume, was calculated using Equation (3).

$$X_{NH_3}(\%) = \frac{\text{moles of } NH_3 \text{ converted}}{\text{moles of initial } NH_3} \times 100 \tag{1}$$

$$E_{H_2} = \frac{3 \cdot X_{NH_3} \cdot F_{NH_3} \times 60}{2 \times 1000 \times 22.4 \times P} \tag{2}$$

E_{H_2}: Energy efficiency of H_2 formation, $mol(kW \cdot h)^{-1}$
X_{NH_3}: Conversion of NH_3 , %
F_{NH_3}: NH_3 flow rate, $mL \cdot min^{-1}$
P: Plasma power, kW

$$\text{SEI} = \frac{\text{energy input (W)}}{\text{discharge volume (mL)}} \tag{3}$$

3.2. Catalyst Preparation

Cobalt nitrate was provided by the Tianjin Kermel Chemical Reagent Co., Ltd (Tianjin, China). Fumed SiO_2 was purchased from the Dalian Luming Nanometer Material Co., Ltd (Dalian, China). Catalysts were synthesized either using the conventional preparation method and improved preparation method. Incipient wetness impregnation was used in this study. Briefly, cobalt nitrate (the theoretical metal loading was 30 wt %) was dissolved in deionized water. The support of fumed SiO_2 was calcined, in advance, at 400 °C for 5 h to remove impurities, such as H_2O, before impregnation, and then the pretreated support was added to the cobalt nitrate solution and stirred until it was thoroughly mixed. For "conventional preparation method", the resulting mixture was kept at room temperature for 3 h and dried in air overnight at 110 °C. The dried sample was finally calcined in air at 540 °C for 5 h. Different from the conventional preparation method, for the "improved preparation method", the resulting mixture was kept at room temperature for 3 h, followed by vacuum-freeze drying overnight at −50 °C before dried in air at 120 °C for 5 h, then the dried sample was calcined in a He-DBD plasma environment at 400 °C for 3 h to obtain the as-prepared catalyst. In addition, all the as-prepared catalysts were treated in NH_3-DBD plasma at 400 °C for 0.3–1.0 h to reduce catalysts before evaluating their activity in NH_3 decomposition.

3.3. Catalyst Characterization

X-ray diffraction (XRD) patterns of as-prepared catalysts were recorded using a Rigaku D-Max 2400 X ray diffractometer with Cu K_α radiation. Transmission electron microcopy (TEM) was used to characterize metal particles formed on the support surface (FEI Tecnai G2 F30 microscope, point resolution 0.2 nm, operated at 300 kV, Utrecht, Netherlands).

The reduction behavior of as-prepared catalyst was evaluated by H_2 temperature-programmed reduction (H_2-TPR) using a Chemisorption instrument (ChemBET 3000, Quantachrome, Boynton Beach, FL, USA). The sample (100 mg) was pretreated at 500 °C for 1 h under He flow (20 mL/min), and then cooled to 50 °C. The pretreated sample was exposed to a H_2/He mixture (10 vol% H_2) and was heated from 150 to 800 °C at a constant heating rate of 14 °C/min to get a H_2-TPR profile. The acid–base properties of the as-prepared catalyst were tested by NH_3 temperature-programmed desorption (NH_3-TPD) using the same Chemisorption instrument with operating H_2-TPR. The sample (140 mg) was pretreated at 500 °C for 1 h under He flow (20 mL·min^{-1}), and then cooled to 150 °C. The pretreated sample was saturated with NH_3 for 30 min, and then purged with He flow for 1 h at 150 °C. The TPD profile was recorded while the sample was heated from 150 to 600 °C at a constant heating rate of 14 °C·min^{-1} under He flow.

The specific surface area (S_g) of fumed SiO_2 support was tested by N_2 physisorption at −196 °C (Micrometrics ASAP 2020, Norcross, GA, USA). Prior to the N_2 physisorption measurement, fumed SiO_2 was degassed at 350 °C for 3 h, and S_g was calculated using the Brunauer−Emmett−Teller (BET) equation.

The metal loading of fumed SiO_2 supported catalyst with different preparation methods was determined using X-ray fluorescence (XRF, SRS-3400, Bruker, Germany).

4. Conclusions

CO_x-free H_2 generation from plasma-catalytic NH_3 decomposition has been significantly promoted over Co/fumed SiO_2 catalyst prepared with an improved preparation method, which featured the use of vacuum-freeze drying and DBD plasma calcination techniques during catalyst preparation. Compared with the activity of the catalyst prepared by the conventional preparation method, the conversion of NH_3 increased by 47% on Co/fumed SiO_2 catalyst prepared by improved method and, correspondingly, the energy efficiency of H_2 production increased from

2.3 to 5.7 $mol(kW \cdot h)^{-1}$. The enhanced activity was mainly attributed to the high dispersion of Co particles on fumed SiO_2 with a narrow particle size distribution (2–3 nm), which brought more active sites, stronger acidity, and a strong metal–support interaction. In addition, the reaction performance was significantly improved with the increase of specific energy input. At 380 °C, the highest energy efficiency of H_2 formation achieved, so far, was 15.9 $mol(kW \cdot h)^{-1}$ over improved prepared Co/fumed SiO_2 catalyst with 98.0% ammonia conversion at the optimal conditions.

Supplementary Materials: The following are available online at http://www.mdpi.com/2073-4344/9/2/107/s1. Figure S1. Peak analysis of H_2-TPR profile obtained over Co_3O_4/fumed SiO_2 catalyst; Table S1. XRF analysis of Co/fumed SiO_2 with improved preparation method; Table S2. XRF analysis of Co/fumed SiO_2 with conventional preparation method.

Author Contributions: Conceptualization, L.W., Y.Y. and H.G.; methodology, L.W. and Y.Y.; validation, L.W., Y.Y. and X.D.; formal analysis, L.W., Y.Y. and B.Z.; investigation, L.W. and Y.Y.; resources, H.C.G and Y.Z.; data curation, L.W. and Y.Y.; writing—original draft preparation, L.W.; writing—review and editing, L.W., Y.Y., B.Z. and Y.Z.; visualization, L.W., Y.Y. and X.D.; supervision, L.W., H.G. and Y.Z.; project administration, L.W., Y.Z. and H.G.; funding acquisition, L.W. and H.G.

Funding: This research was funded by the National Natural Science Foundation of China [20473016 and 20673018], the Liaoning Provincial Natural Science Fund of China [2018011143-301], and the Fundamental Research Funds for the Central Universities of China [DMU20110218002].

Conflicts of Interest: The authors declare no conflict of interest.

References

1. Schüth, F.; Palkovits, R.; Schlögl, R.; Su, D.S. NH_3 as a possible element in an energy infrastructure: Catalysts for NH_3 decomposition. *Energy Environ. Sci.* **2012**, *5*, 6278–6289. [CrossRef]

2. Lan, R.; Irvine, J.T.S.; Tao, S.W. NH_3 and related chemicals as potential indirect H_2 storage materials. *Int. J. Hydrogen Energy* **2012**, *37*, 1482–1494. [CrossRef]

3. García-Bordejé, E.; Armenise, S.; Roldán, L. Toward practical application of H_2 generation from NH_3 decomposition guided by rational catalyst design. *Catal. Rev.* **2014**, *56*, 220–237. [CrossRef]

4. Karim, A.M.; Prasad, V.; Mpourmpakis, G.; Lonergan, W.W.; Frenkel, A.I.; Chen, J.G.; Vlachos, D.G. Correlating particle size and shape of supported Ru/γ-Al_2O_3 Catalysts with NH_3 Decomposition Activity. *J. Am. Chem. Soc.* **2009**, *131*, 12230–12239. [CrossRef] [PubMed]

5. Yin, S.F.; Xu, B.Q.; Wang, S.J.; Ng, C.F.; Au, C.T. Magnesia-carbon nanotubes (MgO-CNTs) nanocomposite: Novel support of Ru catalyst for the generation of CO_x-free hydrogen from ammonia. *Catal. Lett.* **2004**, *96*, 113–116. [CrossRef]

6. Marco, Y.; Roldán, L.; Armenise, S.; García-Bordejé, E. Support-induced oxidation state of catalytic Ru nanoparticles on carbon nanofibers that were doped with heteroatoms (O, N) for the decomposition of NH_3. *ChemCatChem* **2013**, *5*, 3829–3834. [CrossRef]

7. Zheng, W.Q.; Zhang, J.; Ge, Q.J.; Xu, H.Y.; Li, W.Z. Effects of CeO_2 addition on Ni/Al_2O_3 catalysts for the reaction of NH_3 decomposition to H_2. *Appl. Catal. B Environ.* **2008**, *80*, 98–105. [CrossRef]

8. Donald, J.; Xu, C.B.; Hashimoto, H.; Byambajav, E.; Ohtsuka, Y. Novel carbon-based Ni/Fe catalysts derived from peat for hot gas NH_3 decomposition in an inert helium atmosphere. *Appl. Catal. A Gen.* **2010**, *375*, 124–133. [CrossRef]

9. Hansgen, D.A.; Vlachos, D.G.; Chen, J.G. Using first principles to predict bimetallic catalysts for the NH_3 decomposition reaction. *Nat. Chem.* **2010**, *2*, 484–489. [CrossRef]

10. Lu, A.H.; Nitz, J.J.; Comotti, M.; Weidenthaler, C.; Schlichte, K.; Lehmann, C.W.; Terasaki, O.; Schüth, F. Experimental and theoretical investigation of molybdenum carbide and nitride as catalysts for NH_3 decomposition. *J. Am. Chem. Soc.* **2010**, *132*, 14152–14162. [CrossRef]

11. Wang, L.; Zhao, Y.; Liu, C.Y.; Gong, W.M.; Guo, H.C. Plasma driven NH_3 decomposition on a Fe-catalyst: Eliminating surface nitrogen poisoning. *Chem. Commun.* **2013**, *49*, 3787–3789. [CrossRef] [PubMed]

12. Wang, L.; Yi, Y.H.; Zhao, Y.; Zhang, R.; Zhang, J.L.; Guo, H.C. NH_3 decomposition for H_2 generation: Effects of cheap metals and supports on plasma—catalyst synergy. *ACS Catal.* **2015**, *5*, 4167–4174. [CrossRef]

13. Wang, L.; Yi, Y.H.; Guo, Y.J.; Zhao, Y.; Zhang, J.L.; Guo, H.C. Synergy of DBD plasma and Fe-based catalyst in NH_3 decomposition: Plasma enhancing adsorption step. *Plasma Process Polym.* **2017**, *14*, e1600111. [CrossRef]

14. Yi, Y.H.; Wang, L.; Guo, Y.J.; Sun, S.Q.; Guo, H.C. Plasma-assisted ammonia decomposition over Fe-Ni alloy catalysts for CO$_x$-free hydrogen. *AIChE J.* **2019**, *65*, 691–701.

15. Munnik, P.; de Jongh, P.E.; de Jong, K.P. Recent developments in the synthesis of supported catalysts. *Chem. Rev.* **2015**, *115*, 6687–6718. [CrossRef] [PubMed]

16. Muñoz-Flores, B.M.; Kharisov, B.I.; Jiménez-Pérez, V.M.; Martínez, P.E.; López, S.T. Recent advances in the synthesis and main applications of metallic nanoalloys. *Ind. Eng. Chem. Res.* **2011**, *50*, 7705–7721. [CrossRef]

17. Zaera, F. Nanostructured materials for applications in heterogeneous catalysis. *Chem. Soc. Rev.* **2013**, *42*, 2746–2762. [CrossRef]

18. Kogelschatz, U. Dielectric-barrier discharges: Their history, discharge physics, and industrial applications. *Plasma Chem. Plasma Process* **2003**, *23*, 1–45. [CrossRef]

19. Wang, L.; Yi, Y.H.; Guo, H.C.; Tu, X. Atmospheric pressure and room temperature synthesis of methanol through plasma-catalytic hydrogenation of CO$_2$. *ACS Catal.* **2018**, *8*, 90–100. [CrossRef]

20. Kameshima, S.; Tamura, K.; Ishibashi, Y.; Nozaki, T. Pulsed dry methane reforming in plasma-enhanced catalytic reaction. *Catal. Today* **2015**, *256*, 67–75. [CrossRef]

21. Wang, L.; Yi, Y.H.; Wu, C.F.; Guo, H.C.; Tu, X. One-step reforming of CO$_2$ and CH$_4$ into high-value liquid chemicals and fuels at room temperature by plasma-driven catalysis. *Angew. Chem. Int. Ed.* **2017**, *56*, 13679–13683. [CrossRef] [PubMed]

22. Snoeckx, R.; Heijkers, S.; Wesenbeeck, K.V.; Lenaerts, S.; Bogaerts, A. CO$_2$ conversion in a dielectric barrier discharge plasma: N$_2$ in the mix as a helping hand or problematic impurity. *Energy Environ. Sci.* **2016**, *9*, 999–1011. [CrossRef]

23. Gao, Y.; Zhang, S.; Sun, H.; Wang, R.; Tu, X.; Shao, T. Highly efficient conversion of methane using microsecond and nanosecond pulsed spark discharges. *Appl. Energy* **2018**, *226*, 534–545. [CrossRef]

24. Guo, Z.F.; Yi, Y.H.; Wang, L.; Yan, J.H.; Guo, H.C. Pt/TS-1 catalyst promoted C-N coupling reaction in CH$_4$-NH$_3$ plasma for HCN synthesis at low temperature. *ACS Catal.* **2018**, *8*, 10219–10224. [CrossRef]

25. Liu, C.J.; Vissokov, G.P.; Jang, B. Catalyst preparation using plasma technologies. *Catal. Today* **2002**, *72*, 173–184. [CrossRef]

26. Zhang, H.; Chu, W.; Xu, H.Y.; Zhou, J. Plasma-assisted preparation of Fe–Cu bimetal catalyst for higher alcohols synthesis from carbon monoxide hydrogenation. *Fuel* **2010**, *89*, 3127–3131. [CrossRef]

27. Cheng, D.G. Plasma decomposition and reduction in supported metal catalyst preparation. *Catal. Surv. Asia* **2008**, *12*, 145–151. [CrossRef]

28. Hinokuma, S.; Misumi, S.; Yoshida, H.; Machida, M. Nanoparticle catalyst preparation using pulsed arc plasma deposition. *Catal. Sci. Technol.* **2015**, *5*, 4249–4257. [CrossRef]

29. Di, L.B.; Zhang, J.S.; Zhang, X.L. A review on the recent progress, challenges, and perspectives of atmospheric-pressure cold plasma for preparation of supported metal catalysts. *Plasma Process Polym.* **2018**, *15*, e1700234. [CrossRef]

30. Wang, Z.; Zhang, Y.; Neyts, E.C.; Cao, X.X.; Zhang, X.S.; Jang, B.; Liu, C.J. Catalyst Preparation with Plasmas: How Does It Work? *ACS Catal.* **2018**, *8*, 2093–2110. [CrossRef]

31. Wang, N.; Shen, K.; Yu, X.P.; Qian, W.Z.; Chu, W. Preparation and characterization of a plasma treated NiMgSBA-15 catalyst for methane reforming with CO$_2$ to produce syngas. *Catal. Sci. Technol.* **2013**, *3*, 2278–2287. [CrossRef]

32. Yan, X.L.; Zhao, B.R.; Liu, Y.; Li, Y.A. Dielectric barrier discharge plasma for preparation of Ni-based catalysts with enhanced coke resistance: Current status and perspective. *Catal. Today* **2015**, *256*, 29–40. [CrossRef]

33. Guo, Z.L.; Huang, Q.S.; Luo, S.Z.; Chu, W. Atmospheric discharge plasma enhanced preparation of Pd/TiO$_2$ catalysts for acetylene selective hydrogenation. *Top. Catal.* **2017**, *60*, 1009–1015. [CrossRef]

34. Li, Y.A.; Jang, B. Selective hydrogenation of acetylene over Pd/Al$_2$O$_3$ catalysts: Effect of non-thermal RF plasma preparation methodologies. *Top. Catal.* **2017**, *60*, 997–1008. [CrossRef]

35. Liu, Y.; Wang, Z.; Liu, C.J. Mechanism of template removal for the synthesis of molecular sieves using dielectric barrier discharge. *Catal. Today* **2015**, *256*, 137–141. [CrossRef]

36. Zhang, Y.R.; Laerb, K.V.; Neyts, E.C.; Bogaerts, A. Can plasma be formed in catalyst pores? A modeling investigation. *Appl. Catal. B Environ.* **2016**, *185*, 56–67. [CrossRef]

37. Wang, Z.; Wang, W.Z.; Zhang, L.; Jiang, D. Surface oxygen vacancies on Co$_3$O$_4$ mediated catalytic formaldehyde oxidation at room temperature. *Catal. Sci. Technol.* **2016**, *6*, 3845–3853. [CrossRef]

38. Wang, Q.; Peng, Y.; Fu, J.; Kyzasd, G.Z.; Reduwan Billah, S.M.; An, S.Q. Synthesis, characterization, and catalytic evaluation of Co_3O_4/γ-Al_2O_3 as methane combustion catalysts: Significance of Co species and the redox cycle. *Appl. Catal. B Environ.* **2015**, *168–169*, 42–50. [CrossRef]

39. Qazi, S.J.S.; Rennie, A.R.; Cockcroft, J.K.; Vickers, M. Use of wide-angle X-ray diffraction to measure shape and size of dispersed colloidal particles. *J. Colloid Interf. Sci.* **2009**, *338*, 105–110. [CrossRef]

40. Luo, J.Y.; Meng, M.; Li, X.; Li, X.G.; Zha, Y.Q.; Hu, T.D.; Xie, Y.N.; Zhang, J. Mesoporous Co_3O_4–CeO_2 and Pd/Co_3O_4–CeO_2 catalysts: Synthesis, characterization and mechanistic study of their catalytic properties for low-temperature CO oxidation. *J. Catal.* **2008**, *254*, 310–324. [CrossRef]

41. Teng, F.; Chen, M.D.; Li, G.Q.; Teng, Y.; Xu, T.G.; Hang, Y.C.; Yao, W.Q.; Santhanagopalan, S.; Meng, D.D.; Zhu, Y.F. High combustion activity of CH_4 and catalluminescence properties of CO oxidation over porous Co_3O_4 nanorods. *Appl. Catal. B Environ.* **2011**, *110*, 133–140. [CrossRef]

42. Vakros, J.; Kordulis, C.; Lycourghiotis, A. Cobalt oxide supported γ-Alumina catalyst with very high active surface area prepared by equilibrium deposition filtration. *Langmuir* **2002**, *18*, 417–422. [CrossRef]

43. Arnoldy, P.; Moulijn, J.A. Temperature-programmed reduction of CoO/Al_2O_3 catalysts. *J. Catal.* **1985**, *93*, 38–54. [CrossRef]

44. Hong, J.P.; Du, J.; Wang, B.; Zhang, Y.H.; Liu, C.C.; Xiong, H.F.; Sun, F.L.; Chen, S.F.; Li, J.L. Plasma-assisted preparation of highly dispersed Co enhanced Fischer–Tropsch synthesis performance. *ACS Catal.* **2018**, *8*, 6177–6185. [CrossRef]

MDPI
St. Alban-Anlage 66
4052 Basel
Switzerland
Tel. +41 61 683 77 34
Fax +41 61 302 89 18
www.mdpi.com

Catalysts Editorial Office
E-mail: catalysts@mdpi.com
www.mdpi.com/journal/catalysts

www.ingramcontent.com/pod-product-compliance
Lightning Source LLC
Chambersburg PA
CBHW051729210326
41597CB00032B/5653